Otto Endler

Valuation Theory

Springer-Verlag
Berlin Heidelberg New York 1972

Otto Endler
Mathematisches Institut der Universität Bonn
and Instituto de Matemática P. e A., Rio de Janeiro

AMS Subject Classifications (1970)
Primary 12J20, 12J10, 10M10, 13A15
Secondary 13B15, 12B10, 13B20, 10B40, 13F05, 13F15, 13J15, 14A05, 13J05, 13J10, 14H05

ISBN-13: 978-3-540-06070-3 e-ISBN-13: 978-3-642-65505-0
DOI: 10.1007/978-3-642-65505-0

Library of Congress Catalog Card Number 72-92285.

To the Memory of

WOLFGANG KRULL

26 August 1899 – 12 April 1971

Preface

These are the revised notes of a course for graduate students and some seminar talks which I gave at the University of Rochester during Fall Term 1969/70. They would not have been written without the encouragement and the aid which I received, during all stages of the work, by friends from Rochester, Rio de Janeiro, and Bonn. I wish to thank all of them: Barbara Grabkowicz encouraged me to write these notes in English and read carefully parts of a preliminary manuscript, as did Gervásio G. Bastos, Yves A. E. Lequain, Walter Strubel, and Antonio J. Engler. Many valuable suggestions were given me by Yves A. E. Lequain, and several improvements of theorems and proofs are due to him. I am particularly grateful to Linda C. Hill for her criticism in reading the last version and for improving and smoothing many of my formulations. Last but not least I thank Wilson Góes for the excellent typing.

Most of this book was elaborated when I stayed in Rio de Janeiro as a Visiting Professor at IMPA (Institute for Pure and Applied Mathematics) and as a Pesquisador-Conferencista of CNPq (National Research Council). Thanks are also due to these institutions.

Rio de Janeiro, September 1972

Otto Endler

Contents

Symbols Used in Text

$\Rightarrow$	implication
$\Leftrightarrow$	if and only if
$S \to T$	map from the set S into the set T
$x \mapsto y$	map assigning y to x
ι_S	identical map of the set S
$\iota_{S,T}$	identical imbedding of S into T, where S is a subset of T
$\mu \vert S$	restriction of the map μ to the subset S
∇	diagonal map
$\emptyset$	empty set
$\subseteq$	inclusion
$\subset$	proper inclusion
$\in$	is an element of
$S \setminus T$	difference set (set of all $x \in S$ such that $x \notin T$)
$S_1 \times \ldots \times S_n$	product of the sets $S_1, \ldots, S_n$
$\times_{\iota \in I} S_\iota$	product of the family of sets $(S_\iota)_{\iota \in I}$
$\#$	number of elements (cardinality)
$\mathbb{N}$	set of all non-negative rational integers
$\mathbb{Z}$	ring of all rational integers
$\mathbb{Q}$	field of all rational numbers
$\mathbb{R}$	field of all real numbers
$\mathbb{C}$	field of all complex numbers
$\mathbb{F}_p$	prime field of characteristic p $(\neq 0)$
$\hat{\mathbb{Q}}_p$	field of all p-adic numbers
$\mathbb{R}_+$	set of all non-negative real numbers
$\mathbb{R}_+^*$	multiplicative group of all positive real numbers
$\tilde{K} = K \cup \{\infty\}$	projective field obtained from the field K
$ac(K)$	algebraic closure of the field K
$sc(K)$	separable closure of the field K
K_{sep}	maximal separable subextension of a given algebraic field extension $L \vert K$
K_{ins}	maximal purely inseparable subextension of a given normal field extension $N \vert K$
$[L:K]_{sep}$	separability degree of the algebraic extension $L \vert K$

$[L:K]_{insep}$	inseparability degree of the algebraic extension $L\|K$
Char K	characteristic of the field K
$Mon(L\|K)$	set of all K-monomorphisms from L into a given algebraically closed field Ω
$Mon_{\bar{K}}(L\|K)$	set of all classes of $\bar{K}$-conjugate K-monomorphisms from L into Ω
$Aut(N\|K)$	group of all K-automorphisms of N (Galois group of $N\|K$)
$k(H)$	fixed field corresponding to the subgroup H of $Aut(N\|K)$
$g(L)$	subgroup of $Aut(N\|K)$ corresponding to the sub-extension L of $N\|K$
$(G:H)$	group index of H in G
$Hom(A,B)$	group of homomorphisms from A into B (where A and B are abelian groups)
$H_{\Omega}(X)$	p-character group of X (where X is an abelian torsion group and $p = Char\,\Omega$)
$I_S(R)$	integral closure of R in S
$\| \ \|_{\mathbb{C}}$	absolute value of $\mathbb{C}$
$\| \ \|_{\mathbb{R}}$	absolute value of $\mathbb{R}$
φ_{∞}	absolute value of $\mathbb{Q}$
φ_p	p-adic valuation of $\mathbb{Q}$
v_p	p-adic exponential valuation of $\mathbb{Q}$
T_{φ}	topology defined by the valuation φ
(K,φ)	valued field (where φ is a valuation of K)
(K,v)	valued field (where v is an exponential valuation of K)
(K,A)	valued field (where A is a valuation ring of K)
$(\hat{K},\hat{\varphi})$	completion of (K,φ)
$(\hat{K},\hat{v})$	completion of (K,v)
$(\hat{K},\hat{A})$	completion of (K,A)
$\mathcal{B}(L;A)$	set of all valuation rings of L which lie over the valuation ring A
$e_{B\|K}$	ramification index of B over K
$f_{B\|K}$	residue degree of B over K

G^Z	decomposition group
K^Z	decomposition field
G^T	inertia group
K^T	inertia field
G^V	ramification group
K^V	ramification field
deg F	degree of the polynomial F
$P_{x\|K}$	minimal polynomial of x over K
$\delta_{i,j}$	Kronecker symbol

Introduction

Giving a course or writing a book on Valuation Theory, one is faced with the problem whether more emphasis should be given to usual valuations (by some authors called "absolute values"), which include archimedean valuations and in particular the usual absolute value, or to Krull valuations, which correspond to valuation rings of arbitrary rank and generalize non-archimedean valuations only. If one is interested only in applications to Algebraic Number Theory it suffices to consider usual valuations and one may even restrict oneself to archimedean and discrete non-archimedean valuations, since these are the only ones which occur in algebraic number fields (i.e., finite extensions of $\mathbb{Q}$). On the other hand, arbitrary Krull valuations, or rather the corresponding valuation rings, play an important role in Commutative Algebra and its application to Algebraic Geometry, as well as in the theory of Diophantine equations.

In this book, we try to meet both demands. The standard material on usual (archimedean and non-archimedean) valuations is given rather briefly in Chapter I; more can be found in any book on Algebraic Number Theory using valuation-theoretical methods (e.g. Weiss [35]). The basic theory of valuation rings, Krull valuations, and places is given in Chapter II, which includes also some results on Prüfer rings, Krull rings, and Dedekind rings. Chapter III is concerned with extensions of valuation rings; it contains Infinite Ramification Theory and emphasizes the importance of henselizations. Some more advanced topics of Valuation Theory, such as the theory of maximal valued fields and Ribenboim's generalization of the approximation theorem, are not presented in this book (cf Ribenboim [30]). In Chapter IV we consider again usual valuations. Krull's results on the existence of fields with prescribed valuations are presented here in a new and generalized form.

A list of Exercises for each chapter is given at the end of the book.

Numbers occuring in brackets refer to the bibliography.

CHAPTER I

Valuations

§1 Valuations

We denote by $\mathbb{R}_+$ the set of all non-negative real numbers. A <u>valuation</u> [1] of a field K is a mapping $\varphi: K \to \mathbb{R}_+$ which satisfies the following conditions:

(V_1) $\varphi x = 0 \Leftrightarrow x = 0$, for all $x \in K$

(V_2) $\varphi(x \cdot y) = \varphi x \cdot \varphi y$, for all $x,y \in K$ (homomorphy)

(V_3) $\varphi(x+y) \leqq \varphi x + \varphi y$, for all $x,y \in K$ (triangle inequality).

Any field K has the <u>trivial</u> valuation τ, determined by $\tau K = \{0,1\}$. The best known examples of valuations are the absolute values $|\ |_{\mathbb{R}}$ and $|\ |_{\mathbb{C}}$ of $\mathbb{R}$ and $\mathbb{C}$, the fields of real and complex numbers, respectively. The field $\mathbb{Q}$ of rational numbers admits, for every prime number p, the p-adic valuation φ_p, uniquely determined by $\varphi_p p = p^{-1}$ and $\varphi_p q = 1$ for any prime number q, $q \neq p$. We shall see later that the p-adic valuations and the restriction of the absolute value to $\mathbb{Q}$ are essentially the only valuations of $\mathbb{Q}$.

From the axioms (V_1) - (V_3), it follows easily that any valuation φ of K has the following properties:

(1.1) a) $\varphi z = 1$, <u>for any root of unity</u> $z \in K$; <u>in particular</u> $\varphi 1 = \varphi(-1) = 1$.

b) $\varphi(x-y) \leqq \varphi x + \varphi y$, <u>for all</u> $x,y \in K$.

c) $\varphi(x \cdot y^{-1}) = \varphi x \cdot (\varphi y)^{-1}$, <u>for all</u> $x,y \in K$, $y \neq 0$.

d) $|\varphi x - \varphi y| \leqq \varphi(x-y)$, <u>for all</u> $x,y \in K$.

---

[1] Some authors prefer the name "absolute value" and reserve the name "valuation" for (mostly additively written) non-archimedean valuations or, more generally, for Krull valuations.

From (1.1) a) we conclude:

(1.2) <u>If</u> K <u>is finite, then</u> K <u>has no non-trivial valuation.</u>

Let φ be a valuation of K. The mapping $d_\varphi: K \times K \to \mathbb{R}_+$ defined by

$$d_\varphi(x,y) = \varphi(x-y), \text{ for all } x,y \in K,$$

is a metric on K and hence defines on K a Hausdorff topology T_φ. For any $x \in K$, the set of all $\mathcal{U}_\varepsilon(x) = \{y \in K \mid \varphi(y-x) < \varepsilon\}$ $(\varepsilon > 0)$ is a basis of open neighborhoods of x. It is easy to see that T_φ is discrete if and only if φ is trivial. Property (1.1) d) yields:

(1.3) <u>The mapping</u> $\varphi: K \to \mathbb{R}_+$ <u>is uniformly continuous</u> (when K is provided with d_φ and R_+ with the metric defined by the absolute value).

Moreover we prove,

(1.4) <u>The field</u> K <u>with</u> T_φ <u>is a topological field</u>.

<u>Proof</u>: The continuity of subtraction and multiplication follows from $\varphi((x-y) - (x_o-y_o)) \leq \varphi(x-x_o) + \varphi(y-y_o)$ and $\varphi(x\cdot y - x_o\cdot y_o) \leq$ $\leq \varphi(x-x_o)\cdot\varphi(y-y_o) + \varphi(x-x_o)\cdot\varphi y_o + \varphi x_o\cdot\varphi(y-y_o)$. The mapping $x \to x^{-1}$ is continuous at any $x_o \neq 0$ since the inequality $\varphi(x-x_o) <$ $< 1/2 \min\{\varphi x_o, \varepsilon\cdot(\varphi x_o)^2\}$ implies $\varphi(x^{-1} - x_o^{-1}) < \varepsilon$. □

Two valuations φ, ψ of K are called <u>equivalent</u> if $\varphi = \psi^\rho$ for some real number $\rho > 0$. The trivial valuation of K is equivalent to itself only. For non-trivial valuations we prove,

(1.5) <u>For any non-trivial valuations</u> φ, ψ <u>of</u> K <u>the following conditions are equivalent:</u>

(i) φ <u>is equivalent to</u> ψ.

(ii) $T_\varphi = T_\psi$

(iii) T_φ <u>is stronger than</u> T_ψ.

(iv) <u>For any</u> $x \in K$, $\varphi x < 1$ <u>implies</u> $\psi x < 1$.

(v) <u>For any</u> $x \in K$, $\varphi x \leqq 1$ <u>if and only if</u> $\psi x \leqq 1$.

<u>Proof</u>: (i) ⇒ (ii) and (ii) ⇒ (iii) are trivial. (iii) ⇒ (iv):

We have $\{y \in K \mid \varphi y < \epsilon\} \subseteq \{y \in K \mid \psi y < 1\}$ for some $\epsilon > 0$. If $x \in K$ such that $\varphi x < 1$, then $\varphi x^n < \epsilon$ for some $n \in \mathbb{N}$, hence $\psi x^n < 1$, hence $\psi x < 1$. (iv) ⇒ (v): Since φ is non-trivial we have $0 < \varphi y < 1$ for some $y \in K$. If $\varphi x \leqq 1$, then for all $n \in \mathbb{N}$ we have $\varphi(x^n \cdot y) < 1$ hence $\psi(x^n \cdot y) < 1$, $(\psi x)^n < (\psi y)^{-1}$, hence [2] $\psi x \leqq 1$. If $\psi x \leqq 1$, then $\psi x^{-1} \nless 1$, hence $\varphi x^{-1} \nless 1$, $\varphi x \leqq 1$.
(v) ⇒ (i): Since φ is non-trivial, we have $\varphi z > 1$ for some $z \in K$, hence $\varphi z \nleqq 1$, $\psi z \nleqq 1$, $\psi z > 1$. For any non-zero $x \in K$ we have $\frac{\log \varphi x}{\log \varphi z} = \frac{\log \psi x}{\log \psi z}$; in fact, for all $m,n \in \mathbb{Z}$, $n > 0$, we have $\frac{m}{n} \geqq \frac{\log \varphi x}{\log \varphi z} \Leftrightarrow (\varphi z)^m \geqq (\varphi x)^n \Leftrightarrow \varphi(x^n \cdot z^{-m}) \leqq 1 \Leftrightarrow \psi(x^n \cdot z^{-m}) \leqq 1 \Leftrightarrow (\psi z)^m \geqq$ $\geqq (\psi x)^n \Leftrightarrow \frac{m}{n} \geqq \frac{\log \psi x}{\log \psi z}$. Therefore $\log \varphi x = \rho \cdot \log \psi x$, $\varphi x = \psi^\rho x$, for all $x \in K$, where $\rho = \frac{\log \varphi z}{\log \psi z} > 0$. □

From (1.5) we conclude that for any two inequivalent, non-trivial valuations φ, ψ of K the topologies T_φ, T_ψ are incomparable, and that $\varphi x < 1 \leqq \psi x$ for some $x \in K$. The last statement can be strengthened as follows:

(1.6) <u>Let</u> $\varphi_1,\dots,\varphi_n$ $(n \geqq 2)$ <u>be pairwise inequivalent, non-trivial valuations of</u> K. <u>Then there exists some</u> $x \in K$ <u>such that</u> $\varphi_1 x > 1$, $\varphi_2 x < 1,\dots,\varphi_n x < 1$.

<u>Proof</u>: If $n = 2$, let $y,z \in K$ such that $\varphi_1 y < 1 \leqq \varphi_2 y$ and $\varphi_2 z < 1 \leqq \varphi_1 z$; then the element $x = y^{-1} \cdot z$ has the

---

[2] We use here the following lemma: Let $\alpha,\beta,\gamma \in R$ such that $\gamma^n \leqq \alpha + n\beta$ for all $n \in \mathbb{N}$; then $\gamma \leqq 1$. (In fact, $\gamma > 1$ implies $n \cdot (\gamma - 1) > \alpha$ and $\frac{n-1}{2} \cdot (\gamma-1)^2 > \beta$ for sufficiently large $n \in \mathbb{N}$, hence $\gamma^n \geqq 1 + n \cdot (\gamma - 1) + \binom{n}{2} \cdot (\gamma-1)^2 > \alpha + n\beta$).

desired property. We assume (1.6) to be true for $n-1$ valuations, where $n > 2$. Let $y,z \in K$ such that $\varphi_1 y > 1$, $\varphi_2 y < 1,\dots,\varphi_{n-1} y < 1$ and $\varphi_1 z > 1$, $\varphi_n z < 1$. If $\varphi_n y \leqq 1$, then $x = y^m \cdot z$, for sufficiently large $m \in \mathbb{N}$, has the desired properties. Let $\varphi_n y > 1$. Since the sequence $(\varphi_k(y^i \cdot (1+y^i)^{-1}))_{i\in\mathbb{N}}$ converges to 1 for $k \in \{1,n\}$ and to 0 for $k \in \{2,\dots,n-1\}$, the sequence $(\varphi_k(z \cdot y^i \cdot (1+y^i)^{-1}))_{i\in\mathbb{N}}$ converges to $\varphi_1 z > 1$ for $k = 1$, to $\varphi_n z < 1$ for $k = n$, and to 0 for $k \in \{2,\dots,n-1\}$. Therefore, for sufficiently large $m \in \mathbb{N}$, any x of the form $z \cdot y^m \cdot (1+y^m)^{-1}$ has the desired properties. □

We shall use the following Corollary of (1.6):

(1.7) Let $\varphi_1,\dots,\varphi_n$ $(n \geqq 2)$ be as in (1.6). For any $\varepsilon > 0$ there exists some $y \in K$ such that $\varphi_1(y-1) < \varepsilon$, $\varphi_2 y < \varepsilon,\dots,\varphi_n y < \varepsilon$.

Proof: Let $x \in K$ with the properties indicated in (1.6). Then the sequences $(\varphi_1(x^i \cdot (1+x^i)^{-1} - 1))_{i\in\mathbb{N}}$ and $(\varphi_k(x^i \cdot (1+x^i)^{-1}))_{i\in\mathbb{N}}$ for $2 \leqq k \leqq n$ converge to zero. Hence for sufficiently large $m \in \mathbb{N}$, any y of the form $x^m \cdot (1+x^m)^{-1}$ has the desired properties. □

(1.8) APPROXIMATION THEOREM - Let $\varphi_1,\dots,\varphi_n$ be pairwise inequivalent non-trivial valuations of K. For any $x_1,\dots,x_n \in K$ and $\varepsilon > 0$ there exists an $x \in K$ such that $\varphi_k(x-x_k) < \varepsilon$ for $1 \leqq k \leqq n$.

Proof: Let $\rho > \max\{\varphi_k x_j \mid j,k = 1,\dots,n\}$. By (1.7), there exist elements $y_1,\dots,y_n$ such that $\varphi_k(y_j - \delta_{jk}) < \varepsilon \cdot (n \cdot \rho)^{-1}$ for $j,k \in \{1,\dots,n\}$ ($\delta_{j,k}$ Kronecker index). Let $x = \sum_{j=1}^{n} x_j \cdot y_j$; thus $\varphi_k(x-x_k) \leqq \sum_{j=1}^{n} \varphi_k(x_j \cdot (y_j - \delta_{j,k})) < n \cdot \rho \cdot \varepsilon \cdot (n \cdot \rho)^{-1} = \varepsilon$. □

The approximation theorem can also be formulated in a purely topological way. In fact, the following statement is equivalent to (1.8).

(1.9) Let K_k be the field K provided with the topology T_{φ_k} $(k=1,\dots,n)$. The image of K under the diagonal mapping

$\nabla: K \to K_1 \times \ldots \times K_n$ <u>is dense in the product space</u> $K_1 \times \ldots \times K_n$.

A valuation φ of K is called <u>archimedean</u> if $\varphi(m \cdot 1) > 1$ for some $m \in \mathbb{N}$ (where $m \cdot 1$ represents m times the unity of K), otherwise, <u>non-archimedean</u>. The absolute values $|\ |_{\mathbb{C}}$ and $|\ |_{\mathbb{R}}$ are archimedean. The p-adic valuations of $\mathbb{Q}$ are non-archimedean, and so is the trivial valuation of any field K.

If K_o is a subfield of K, we denote by $\varphi|K_o$ the restriction of φ to K_o; this is a valuation of K_o. If $\varphi|K_o$ is trivial, φ is called a <u>valuation of</u> $K|K_o$. The following statements are trivial.

(1.10) <u>If</u> φ <u>is archimedean, then so is any valuation equivalent to</u> φ.

(1.11) φ <u>is archimedean if and only if its restriction to the prime field is archimedean</u>.

(1.12) <u>If</u> $\varphi|K_o$ <u>is trivial for some subfield</u> K_o <u>of</u> K, <u>then</u> φ <u>is non-archimedean</u>.

(1.13) <u>If</u> K <u>has prime characteristic, then</u> K <u>has no archimedean valuation</u>.

Strengthening (1.13), we shall see later that the only archimedean valuations are essentially the absolute value $|\ |_{\mathbb{C}}$ and its restrictions to subfields of $\mathbb{C}$. We shall be interested chiefly in non-archimedean valuations and their generalizations, the so-called "general valuations" or "Krull valuations."

There are several equivalent definitions of non-archimedean valuations:

(1.14) <u>Let</u> $\varphi: K \to \mathbb{R}_+$ <u>be a mapping satisfying</u> (V_1) <u>and</u> (V_2). <u>The following conditions are equivalent</u>:

(i) φ <u>is a non-archimedean valuation</u>.

(ii) φ <u>satisfies</u> (V_3), <u>and</u> $\{\varphi(m\cdot 1) \mid m\in\mathbb{N}\}$ <u>is bounded</u>.

(iii) <u>For all</u> $x \in K$, $\varphi x \leqq 1$ <u>implies</u> $\varphi(x+1) \leqq 1$.

(iv) $\varphi(x+y) \leqq \max\{\varphi x,\varphi y\}$ <u>for all</u> $x,y \in K$.

(v) <u>For any</u> $\rho > 0$, φ^ρ <u>is a valuation of</u> K.

<u>Proof</u>: (i) $\Rightarrow$ (ii) is trivial. (ii) $\Rightarrow$ (iii): If $\varphi x \leqq 1$, then

$$(\varphi(x+1))^n = \varphi\left(\sum_{i=o}^{n} \binom{n}{i}\cdot x^i\right) \leqq \sum_{i=o}^{n} \varphi\left(\binom{n}{i}\cdot 1\right)\cdot(\varphi x)^i \leqq (n+1)\cdot$$

$\cdot \sup\{\varphi(m\cdot 1) \mid m\in\mathbb{N}\}$, hence $\overset{2}{=}$ $\varphi(x+1) \leqq 1$. (iii) $\Rightarrow$ (iv): We may assume $\varphi x \leqq \varphi y \neq 0$. Then $\varphi(\frac{x}{y}) \leqq 1$, $\varphi(x+y) = \varphi y\cdot\varphi(\frac{x}{y}+1) \leqq \varphi y =$ $= \max\{\varphi x,\varphi y\}$. (iv) $\Rightarrow$ (v): For any $\rho > 0$, φ^ρ satisfies (V_1) and (V_2). It satisfies (V_3), too, since $\varphi^\rho(x+y) \leqq (\max\{\varphi x,\varphi y\})^\rho =$ $= \max\{\varphi^\rho x,\varphi^\rho y\} \leqq \varphi^\rho x + \varphi^\rho y$. (v) $\Rightarrow$ (i): For all $m,n \in \mathbb{N}$ we have $(\varphi(m\cdot 1))^n = \varphi^n(m\cdot 1) \leqq m\cdot\varphi^n 1 = m$, hence $\overset{2}{=}$ $\varphi(m\cdot 1) \leqq 1$. □

Condition (iv) is sometimes called the "ultrametric triangle inequality" because it implies that d_φ is an ultrametric (i.e., satisfies $d_\varphi(x,y) \leqq \max\{d_\varphi(x,z),d_\varphi(z,y)\}$ for all $x,y,z \in K$). Therefore non-archimedean valuations are also called ultrametric valuations.

From (iv) it follows immediately:

(1.15) <u>For all</u> $x,y \in K$, $\varphi x \neq \varphi y$ <u>implies</u> $\varphi(x+y) = \max\{\varphi x,\varphi y\}$.

We want to mention that some authors prefer a slightly more general definition of valuation, replacing the axiom (V_3) by one of the following equivalent conditions:

(V_3') There exists some $\gamma \in \mathbb{R}_+$ such that $\varphi(x+y) \leqq \gamma\cdot\max\{\varphi x,\varphi y\}$ for all $x,y \in K$.

(V_3'') There exists some $\gamma \in \mathbb{R}_+$ such that for every $x \in K$, $\varphi x \leqq 1$ implies $\varphi(x+1) \leqq \gamma$.

It can be shown that the mappings $\varphi\colon K \to \mathbb{R}_+$ which satisfy (V_1), (V_2) and (V_3') are exactly the positive real powers of valuations, and that φ is a valuation (resp. non-archimedean valuation) if and only if $\|\varphi\| \leqq 2$ (resp. $= 1$), where $\|\varphi\|$, the "norm of φ", is the infimum of the set of γ's satisfying (V_3'). Moreover, $\|\varphi\| = \max\{\varphi 1, \varphi 2\}$. (Cf. Weiss [35], Chapter 1.)

We finish this section proving the following theorem, due to Ostrowski, which determines all valuations of $\mathbb{Q}$. Let $\mathcal{P}$ be the set of all prime numbers and let φ_∞ be the restriction to $\mathbb{Q}$ of the absolute value $|\ |_{\mathbb{R}}$.

(1.16) THEOREM - <u>Any non-trivial valuation of</u> $\mathbb{Q}$ <u>is equivalent to</u> φ_p <u>for exactly one</u> $p \in \mathcal{P} \cup \{\infty\}$. <u>For every non-zero</u> $x \in \mathbb{Q}$, $\{p \in \mathcal{P} \mid \varphi_p x \neq 1\}$ <u>is finite and</u> $\prod_{p \in \mathcal{P} \cup \{\infty\}} \varphi_p x = 1$.

<u>Proof</u>: 1) For any $p \in \mathcal{P}$, φ_p is inequivalent to the archimedean valuation φ_∞. If $p, q \in \mathcal{P}$, $p \neq q$, then $\varphi_p p \neq 1 = \varphi_p p$, hence φ_p, φ_q are inequivalent.

2) Let φ be a non-trivial, non-archimedean valuation of $\mathbb{Q}$; then $\mathfrak{P} = \{a \in \mathbb{Z} \mid \varphi a < 1\}$ is a non-zero prime ideal of $\mathbb{Z}$, hence $\mathfrak{P} = p \cdot \mathbb{Z}$ for some prime number p. There exists some $\rho > 0$ such that $\varphi p = p^{-\rho} = \varphi_p^\rho\, p$. Moreover, φ and φ_p^ρ coincide on $\mathbb{Z} \setminus \mathfrak{P} = \{a \in \mathbb{Z} \mid \varphi a = 1\}$. Since φ and φ_p^ρ are multiplicative homomorphisms, and any non-zero $x \in \mathbb{Q}$ is of the form $p^m \cdot \frac{a}{b}$ with $a, b \in \mathbb{Z} \setminus \mathfrak{P}$, $m \in \mathbb{Z}$, we have $\varphi = \varphi_p^\rho$.

3) Let φ be an archimedean valuation of $\mathbb{Q}$. For all integers $m > 1$, $n > 1$ and $t \geqq 1$ there exist $s \geqq 0$ and $a_o, \ldots, a_s \in \{0, \ldots, n-1\}$ such that $a_s \neq 0$ and $m^t = a_o + a_1 \cdot n + \ldots + a_s \cdot n^s$. Since $m^t \geqq n^s$, we have $s \leqq t \cdot \frac{\log m}{\log n}$ and $(\varphi m)^t \leqq \sum_{i=o}^{s} \varphi a_i \cdot \varphi n^i \leqq n \cdot (\sum_{i=o}^{s} \varphi n^i) \leqq n \cdot (s+1) \cdot \max\{1, \varphi n^s\} \leqq n \cdot (\frac{\log m}{\log n} \cdot t + 1) \cdot (\max\{1, (\varphi n)^{\frac{\log m}{\log n}}\})^t$. It follows [2] that $\varphi m \leqq \max\{1, (\varphi n)^{\frac{\log m}{\log n}}\}$. We have $\varphi n > 1$ for all $n > 1$, since otherwise $\varphi m \leqq 1$ for all $m > 1$, contrary to the hypothesis that φ be archimedean. Therefore, $1 < (\varphi n)^{\frac{\log m}{\log n}}$, $\varphi m \leqq (\varphi n)^{\frac{\log m}{\log n}}$, $(\varphi m)^{\frac{1}{\log m}} \leqq (\varphi n)^{\frac{1}{\log n}}$. This in-

equality being true for every $m > 1$ and $n > 1$, we see that there exists some $\rho > 0$ such that for every $m > 1$ we have $(\varphi m)^{\frac{1}{\log m}} = e^{\rho}$, hence $\varphi m = m^{\rho} = \varphi_{\infty}^{\rho} m$. This equality holds also for $m \in \{0,1\}$, hence for all $m \in \mathbb{Z}$. From multiplicativity it follows that $\varphi = \varphi_{\infty}^{\rho}$.

4) Any non-zero $x \in \mathbb{Q}$ is of the form $\pm \prod_{p \in \mathcal{P}} p^{n_p}$, with $n_p \in \mathbb{Z}$ for all $p \in \mathcal{P}$ and $n_p = 0$ for almost all $p \in \mathcal{P}$. Hence $\varphi_p x = 1$ for almost all $p \in \mathcal{P}$. For any $q \in \mathcal{P}$, we have $\prod_{p \in \mathcal{P}} \varphi_p q = \varphi_q q = \frac{1}{q} = (\varphi_{\infty} q)^{-1}$ hence, $\prod_{p \in \mathcal{P} \cup \{\infty\}} \varphi_p q = 1$. By multiplicativity, this equation holds for any non-zero $x \in \mathbb{Q}$. □

§2 Completions and extensions of valuations

It is well known that the field $\mathbb{R}$ of real numbers is the completion of $\mathbb{Q}$ with respect to the absolute value. Similarly any valued field can be imbedded in a completion as we show in the following.

Let φ be a non-trivial valuation of K. A sequence $(x_i)_{i \in \mathbb{N}}$ of elements of K is called φ-convergent to x, if it converges to x with respect to the topology T_{φ}. We write in this case $\lim_{\varphi}(x_i)_{i \in \mathbb{N}} = x$ and note that this occurs if and only if $(\varphi(x_i - x))_{i \in \mathbb{N}}$ converges to 0 in $\mathbb{R}$ (with respect to the natural topology, defined by $| \ |_{\mathbb{R}}$). In this case $(\varphi x_i)_{i \in \mathbb{N}}$ is also convergent and we have $\lim(\varphi x_i)_{i \in \mathbb{N}} = \varphi(\lim_{\varphi}(x_i)_{i \in \mathbb{N}})$. [3]

A sequence $(x_i)_{i \in \mathbb{N}}$ is called φ-Cauchy, if it is a Cauchy sequence with respect to d_{φ} (or, more precisely, to the uniform structure defined by d_{φ}), i.e., if for any $\varepsilon > 0$ there is some

---

[3] Note that the convergence of $(\varphi x_i)_{i \in \mathbb{N}}$ does not imply the φ-convergence of $(x_i)_{i \in \mathbb{N}}$.

$n_o \in \mathbb{N}$ such that $\varphi(x_n - x_m) < e$ for all $m,n \geq n_o$. In this case, $(\varphi x_i)_{i\in\mathbb{N}}$ is a Cauchy sequence and hence convergent.

Obviously any φ-convergent sequence is φ-Cauchy. If any φ-Cauchy sequence is φ-convergent, then K is called φ-complete. We note that φ-convergence, φ-Cauchy and φ-complete depend only on the equivalence class of φ.

For abbreviation, we write X,Y etc., instead of $(x_i)_{i\in\mathbb{N}}$, $(y_i)_{i\in\mathbb{N}}$ etc. We shall make use of the following lemma.

(2.1) Let φ be a valuation of K. Then:

a) The set of all φ-Cauchy sequences $(x_i)_{i\in\mathbb{N}}$ of elements of K form a ring $\mathcal{C}$ (with respect to componentwise addition and multiplication).

b) The set of all φ-null sequences (i.e., sequences which are φ-convergent to 0) forms a maximal ideal $\mathfrak{n}$ of $\mathcal{C}$.

c) The mapping $\Phi: \mathcal{C} \to \mathbb{R}_+$ defined by $\Phi X = \lim(\varphi x_i)_{i\in\mathbb{N}}$ has the following properties:

$$\Phi(X+Y) \leq \Phi X + \Phi Y; \quad \Phi(X\cdot Y) = \Phi X \cdot \Phi Y; \quad \Phi X = 0 \Leftrightarrow X \in \mathfrak{n}.$$

Proof: a) and c) are obvious. b) $\mathfrak{n}$ is an ideal of $\mathcal{C}$, since if $X,Y \in \mathfrak{n}$, $Z \in \mathcal{C}$, then $\Phi(X+Y) \leq \Phi X + \Phi Y = 0$ and $\Phi(X\cdot Z) = \Phi X \cdot \Phi Z = 0$. To prove that $\mathfrak{n}$ is a maximal ideal, it suffices to show that for every $X \in \mathcal{C} \setminus \mathfrak{n}$ there is some $Y \in \mathcal{C}$ such that $(1)_{n\in\mathbb{N}} - X\cdot Y \in \mathfrak{n}$. In fact, we have $\xi = \lim(\varphi x_i)_{i\in\mathbb{N}} > 0$, and choosing $n_o \in \mathbb{N}$ such that $|\varphi x_n - \xi| < \frac{\xi}{2}$ for all $n \geq n_o$, we have $\varphi x_n \geq \xi - |\varphi x_n - \xi| > \frac{\xi}{2}$ for all $n \geq n_o$. Let $y_i = 0$ for $i < n_o$ and $y_i = x_i^{-1}$ for $i \geq n_o$. Then $Y = (y_i)_{i\in\mathbb{N}}$ is φ-Cauchy since $\varphi(y_i - y_j) = \varphi(\frac{x_j - x_i}{x_i \cdot x_j}) \leq \frac{4}{\xi^2} \cdot \varphi(x_j - x_i)$ for $i,j \geq n_o$. It is clear that $(1)_{i\in\mathbb{N}} - X\cdot Y \in \mathfrak{n}$. □

It is practical to use the following definitions. A valued field is a couple (K,φ) consisting of a field K and a valuation

φ of K. We write $(K,\varphi) \subseteq (L,\psi)$ if K is a subfield of L and $\varphi = \psi|K$. More generally, a monomorphism (i.e., an injective homomorphism) λ from K into a field L is called an <u>imbedding</u> of (K,φ) in (L,ψ) if $\psi \circ \lambda = \varphi$; in this case we write $\lambda: (K,\varphi) \to (L,\psi)$. It is obvious that λ is a continuous mapping, when K (resp. L) is endowed with T_φ (resp. T_ψ). A bijective imbedding is called an isomorphism; its inverse is also an imbedding.

We say that (K,φ) is <u>complete</u> if K is φ-complete. A <u>completion</u> of (K,φ) is a complete valued field $(L,\psi) \supseteq (K,\varphi)$ such that K is dense in L with respect to T_ψ. Often it is practical to use a slightly more general definition: A completion of (K,φ) is an imbedding $\lambda: (K,\varphi) \to (L,\psi)$ such that (L,ψ) is complete and λK is dense in L. It is clear that in this case there exists an isomorphism $\iota: (L,\psi) \to (L',\psi')$ such that $(L',\psi') \supseteq (K,\varphi)$ is a completion of (K,φ), in the usual sense.

We first prove the existence of completions, by means of (2.1).

(2.2) THEOREM - <u>Let (K,φ) be a valued field and let $\mathcal{C}$, $\mathfrak{n}$, Φ be as in (2.1). Then Φ induces a valuation $\hat{\varphi}$ of the field $\hat{K} = \mathcal{C}/\mathfrak{n}$, and $\mu: (K,\varphi) \to (\hat{K},\hat{\varphi})$ is a completion, where $\mu x = (x)_{i\in\mathbb{N}} + \mathfrak{n}$ for all $x \in K$.</u>

<u>Proof</u>: 1) For any $X \in \mathcal{C}$ and $Z \in \mathfrak{n}$ we have $\Phi X = \Phi(X+Z-Z) \leq \Phi(X+Z) + \Phi Z = \Phi(X+Z) \leq \Phi X + \Phi Z = \Phi X$, hence $X + \mathfrak{n} = Y + \mathfrak{n}$ implies $\Phi X = \Phi Y$. Therefore Φ induces a mapping $\hat{\varphi}: \mathcal{C}/\mathfrak{n} \to \mathbb{R}_+$, and from the properties of Φ it follows that $\hat{\varphi}$ is a valuation of $\hat{K} = \mathcal{C}/\mathfrak{n}$.

2) μ is clearly a monomorphism such that $\hat{\varphi}(\mu x) = \Phi((x)_{i\in\mathbb{N}}) = \varphi x$ for all $x \in K$.

3) To prove that μK is dense in K it suffices to show that $\lim_{\hat{\varphi}} (\mu x_i)_{i\in\mathbb{N}} = \hat{x}$ for any $\hat{x} = X + \mathfrak{n} \in \hat{K}$. In fact, for every $n \in \mathbb{N}$ we have $\hat{x} - \mu x_n = (x_i - x_n)_{i\in\mathbb{N}} + \mathfrak{n}$, hence $\hat{\varphi}(\hat{x} - \mu x_n) =$

$= \lim(\varphi(x_i - x_n))_{i\in\mathbb{N}} = \alpha_n$, say, and as $X \in C$, we have $(\hat{\varphi}(\hat{x}-\mu x_n))_{n\in\mathbb{N}} =$ $= \lim(\alpha_n)_{n\in\mathbb{N}} = 0.$

4) Let $(\hat{x}_n)_{n\in\mathbb{N}}$ be a $\hat{\varphi}$-Cauchy sequence. Since μK is dense in $\hat{K}$, there exists a sequence $Y = (y_n)_{n\in\mathbb{N}}$ of elements of K such that $\lim_{\hat{\varphi}} (\hat{x}_n - \mu y_n)_{n\in\mathbb{N}} = 0$; hence $(\mu y_n)_{n\in\mathbb{N}}$ is $\hat{\varphi}$-Cauchy, hence $Y \in C$. Let $\hat{y} = (y_n)_{n\in\mathbb{N}} + \mathfrak{n}$; then $\lim_{\hat{\varphi}} (\mu y_n)_{n\in\mathbb{N}} = \hat{y}$ as was proven in 3), hence $\lim_{\hat{\varphi}}(\hat{x}_n)_{n\in\mathbb{N}} = \lim_{\hat{\varphi}}(\hat{x}_n - \mu y_n)_{n\in\mathbb{N}} + \lim_{\hat{\varphi}}(\mu y_n)_{n\in\mathbb{N}} =$ $= \hat{y}$. □

For example, in the case $(K,\varphi) = (\mathbb{Q},\varphi_\infty)$, one gets the valued field $(\mathbb{R}, |\ |_{\mathbb{R}})$ as constructed by Cantor, the rational numbers $x \in \mathbb{Q}$ being identified with the corresponding real numbers $\mu x = (x)_{n\in\mathbb{N}} + \mathfrak{n}$. The completion $(\hat{\mathbb{Q}}_p, \hat{\varphi}_p)$ of $(\mathbb{Q},\varphi_p)$ is called the (valued) <u>field of p-adic numbers</u>; we shall study it in more detail in §5.

We shall prove now a universal property of completions:

(2.3) THEOREM - <u>Let</u> $\mu: (K,\varphi) \to (\hat{K},\hat{\varphi})$ <u>be a completion</u> [4] <u>and</u> $\lambda: (K,\varphi) \to (L,\psi)$ <u>an imbedding of</u> (K,φ) <u>in a complete valued field</u> (L,ψ) . <u>Then there exists one and only one imbedding</u> $\hat{\lambda}: (\hat{K},\hat{\varphi}) \to (L,\psi)$ <u>such that</u> $\lambda = \hat{\lambda}\circ\mu$.

<u>Proof</u>: 1) (Uniqueness). Suppose $\hat{\lambda}_i: (\hat{K},\hat{\varphi}) \to (L,\psi)$ are imbeddings such that $\lambda = \hat{\lambda}_i \circ \mu$ $(i=1,2)$; then $\hat{\lambda}_1$ and $\hat{\lambda}_2$ coincide on the dense subset μK of $\hat{K}$. It follows that $\hat{\lambda}_1 = \hat{\lambda}_2$, since T_ψ is Hausdorff and $\hat{\lambda}_1$, $\hat{\lambda}_2$ are continuous mappings from $\hat{K}$ to L, provided with the topology $T_{\hat{\varphi}}$, T_ψ , respectively.

2) (Existence). For every $\hat{x} \in \hat{K}$ there exists a sequence $X = (x_n)_{n\in\mathbb{N}}$ of elements of K such that $\lim_{\hat{\varphi}}(\mu x_n)_{n\in\mathbb{N}} =$ $= \hat{x}$. So $(\mu x_n)_{n\in\mathbb{N}}$ is $\hat{\varphi}$-Cauchy, hence X is φ-Cauchy, hence $(\lambda x_n)_{n\in\mathbb{N}}$ is ψ-Cauchy. It is easy to see that $\lim_\psi(\lambda x_n)_{n\in\mathbb{N}}$ depends

---

[4] This need not be the completion constructed in (2.3).

only on $\hat{x}$. Therefore a mapping $\hat{\lambda}: \hat{K} \to L$ is well defined by $\hat{\lambda}\hat{x} = \lim_{\psi}(\lambda x_n)_{n\in\mathbb{N}}$. It is clear that it is a monomorphism and satisfies $\lambda = \hat{\lambda}\circ\mu$. Moreover, $\hat{\lambda}: (\hat{K},\hat{\varphi}) \to (L,\psi)$ is an imbedding, since $\psi(\hat{\lambda}\hat{x}) = \psi(\lim_{\psi}(\lambda x_n)_{n\in\mathbb{N}}) = \lim(\psi(\lambda x_n)_{n\in\mathbb{N}}) = \lim(\varphi(x_n))_{n\in\mathbb{N}} = \lim(\hat{\varphi}(\mu x_n)_{n\in\mathbb{N}}) = \hat{\varphi}(\lim_{\hat{\varphi}}(\mu x_n)_{n\in\mathbb{N}}) = \hat{\varphi}\hat{x}$ for all $\hat{x} \in \hat{K}$. □

The universal property yields the uniqueness of a completion up to an isomorphism. In fact, the next two corollaries follow immediately from (2.3).

(2.4) COROLLARY 1 - Let $\mu_i: (K,\varphi) \to (\hat{K}_i,\hat{\varphi}_i)$ be completions $(i=1,2)$. Then there exists exactly one imbedding $\lambda: (\hat{K}_1,\hat{\varphi}_1) \to (\hat{K}_2,\hat{\varphi}_2)$ such that $\mu_2 = \lambda\circ\mu_1$, and this is an isomorphism.

(2.5) COROLLARY 2 - Let $(K,\varphi) \subseteq (\hat{K},\hat{\varphi})$ be a completion of (K,φ) . If $\lambda: (\hat{K},\hat{\varphi}) \to (\hat{K},\hat{\varphi})$ is an imbedding such that $\lambda|K = \iota_K$ then $\lambda = \iota_{\hat{K}}$.[5]

Restricting ourselves to subfields of a given complete valued field (L,ψ) containing (K,φ), the completion of (K,φ) is even unique in the strict sense, as we shall show now:

(2.6) COROLLARY 3 - Let $(K,\varphi) \subseteq (L,\psi)$ and let (L,ψ) be complete. Then the topological closure of K in L (with respect to T_ψ) is the only subfield $\bar{K}$ of L such that $(\bar{K},\psi|\bar{K})$ is a completion of (K,φ).

Proof: The topological closure $\bar{K}$ of K in L is a field, as follows from (1.4). $(\bar{K},\psi|\bar{K})$ is complete, since (L,ψ) is complete and $\bar{K}$ is closed in L . Since K is dense in $\bar{K}$, $(\bar{K},\psi|\bar{K})$ is in fact a completion of (K,φ) . On the other hand, let $\mu: (K,\varphi) \to (\hat{K},\hat{\varphi})$ be any completion of (K,φ) and let $\hat{\lambda}: (\hat{K},\hat{\varphi}) \to (L,\psi)$ be

---

[5] For any sets S, T such that $S \subseteq T$, we write $\iota_{S,T}$ for the identical imbedding $S \to T$. In particular, we set $\iota_S = \iota_{S,S}$.

determined by (2.3) (where $\lambda = \iota_{K,L}$). Then there is only one subfield $\widetilde{K}$ of L such that $(\widetilde{K},\psi|\widetilde{K})$ is a completion of (K,φ), namely $\widetilde{K} = \hat{\lambda}\hat{K}$. □

We are going to investigate the valuations of a field extension L of K which extend a given non-trivial valuation φ of K. We note that if ψ_1 , ψ_2 are equivalent valuations of L which extend φ then $\psi_1 = \psi_2$ (in fact, $\psi_2 = \psi_1^\rho$, $\psi_2|K = \varphi = \psi_1|K$ imply $\rho = 1$).

We first consider the case in which (K,φ) is complete and L a finite extension of K .

(2.7) THEOREM - <u>Let</u> (K,φ) <u>be complete and</u> $L|K$ <u>be a finite field extension. Then there is at most one valuation</u> ψ <u>of</u> L <u>which extends</u> φ, <u>and</u> (L,ψ) <u>is complete</u>.

<u>Proof</u>: Let $y_1,\dots,y_n$ be a basis of $L|K$. Any $z \in L$ will be written as $z = c_1(z)\cdot y_1 +\dots+ c_n(z)\cdot y_n$ where $c_i(z) \in K$ $(i=1,\dots,n)$. Assume that there is a valuation ψ of L which extends φ . Obviously $\psi z \leqq \varphi(c_1(z))\cdot\psi y_1 +\dots+ \varphi(c_n(z))\cdot\psi y_n \leqq$ $\leqq \rho\cdot\max\{\varphi(c_1(z)),\dots,\varphi(c_n(z))\}$ for all $z \in L$, where $\rho = \psi y_1 +\dots+ \psi y_n > 0$. On the other hand, we prove the existence of a real number $\tau > 0$ such that $\max\{\varphi(c_1(z)),\dots,\varphi(c_n(z))\} \leqq \tau\cdot\psi z$ for all $z \in L$. For any subset I of $\{1,\dots,n\}$, let L_I be the set of all $z \in L$ such that $c_j(z) = 0$ for all $j \in \{1,\dots,n\}\setminus I$, and let $V_I = \{\psi(c_i(z)\cdot z^{-1}) \mid z \in L_I\setminus\{0\},\ i \in I\}$. We assume by induction that, for any I with less than r elements, V_I is bounded by some real number $\tau_I > 0$. Suppose that $V_{\{1,\dots,r\}}$ is not bounded. Then there is a sequence $(z_k)_{k\in\mathbb{N}}$ of non-zero elements $z_k \in L_{\{1,\dots,r\}}$ such that $\lim(\psi(c_j(z_k)\cdot z_k^{-1}))_{k\in\mathbb{N}} = \infty$ for some $j \in \{1,\dots,r\}$, say, for $j = r$. Let $J = \{1,\dots,r-1\}$. Since V_J is bounded, by hypothesis, we have $c_r(z_k) \neq 0$ for almost all $k \in \mathbb{N}$. Replacing $(z_k)_{k\in\mathbb{N}}$ by an appropriate subsequence, we may assume that $c_r(z_k) \neq 0$ for all

$k \in \mathbb{N}$. Let $u_k = z_k \cdot c_r(z_k)^{-1}$ for any $k \in \mathbb{N}$. Then $\lim(\psi u_k)_{k\in\mathbb{N}} = 0$, hence $(u_k)_{k\in\mathbb{N}}$ is a ψ-null sequence. For all $k \in \mathbb{N}$ we have $c_r(u_k) = 1$, $c_{r+1}(u_k) = \ldots = c_n(u_k) = 0$, hence $c_j(u_k - u_h) = c_j(u_k) - c_j(u_h) = 0$ for all $k, h \in \mathbb{N}$, $j \in \{r,\ldots,n\}$. Therefore $u_k - u_h \in L_J$ for all $k, h \in \mathbb{N}$ and, by hypothesis, $\varphi(c_i(u_k - u_h)) \leqq \tau_J \cdot \psi(u_k - u_h)$ for all $i \in J$. We conclude that, for any $i \in J$, the sequence $(c_i(u_k))_{k\in\mathbb{N}}$ is φ-Cauchy, with φ-limit $d_i \in K$ (say). It follows that the non-zero element $d_1 \cdot y_1 + \ldots + d_{r-1} \cdot y_{r-1} + y_r$ of L is a ψ-limit of the ψ-null sequence $(u_k)_{k\in\mathbb{N}}$, a contradiction. Therefore $V_{\{1,\ldots,r\}}$ is bounded and so is $V_{J'}$, for any subset J' of $\{1,\ldots,n\}$ consisting of r elements, as is proven similarly. For $r = n$ we conclude that $V_{\{1,\ldots,n\}}$ is bounded by τ (say); therefore $\max\{\varphi(c_1(z)),\ldots,\varphi(c_n(z))\} \leqq \tau \cdot \psi z$ for any $z \in L$.

Let $(z_k)_{k\in\mathbb{N}}$ be any ψ-Cauchy sequence of elements $z_k \in L$. For any $k, h \in \mathbb{N}$ and $i \in \{1,\ldots,n\}$ we have $\varphi(c_i(z_k) - c_i(z_h)) \leqq \tau \cdot \psi(z_k - z_h)$; therefore, for any $i \in \{1,\ldots,n\}$, $(c_i(z_k))_{k\in\mathbb{N}}$ is a φ-Cauchy sequence, with φ-limit $d_i \in K$ (say). It follows that $d_1 \cdot y_1 + \ldots + d_n \cdot y_n$ is a ψ-limit of $(z_k)_{k\in\mathbb{N}}$. Therefore (L,ψ) is complete.

Assume that ψ_1, ψ_2 are valuations of L which extend φ. We have just proved the existence of positive real numbers ρ_1, τ_2 such that $\psi_1 z \leqq \rho_1 \cdot \max\{\varphi(c_1(z)),\ldots,\varphi(c_n(z))\} \leqq \rho_1 \cdot \tau_2 \cdot \psi_2 z$ for any $z \in L$; therefore the topology of L defined by ψ_2 is stronger than that defined by ψ_1. From (1.5) we conlcude that ψ_1, ψ_2 are equivalent, hence $\psi_1 = \psi_2$. □

Part of (2.7) can be generalized to infinite algebraic extensions. In fact, since any algebraic extension is the union of its finite subextensions, we get as an immediate consequence of (2.7):

(2.8) COROLLARY - Let (K,φ) be complete and $L|K$ an algebraic extension. Then there exists at most one valuation ψ of L which extends φ.

Note, however, that (L,ψ) is not complete, in general.

In theorem (2.7) and its corollary (2.8), "at most" may be replaced by "exactly". In fact, the existence of a valuation of L which extends φ will be proven in §13 in the case of non-archimedean (and, more generally, in the case of Krull) valuations, even without assuming the completeness of (K,φ) [6]. In the archimedean complete case, the existence will follow immediately from Ostrowski's theorem, which in turn is a consequence of the following lemma.

(2.9) Let (L,ψ) be a valued field such that $(\mathbb{R}, |\ |_{\mathbb{R}}) \subseteq (L,\psi)$. Then $L|\mathbb{R}$ is algebraic (hence $L = \mathbb{R}$ or $L = \mathbb{C}$).

For the proof of this lemma we refer to Weiss [35], lemma 1-8-2.

(2.10) THEOREM (Ostrowski) - Any complete valued field (K,φ) such that φ is archimedean is isomorphic to $(\mathbb{R}, |\ |_{\mathbb{R}}^{\rho})$ or $(\mathbb{C}, |\ |_{\mathbb{C}}^{\rho})$ for some $\rho > 0$.

Proof: By (1.13), (1.11) and (1.16), there exists an imbedding $(\mathbb{Q},\varphi_{\infty}^{\rho}) \to (K,\varphi)$, for some $\rho > 0$. Since $(\mathbb{R},|\ |_{\mathbb{R}}^{\rho})$ is the completion of $(\mathbb{Q},\varphi_{\infty}^{\rho})$, this imbedding extends to an imbedding $(\mathbb{R},|\ |_{\mathbb{R}}^{\rho}) \to$ $\to (K,\varphi)$, by (2.3), which in turn extends to an isomorphism $(L,\psi) \to$ $\to (K,\varphi)$ for some valued field $(L,\psi) \supseteq (\mathbb{R},|\ |_{\mathbb{R}}^{\rho})$. By (2.9), (L,ψ) equals $(\mathbb{R},|\ |_{\mathbb{R}}^{\rho})$ or $(\mathbb{C},|\ |_{\mathbb{C}}^{\rho})$. □

Another proof of Ostrowski's theorem, can be found in Bachman [1]. We get as an immediate consequence of (2.10):

(2.11) COROLLARY - Let (K,φ) be any valued field such that φ is archimedean. Then there exists either a completion $(K,\varphi) \to$ $(\mathbb{R},|\ |_{\mathbb{R}}^{\rho})$ or a completion $(K,\varphi) \to (\mathbb{C},|\ |_{\mathbb{C}}^{\rho})$, for some $\rho > 0$.

In the first case, φ is called real-archimedean, in the

[6] Another proof in the non-archimedean complete case uses Hensel's lemma.

second case, complex-archimedean. For example, the valuations φ_∞ of $\mathbb{Q}$ and $|\ |_{\mathbb{R}}$ of $\mathbb{R}$ are real-archimedean, whereas $|\ |_{\mathbb{C}}$ is complex-archimedean.

For an arbitrary valued field (K,φ) and any finite separable (hence simple) extension L of K, the determination of the valuations of L which extend φ can be reduced to the complete case, as will be shown in the following theorem. In fact, let $(\hat{K},\hat{\varphi})$ be a completion of (K,φ), Ω a separable closure of $\hat{K}$, and ω the unique valuation of Ω which extends $\hat{\varphi}$ (cf. (2.8) and the remark following it). In particular, we have $\omega\circ\hat{\sigma} = \omega$ for any $\hat{K}$-automorphism $\hat{\sigma}$ of Ω. For any subfield $\hat{L}$ of Ω, we write $(\hat{L},\hat{\varphi})$ instead of $(\hat{L}, \omega|\hat{L})$.

(2.12) THEOREM - *Let $L = K(y)$ be a finite separable extension of K, of degree n, and let $P_{y|K} = \hat{P}_1 \cdot \ldots \cdot \hat{P}_r$ be the factorization of the minimal polynomial $P_{y|K}$ of y over K in (necessarily distinct) irreducible monic polynomials in $\hat{K}[X]$. For any $i \in \{1,\ldots,r\}$, let $\hat{y}_i \in \Omega$ be a root of $\hat{P}_i$, and let $\lambda_i: L \to \hat{K}(\hat{y}_i)$ be the K-monomorphism determined by $\lambda_i y = \hat{y}_i$.*

Then there exist exactly r valuations of L which extend φ, namely $\psi_i = \omega\circ\lambda_i$ $(i=1,\ldots,r)$, and $\lambda_i: (L,\psi_i) \to (\hat{K}(\hat{y}_i),\hat{\varphi})$ is a completion for any $i \in \{1,\ldots,r\}$. Moreover, $\sum_{i=1}^{r} [\hat{K}(\hat{y}_i):\hat{K}] = n$.

Proof: Let $\mathfrak{Y}$ be the set of all roots $\hat{y} \in \Omega$ of $P_{y|K}$ and, for any $\hat{y} \in \mathfrak{Y}$, let $\lambda_{\hat{y}}: L \to \hat{K}(\hat{y})$ be the K-monomorphism determined by $\lambda_{\hat{y}} y = \hat{y}$. Then $\psi_{\hat{y}} = \omega\circ\lambda_{\hat{y}}$ is a valuation of L which extends φ, $\lambda_{\hat{y}}: (L,\psi_{\hat{y}}) \to (\hat{K}(\hat{y}),\hat{\varphi})$ is an imbedding and, by (2.7), $(\hat{K}(\hat{y}),\hat{\varphi})$ is complete. Moreover, $\lambda_{\hat{y}}L = K(\hat{y})$ is dense in $\hat{K}(\hat{y})$ since $\hat{K}$ is the topological closure of K in $\hat{K}(\hat{y})$, by (2.6); therefore $\lambda_{\hat{y}}: (L,\psi_{\hat{y}}) \to (\hat{K}(\hat{y}),\hat{\varphi})$ is a completion. We show that any extension ψ of φ to L is equal to $\psi_{\hat{y}}$ for some $\hat{y} \in \mathfrak{Y}$. In fact, let $(\tilde{L},\tilde{\psi})$ be a completion of (L,ψ). By (2.3) there is an imbedding $\mu: (\hat{K},\hat{\varphi}) \to (\tilde{L},\tilde{\psi})$ such that $\mu|K = \iota_K$. Let $L' = (\mu\hat{K})(y)$; since $[L':\mu\hat{K}] \leqq [L:K] = n < \infty$, $(L',\tilde{\psi}|L')$ is complete, by (2.7). Since $(L,\psi) \subseteq (L',\tilde{\psi}|L') \subseteq (\tilde{L},\tilde{\psi})$, we conclude that $L' = \tilde{L}$; hence $\tilde{L}|\mu\hat{K}$ is a finite extension. The K-isomorphism $\mu^{-1}:\mu\hat{K} \to \hat{K}$ extends to a K-isomorphism $\tilde{\lambda}$ from $\tilde{L} = (\mu\hat{K})(y)$ onto a subfield $\hat{K}(\hat{y})$ of Ω, where $\hat{y} = \tilde{\lambda}y \in \Omega$, and obviously $\hat{y} \in \mathcal{Y}$. The valuation $\tilde{\psi}\circ\tilde{\lambda}^{-1}$ of $\hat{K}(\hat{y})$ extends $\hat{\varphi}$ and therefore coincides with $\omega|\hat{K}(\hat{y})$; hence $\tilde{\psi} = \omega\circ\tilde{\lambda}$ and $\psi = \tilde{\psi}|L = (\omega\circ\tilde{\lambda})|L = \omega\circ\lambda_{\hat{y}}$.

Let $\hat{y}_i \in \mathcal{Y}$, $\lambda_i = \lambda_{\hat{y}_i}$, $\psi_i = \psi_{\hat{y}_i}$ (i=1,2). If $\hat{y}_1$, $\hat{y}_2$ are roots of the same irreducible factor $\hat{P} \in \hat{K}[X]$ of $P_{y|K}$, then there is a $\hat{K}$-automorphism $\hat{\sigma}$ of Ω such that $\hat{\sigma}\hat{y}_1 = \hat{y}_2$, hence $\hat{\sigma}\circ\lambda_1 = \lambda_2$, $\psi_1 = \omega\circ\lambda_1 = \omega\circ\hat{\sigma}\circ\lambda_1 = \omega\circ\lambda_2 = \psi_2$. On the other hand, if $\psi_1 = \psi_2$ then, by (2.4), there exists an isomorphism $\mu: (\hat{K}(\hat{y}_1),\hat{\varphi}) \to (\hat{K}(\hat{y}_2),\hat{\varphi})$ such that $\mu\circ\lambda_1 = \lambda_2$; in particular, μ is a $\hat{K}$-isomorphism and $\mu\hat{y}_1 = \hat{y}_2$. Since K is dense in $\hat{K}$, μ is even a $\hat{K}$-isomorphism; therefore $\hat{y}_1$, $\hat{y}_2$ are roots of the same irreducible factor $\hat{P} \in \hat{K}[X]$ of $P_{y|K}$. This completes the proof of the first statements of the theorem. Its last statement is trivial. □

For non-archimedean (and, more generally, for Krull) valuations, this theorem will be generalized to arbitrary algebraic field extensions in § 7.

In the case of an archimedean valuation φ of K, the hypothesis of separability is superfluous, since K has characteristic zero, by (1.13). In this case, we get the following corollary:

(2.13) COROLLARY - Let φ be an archimedean valuation of K, $L|K$ be a finite extension of degree n, and let r be the number of valuations of L which extend φ.

a) If φ is complex-archimedean, then $r = n$ and all valuations of L which extend φ are complex-archimedean.

b) If φ is real-archimedean, then $\frac{1}{2}n \leqq r \leqq n$, and $2r - n$ of

<u>the valuations of</u> L <u>lying over</u> φ <u>are real-archimedean, whereas the other</u> n-r <u>ones are complex-archimedean.</u>

<u>Proof</u>: We may assume $L = K(y)$ and use the notations of (2.12).

a) If φ is complex-archimedean then $\hat{K} \cong \mathbb{C}$, hence $\hat{P}_1,\dots,\hat{P}_r$ have degree 1, n=r, and $\hat{K}(\hat{y}_1) = \dots = \hat{K}(\hat{y}_r) = \hat{K} \cong \mathbb{C}$.

b) If φ is real-archimedean then $\hat{K} \cong \mathbb{R}$, and we may assume that $\hat{P}_1,\dots,\hat{P}_s$ have degree 1, whereas $\hat{P}_{s+1},\dots,\hat{P}_r$ have degree 2, for some $s \in \{0,\dots,r\}$. Therefore $2r \geqq n = s + 2(r-s) = 2r - s \geqq r$, hence $\frac{1}{2} n \leqq r \leqq n$, $s = 2r - n$, $r-s = n-r$; moreover $\hat{K}(\hat{y}_1) = \dots = \hat{K}(\hat{y}_2) = \hat{K} \cong \mathbb{R}$, $\hat{K} \subset \hat{K}(\hat{y}_{s+1}) = \dots = \hat{K}(\hat{y}_r) \cong \mathbb{C}$. □

As to the question of extending valuations to transcendental field extensions, we want to mention (without proof) that any non-archimedean valuation φ of K extends to any field extension L of K. (A similar statement will be proven for Krull valuations in §13.) For example, if $L = K(x)$, x transcental over K, then one of the valuations ψ of L which extend φ is defined by

$$\psi\left(\frac{a_o + a_1\cdot x + \dots + a_i\cdot x^i}{b_o + b_1\cdot x + \dots + b_j\cdot x^j}\right) = \frac{\max\{\varphi a_o,\dots,\varphi a_i\}}{\max\{\varphi b_o,\dots,\varphi b_j\}}, (a_o,\dots,a_i;b_o,\dots,b_j \in K).$$

On the other hand, archimedean valuations do not extend, in general, to transcendental extensions, as follows from (2.9).

§3 Non-archimedean valuations

In this section, φ is always a non-archimedean valuation of a field K.

We consider first the topology T_φ of K. Let $\mathcal{U}'_\epsilon(x) = \{y \in K \mid \varphi(y-x) \leqq \epsilon\}$. It is clear that $\mathcal{U}_\epsilon(x) \subseteq \mathcal{U}'_\epsilon(x)$ for any $\epsilon > 0$ and that the equality holds if and only if $\epsilon \notin \varphi K$.

(3.1) <u>For any</u> $\epsilon > 0$, $\mathcal{U}_\epsilon(0)$ <u>and</u> $\mathcal{U}'_\epsilon(0)$ <u>are open and closed sub-</u>

<u>groups of the additive group</u> K^+ <u>of</u> K. <u>For any</u> $\varepsilon > 0$ <u>and</u> $x \in K$, $\mathcal{U}_\varepsilon(x)$ <u>and</u> $\mathcal{U}'_\varepsilon(x)$ <u>are open and closed neighborhoods of</u> x.

<u>Proof</u>: $\mathcal{U}_\varepsilon(0)$ and $\mathcal{U}'_\varepsilon(0)$ are subgroups of K^+ because of (1.14) iv).

For any $x \in \mathcal{U}'_\varepsilon(0)$ we have $\mathcal{U}_\varepsilon(x) \subseteq \mathcal{U}'_\varepsilon(0)$, since $\varphi x \leqq \varepsilon$, $\varphi(y-x) < \varepsilon$ imply $\varphi y \leqq \varepsilon$, hence $\mathcal{U}'_\varepsilon(0)$ is open. We know already that $\mathcal{U}_\varepsilon(0)$ is open, too. $\mathcal{U}_\varepsilon(0)$ and $\mathcal{U}'_\varepsilon(0)$ are also closed, since any open subgroup of a topological group is closed [7]. The rest follows from the fact that the translation $y \to x+y$ is a homeomorphism and maps $\mathcal{U}_\varepsilon(0)$, $\mathcal{U}'_\varepsilon(0)$ onto $\mathcal{U}_\varepsilon(x)$, $\mathcal{U}'_\varepsilon(x)$, respectively. □

Since the connected component of any $x \in K$ is contained in any open and closed neighborhood of x, we conclude from (3.1):

(3.2) <u>The topological field</u> K (<u>with respect to</u> T_φ) <u>is totally disconnected</u>.

Let $\mathbb{R}_{disc}$ be the field of real numbers endowed with the discrete topology. From (1.15) it follows that $\varphi(\mathcal{U}_{\varphi x}(x)) = \{\varphi x\}$ for any non-zero $x \in K$, hence:

(3.3) <u>The restriction of</u> φ <u>to</u> $K \setminus \{0\}$ <u>is a continuous mapping into</u> $\mathbb{R}_{disc}$.

Let $(\hat{K},\hat{\varphi})$ be a completion of (K,φ) and let $\hat{\mathcal{U}}_\varepsilon$, $\hat{\mathcal{U}}'_\varepsilon$ refer to the valuation $\hat{\varphi}$, which of course is non-archimedean, too. We show,

(3.4) <u>For any</u> $\varepsilon > 0$, $\hat{\mathcal{U}}_\varepsilon(0)$ (<u>resp.</u> $\hat{\mathcal{U}}'_\varepsilon(0)$) <u>is the topological closure of</u> $\mathcal{U}_\varepsilon(0)$ (<u>resp.</u> $\mathcal{U}'_\varepsilon(0)$) <u>in the topological field</u> $\hat{K}$.

<u>Proof</u>: Since $\mathcal{U}_\varepsilon(0) \subseteq \hat{\mathcal{U}}'_\varepsilon(0)$ and $\hat{\mathcal{U}}_\varepsilon(0)$ is closed in $\hat{K}$, $\hat{\mathcal{U}}_\varepsilon(0)$ contains the topological closure $\overline{\mathcal{U}_\varepsilon(0)}$ of $\mathcal{U}_\varepsilon(0)$ in $\hat{K}$. On the other hand, let $\hat{x} \in \hat{\mathcal{U}}_\varepsilon(0)$, $\hat{x} \neq 0$. There exists a sequence

---

[7] In fact, if H is an open subgroup of the topological group G, then $G \setminus H$ is equal to the union of open co-sets $\sigma \cdot H$, where $\sigma \in G \setminus H$. Therefore $G \setminus H$ is open and H is closed.

$(x_i)_{i\in\mathbb{N}}$ of elements of K which is $\hat{\varphi}$-convergent to $\hat{x}$, and from (3.3) we conclude that for sufficiently large $n \in \mathbb{N}$ we have $\varphi x_n = \hat{\varphi}\hat{x}$, hence $x_n \in \mathfrak{U}_e(0)$; therefore, $\hat{x} \in \overline{\mathfrak{U}_e(0)}$. The statement for $\mathfrak{U}_e'(0)$, $\hat{\mathfrak{U}}_e'(0)$ is proven similarly. □

It is usual and sometimes a little more convenient to write non-archimedean valuations additively or, more precisely, to replace them by "exponential valuations." This is possible because non-archimedean valuations can be defined by (V_1), (V_2), and (1.14) iv), in which no addition of values is needed [8]. Thus $\mathbb{R}_+\setminus\{0\}$ may be considered merely as multiplicative group, totally ordered in the natural manner, and this group is order-isomorphic to the additive group $\mathbb{R}^+$ of the field $\mathbb{R}$ by means of the logarithm function.

We define an <u>exponential valuation</u> v of K as a mapping $v\colon K \to \mathbb{R} \cup \{\infty\}$ satisfying the following conditions:

(E_1) $vx = \infty \Leftrightarrow x = 0$, for all $x \in K$

(E_2) $v(x\cdot y) = vx + vy$, for all $x,y \in K$

(E_3) $v(x+y) \geqq \min\{vx,vy\}$, for all $x,y \in K$

(with the usual rules for the symbol ∞), and we see immediately:

(3.5) <u>A 1-1 correspondence between the non-archimedean valuations φ of K and the exponential valuations v of K is given by</u>

$$v \mapsto \varphi = e^{-v} , \quad \varphi \mapsto v = -\log \varphi$$

(<u>with the convention</u> $e^{-\infty} = 0$, $-\log 0 = \infty$).

It is clear that under this 1-1 correspondence the trivial valuation τ corresponds to the trivial exponential valuation which maps all non-zero elements of K onto 0. Moreover, considering two

---

[8] The same argument can be used for the more general valuation defined by (V_1), (V_2) and (V_3'). These also can be written additively.

exponential valuations v, w of K equivalent if $v = \rho \cdot w$ for some $\rho > 0$, it is clear that equivalent non-archimedean valuations correspond to equivalent exponential valuations. Finally we shall use for exponential valuations v the notions we have introduced for valuations; in particular, we shall speak of v-convergent, v-Cauchy etc. Moreover, setting $\mathfrak{w}_\gamma(x) = \{y \in K \mid v(y-x) > \gamma\}$ and $\mathfrak{w}'_\gamma(x) = \{y \in K \mid v(y-x) \geqq \gamma\}$, we have $\mathfrak{w}_\gamma(x) = \mathfrak{u}_\varepsilon(x)$ and $\mathfrak{w}'_\gamma(x) = \mathfrak{u}'_\varepsilon(x)$ for $\gamma = -\log \varepsilon$.

For example, we mention the <u>p-adic exponential valuation</u> v_p of $\mathbb{Q}$ (p a prime number), uniquely determined by $v_p p = 1$ and $v_p q = 0$ for any prime number q, $q \neq p$. To any non-zero $x = \pm \prod_{q \in \mathcal{P}} q^{n_q} \in \mathbb{Q}$, v_p assigns its p-exponent $n_p \in \mathbb{Z}$ (this explains the name "exponential valuation"). Note that v_p does not correspond to the p-adic valuation φ_p under the correspondence indicated in (3.5), but to some valuation equivalent to φ_p.

A subring A of a field K is called a <u>valuation ring</u> of K, if $x \in A$ or $x^{-1} \in A$ for any non-zero $x \in K$. It is clear that K is the quotient field of A. We shall study valuation rings in §6. Here we show only:

(3.6) <u>Let</u> v <u>be an exponential valuation of</u> K. <u>Then:</u>

a) $A_v = \mathfrak{w}'_0(0) = \{x \in K \mid vx \geqq 0\}$ <u>is a valuation ring of</u> K.

b) $\mathfrak{M}_v = \mathfrak{w}_0(0) = \{x \in K \mid vx > 0\}$ <u>is the only maximal ideal of</u> A_v.

c) $U_v = A_v \setminus \mathfrak{M}_v$ <u>is the multiplicative group of all units of</u> A_v.

d) v <u>is trivial</u> $\Leftrightarrow A_v = K \Leftrightarrow \mathfrak{M}_v = \{0\} \Leftrightarrow U_v = K \setminus \{0\}$.

<u>Proof</u>: a) and c) are trivial. b) Obviously $\mathfrak{M}_v$ is an ideal of A_v and consists of all non-units of A_v; hence $\mathfrak{M}_v$ is the only maximal ideal of A_v. d) is trivial. □

We call A_v <u>the ring of</u> v and $\mathfrak{M}_v$ the <u>ideal of</u> v. It is obvious that $\{\mathfrak{w}_\gamma(0) \mid \gamma \in \mathbb{R}_+\} \cup \{\mathfrak{w}'_\gamma(0) \mid \gamma \in \mathbb{R}_+\}$ is the set of all non-zero ideals of A_v and $\{\mathfrak{w}_\gamma(0) \mid \gamma \in \mathbb{R}\} \cup \{\mathfrak{w}'_\gamma(0) \mid \gamma \in \mathbb{R}\}$ the set of all fractional ideals of A_v [2]. The field $\mathcal{K}_v = A_v/\mathfrak{M}_v$ is called the <u>residue field</u> of v. We denote by $\kappa_v\colon A_v \to \mathcal{K}_v$ the canonical homomorphism; κ_v is surjective and has kernel $\mathfrak{M}_v$. Note that two exponential valuations v, w of K are equivalent if and only if $A_v = A_w$, if and only if $\mathfrak{M}_v = \mathfrak{M}_w$, if and only if $U_v = U_w$, as follows easily from (1.5); in this case we have also $\mathcal{K}_v = \mathcal{K}_w$ and $\kappa_v = \kappa_w$.

The image vK^* of the multiplicative group K^* of K is denoted by Γ_v and called the <u>value group</u> of v; it is a subgroup of the additive group $\mathbb{R}^+$ of the field $\mathbb{R}$. If v, w are equivalent, say $w = \rho \cdot v$, where $\rho > 0$, then the multiplication by ρ maps Γ_v isomorphically onto Γ_w.

In general, there exist valuation rings of K which are not rings of exponential valuations, as we shall see in §6.

The next two statements are obvious:

(3.7) <u>For all</u> $x,y \in A_v$ <u>we have</u> $v(x-y) > 0 \Leftrightarrow x-y \in \mathfrak{M}_v \Leftrightarrow \kappa_v(x-y) = 0 \Leftrightarrow \kappa_v x = \kappa_v y$.

(3.8) <u>Let</u> K_o <u>be a subfield of</u> K. <u>Then</u> $v|K_o$ <u>is trivial if and only if</u> $K_o \subseteq A_v$, <u>if and only if</u> $K_o \cap \mathfrak{M}_v = \{0\}$. <u>In this case, the restriction of</u> κ_v <u>to</u> K_o <u>is a monomorphism.</u>

In §5 we shall use the following proposition.

(3.9) <u>Let</u> $\mathcal{K}_v$ <u>have characteristic</u> $p \neq 0$. <u>Then for all</u> $x,y \in A_v$, $v(x-y) > 0$ <u>implies</u> $v(x^p-y^p) > v(x-y)$.

---

[2] A fractional ideal of a domain R is a non-zero R-submodule $\mathfrak{A}$ of its quotient field K such that $x \cdot \mathfrak{A} \subseteq R$ for some non-zero $x \in K$.

<u>Proof</u>: $x^p = (y + (x-y))^p = y^p + (p\cdot c + (x-y)^{p-1})\cdot(x-y)$ for some $c \in A_v$. Since $\kappa_v(p\cdot c) = p\cdot\kappa_v c = 0$ we have $v(p\cdot c + (x-y)^{p-1}) \geq \min\{v(p\cdot c),(p-1)\cdot v(x-y)\} > 0$ and therefore $v(x^p-y^p) > v(x-y)$. □

The characteristic of $\mathcal{K}_v$ may or may not coincide with the characteristic of K. In fact:

(3.10) <u>The following conditions are equivalent:</u>

(i) Char $\mathcal{K}_v \neq$ Char K. [10]

(ii) <u>The restriction of</u> v <u>to the prime field of</u> K <u>is not trivial.</u>

(iii) <u>For some prime</u> p <u>and some</u> $\rho > 0$ <u>there is an imbedding</u> $(\mathbb{Q},\rho\cdot v_p) \to (K,v)$.

<u>In this case</u> Char $K = 0$ <u>and</u> Char $\mathcal{K}_v = p$.

<u>Proof</u>: (i) ⇒ (ii) follows from (3.8). (ii) ⇒ (iii): By (1.2) there is an isomorphism λ from $\mathbb{Q}$ onto the prime field of K, and by (1.16), $v\circ\lambda = \rho\cdot v_p$ for some $\rho > 0$ and some prime number p. Hence $(\mathbb{Q},\rho\cdot v_p) \to (K,v)$ is an imbedding. (iii) ⇒ (i): We have Char K = Char $\mathbb{Q} = 0$. Since $v(p\cdot 1) = v(\lambda p) = \rho\cdot v_p p > 0$ we have $p\cdot\kappa_v 1 = \kappa_v(p\cdot 1) = 0$, hence Char $\mathcal{K}_v = p$. □

Let K_o be a subfield of K and v_o the restriction $v|K_o$ of the exponential valuation v of K. It is clear that:

(3.11) $A_{v_o} = A_v \cap K_o$, $U_{v_o} = U_v \cap K_o$, $\mathfrak{M}_{v_o} = \mathfrak{M}_v \cap K_o$, <u>and there is one and only one monomorphism</u> $\iota: \mathcal{K}_{v_o} \to \mathcal{K}_v$ <u>such that</u> $\kappa_v = \iota\circ\kappa_{v_o}$.

Often $\mathcal{K}_{v_o}$ is identified with its image $\iota\mathcal{K}_{v_o}$, i.e. $\mathcal{K}_{v_o}$ is considered as a subfield of $\mathcal{K}_v$. If $\mathcal{K}_{v_o} = \mathcal{K}_v$ (or more precisely, $\iota\mathcal{K}_{v_o} = \mathcal{K}_v$) then we say that v and v_o have the same residue field. Let Γ_v (resp. Γ_{v_o}) be the value group of v (resp. v_o); it is

---

[10] We write Char K for the characteristic of a field K.

clear that $\Gamma_{v_o} \subseteq \Gamma_v$. We say that (K,v) is an <u>immediate extension</u> of (K_o,v_o) if v and v_o have the same value group (i.e., $\Gamma_{v_o} = \Gamma_v$) and the same residue field. We show:

(3.12) THEOREM - *Let $(\hat{K},\hat{v})$ be a completion of (K,v). Then $(\hat{K},\hat{v})$ is an immediate extension of (K,v), and $A_{\hat{v}}$ (resp. $\mathfrak{M}_{\hat{v}}$) is the topological closure of A_v (resp. $\mathfrak{M}_v$) in $\hat{K}$.*

Proof:[11] For any non-zero $\hat{x} \in \hat{K}$ there exists some $x \in K$ such that $\hat{v}(\hat{x}-x) > \hat{v}\hat{x}$, hence $vx = \hat{v}(\hat{x} - (\hat{x}-x)) = \hat{v}\hat{x}$. For any $\hat{x} \in A_{\hat{v}}$ there exists some $x \in K$ such that $\hat{x} - x \in \mathfrak{M}_{\hat{v}}$, hence $x \in A_{\hat{v}} \cap K = A_v$ and $\kappa_v x = \kappa_{\hat{v}} x = \kappa_{\hat{v}}\hat{x}$. Therefore the first statement holds. The second statement follows from (3.6) and (3.4). □

We finish this section giving a stronger version of the approximation theorem (1.8) for exponential valuations.

(3.13) THEOREM - *Let $v_1,\dots,v_n$ be pairwise inequivalent exponential valuations of K. For any $x_1,\dots,x_n \in K$ and $\gamma_1 \in \Gamma_{v_1},\dots, \gamma_n \in \Gamma_{v_n}$ there exists an $x \in K$ such that $v_i(x-x_i) = \gamma_i$ for $i = 1,\dots,n$.*

Proof: There is at most one trivial exponential valuation in $\{v_1,\dots,v_n\}$. We may assume that v_1 is trivial, hence $\gamma_1 = 0$. We choose $y_2,\dots,y_n \in K$ such that $v_iy_i = \gamma_i$ for $i = 2,\dots,n$. By (1.8), there exist $z,y \in K$ such that $v_i(z-x_i) > \gamma_i$ and $v_i(y-y_i) > \gamma_i$ for $i = 2,\dots,n$, and clearly z and y can be chosen such that $y + z \neq x_1$; hence $v_1((y+z) - x_1) = 0 = \gamma_1$. For $i = 2,\dots,n$ we have $v_iy = v_i((y-y_i) + y_i) = \gamma_i$ and $v_i(y + (z-x_i)) = \gamma_i$. Therefore $x = y + z$ has the desired property. □

The following corollary is sometimes called the "Independence Theorem".

---

[11] (3.12) follows also immediately from (3.3) and (3.4).

(3.14) COROLLARY - Let $v_1,\dots,v_n$ be as in (3.13). For any $\gamma_1 \in \Gamma_{v_1},\dots,\gamma_n \in \Gamma_{v_n}$ there exists some $x \in K$ such that $v_i x = \gamma_i$ for $i = 1,\dots,n$.

§4 Discrete exponential valuations

An exponential valuation v of K is called discrete, if it is non-trivial and its value group $\Gamma_v = vK^*$ is a discrete subspace of $\mathbb{R}$ (endowed with its usual topology). The following proposition shows that the discrete exponential valuations are those with value groups of the form $\rho\cdot\mathbb{Z}$ for some $\rho > 0$.

(4.1) Let Γ be a non-trivial subgroup of $\mathbb{R}$. The following conditions are equivalent:

(i) Γ is a discrete subspace of $\mathbb{R}$.

(ii) Γ is not dense in $\mathbb{R}$.

(iii) $\{\gamma \in \Gamma \mid \gamma > 0\}$ has a least element.

(iv) $\Gamma = \rho\cdot\mathbb{Z}$ for some $\rho > 0$.

Proof: (i) ⇒ (ii) is trivial. (ii) ⇒ (iii): If (iii) does not hold, then for every $\varepsilon > 0$ there is some $\gamma \in \Gamma$ such that $0 < \gamma \leqq \varepsilon$, and for every $\rho \in \mathbb{R}$ there exists an $n \in \mathbb{Z}$ such that $n\cdot\gamma \leqq \rho < (n+1)\cdot\gamma$, hence $0 \leqq \rho - n\cdot\gamma < \gamma \leqq \varepsilon$. Therefore Γ is dense in $\mathbb{R}$. (iii) ⇒ (iv): Let γ_o be the least element of $\{\gamma \in \Gamma \mid \gamma > 0\}$. For any $\gamma \in \Gamma$ there is an $n \in \mathbb{Z}$ such that $n\cdot\gamma_o \leqq \gamma < (n+1)\cdot\gamma_o$, hence $0 \leqq \gamma - n\cdot\gamma_o < \gamma_o$, hence $\gamma = n\cdot\gamma_o$ because of the choice of γ_o . Hence $\Gamma = \gamma_o\cdot\mathbb{Z}$. (iv) ⇒ (i): For any $\gamma \in \Gamma$, the intersection of the open interval $(\gamma-\rho, \gamma+\rho)$ with Γ is equal to $\{\gamma\}$. □

An exponential valuation v of K is called (discrete and) normalized, if its value group is equal to $\mathbb{Z}$. From (4.1) we conclude:

(4.2) Any discrete exponential valuation is equivalent to exactly one normalized exponential valuation.

Since equivalent exponential valuations are "essentially equal", the study of discrete valuations can be reduced to that of normalized exponential valuations.

As an example of a normalized exponential valuation we recall the p-adic exponential valuation v_p of $\mathbb{Q}$, where p is a prime number. We know also from Theorem (1.16) that any non-trivial exponential valuation of $\mathbb{Q}$ is equivalent to a p-adic valuation. We generalize now this statement to any field K which is the quotient field of some PID (principle ideal domain). More generally, we prove for any UFD (unique factorization domain):

(4.3) THEOREM - Let R be a UFD with quotient field K and let $\mathcal{P}$ be a set of representatives of the irreducible elements of R (i.e., any irreducible element of R is associate to exactly one $p \in \mathcal{P}$). Then:

a) For any $p \in \mathcal{P}$, the mapping $v_p: K \to \mathbb{R} \cup \{\infty\}$ defined by $v_p 0 = \infty$ and $v_p(u\cdot \prod_{q\in\mathcal{P}} q^{n_q}) = n_p$ (where $n_q \in \mathbb{Z}$, u a unit of R) is a normalized exponential valuation of K , with $A_v =$ $= R_{p\cdot R} \supseteq R$, $\mathfrak{M}_{v_p} = p\cdot R_{p\cdot R}$, $p\cdot R = \mathfrak{M}_{v_p} \cap R$, $R/p\cdot R \cong \kappa_{v_p} R \subseteq \kappa_{v_p}$.

b) $R = \bigcap_{p\in\mathcal{P}} A_{v_p}$.

c) For any non-zero $x \in K$, the set $\{p \in \mathcal{P} \mid v_p x \neq 0\}$ is finite.

d) If R is a PID then $\kappa_{v_p} R = \kappa_{v_p}$ for all $p \in \mathcal{P}$, and any non-trivial exponential valuation v of K such that $R \subseteq A_v$ is equivalent to v_p for exactly one $p \in \mathcal{P}$ and is therefore discrete.

Proof: a), b), and c) are obvious. d) Assume that R is a PID. For any $\alpha \in \mathcal{K}_{v_p}$ there exist $a,b \in R$, $b \notin p\cdot R$, such that $\alpha = \kappa_{v_p}(a\cdot b^{-1})$, and we have $c\cdot p + d\cdot b = 1$ for appropriate c, $d \in R$; therefore $a\cdot b^{-1} - a\cdot d = (a\cdot b^{-1})\cdot c\cdot p \in \mathfrak{M}_{v_p}$, hence $\alpha = \kappa_{v_p}(a\cdot d) \in \kappa_{v_p} R$. Let v be any non-trivial exponential valuation of K such that $R \subseteq A_v$. Then $\mathfrak{M}_v \cap R$ is a non-zero prime ideal of R and therefore equals $p\cdot R$ for some $p \in \mathcal{P}$; hence $vp = \rho = \rho\cdot v_p p$ for some $\rho > 0$. Since $va = \rho\cdot v_p a = 0$ for all $a \in R \setminus p\cdot R$ and since any non-zero element $x \in K$ is of the form $x = p^m\cdot a\cdot b^{-1}$ with $m \in \mathbb{Z}$ and $a,b \in R \setminus p\cdot R$, we have $v = \rho\cdot v_p$, i.e. v is equivalent to v_p. The uniqueness of p follows from the fact that v_p, v_q are inequivalent for all $p, q \in \mathcal{P}$, $p \neq q$. □

For $R = \mathbb{Z}$, $K = \mathbb{Q}$, $\mathcal{P}$ set of prime numbers, we conclude that the p-adic valuation v_p has the ring $A_{v_p} = \mathbb{Z}_{p\cdot\mathbb{Z}}$, the ideal $\mathfrak{M}_{v_p} = p\cdot\mathbb{Z}_{p\cdot\mathbb{Z}}$ and that its residue field $\mathcal{K}_{v_p}$ is isomorphic to $\mathbb{F}_p = \mathbb{Z}/p\cdot\mathbb{Z}$, the prime field of characteristic p.

In §11, statement d) of theorem (4.3) will be generalized to Prüfer rings. Moreover, in case R is a PID it will be shown that any valuation ring of K which contains R is of the form A_{v_p}, for some $p \in \mathcal{P}$.

For another application of theorem (4.3), let $K = K_o(z)$, where z is transcendental over the subfield K_o, and let $\mathfrak{J}$ be the set of all monic irreducible polynomials $P \in K_o[X]$. Since $R = K_o[z]$ is a PID and K its quotient field, we conclude from (4.3) that, for any $P \in \mathfrak{J}$, the mapping $v_{z;P}: K \to \mathbb{R} \cup \{\infty\}$ defined by $v_{z;P}\, 0 = \infty$ and $v_{z;P}\, (u\cdot \prod_{Q\in\mathfrak{J}} Q(z)^{n_Q}) = n_P$ (where $n_Q \in \mathbb{Z}$, $u \in K_o^*$) is a normalized exponential valuation of $K|K_o$ with $A_{v_{z;P}} \supseteq R$, and any exponential valuation v of K such that $A_v \supseteq R$ is of the form $v_{z;P}$ for some $P \in \mathfrak{J}$. Furthermore, we consider the mapping $v_{z;\infty}: K \to \mathbb{R} \cup \{\infty\}$ defined by $v_{z;\infty}\, 0 = \infty$ and $v_{z;\infty}(F(z)\cdot G(z)^{-1}) = \deg G - \deg F$ (F, $G \in K_o[X]\setminus\{0\}$); it is obvious that $v_{z;\infty}$ is also a normalized exponential valuation of $K|K_o$. Setting $\deg \infty = 1$ we prove:

(4.4) COROLLARY - Let $K = K_o(z)$ be a transcendental extension of the field K_o. Then:

a) $P \mapsto v_{z;P}$ $(P \in \mathfrak{J} \cup \{\infty\})$ is a bijective mapping from $\mathfrak{J} \cup \{\infty\}$ onto the set of all normalized exponential valuations of $K|K_o$. Any non-trivial exponential valuation of $K|K_o$ is equivalent to $v_{z;P}$ for exactly one $P \in \mathfrak{J} \cup \{\infty\}$ and is therefore discrete.

b) $\bigcap_{P\in\mathfrak{J}\cup\{\infty\}} A_{v_{z;P}} = K_o$.

c) For any $P \in \mathfrak{J} \cup \{\infty\}$, the residue field $\mathcal{K}_{v_{z;P}}$ is a simple extension of the isomorphic image $\kappa_{v_{z;P}} K_o$ of K_o, and $[\mathcal{K}_{v_{z;P}} : \kappa_{v_{z;P}} K_o] = \deg P$.

d) For any $x \in K$, the set $\{P \in \mathfrak{J} \cup \{\infty\} \mid v_{z;P}\, x \neq 0\}$ is finite, and we have $\sum_{P\in\mathfrak{J}\cup\{\infty\}} \deg P \cdot v_{z;P}x = 0$.

Proof: a) Because of (4.2) and (4.3), it suffices to prove that if v is an exponential valuation of $K|K_o$ such that $K[z] \not\subseteq A_v$ then v is equivalent to $v_{z;\infty}$. In fact, we have $z \notin A_v$ hence $vz = -\rho$ for some $\rho > 0$; therefore, for all $F = a_n \cdot X^n + \dots + a_o \in K_o[X]$ of degree $n \geq 0$ we have $v(F(z)) = -n\cdot\rho$, since $\min\{v(a_i \cdot z^i) \mid 0 \leq i < n\} > v(a_n \cdot z^n) = -n \cdot \rho$. It follows that $v = \rho \cdot v_{z;\infty}$.

b) It is obvious that $\bigcap_{P\in\mathfrak{J}\cup\{\infty\}} A_{v_{z;P}} = K_o[z] \cap A_{v_{z;\infty}} = K_o$.

c) Let $P \in \mathfrak{J}$, $v = v_{z;P}$. By (3.8), κ_v maps K_o isomorphically onto $\kappa_v K_o$, and by (4.3 d) we have $\mathcal{K}_v = \kappa_v(K_o[z]) = (\kappa_v K_o)[\zeta_v]$, where $\zeta_v = \kappa_v z$. Since $\kappa_v P$ is the minimal polynomial of ζ_v over $\kappa_v K_o$, we have $[\mathcal{K}_v : \kappa_v K_o] = \deg(\kappa_v P) = \deg P$. In case $P = \infty$, the statement follows from the equality $v_{z;\infty} = v_{z^{-1};X}$, which is checked easily.

d) The first statement follows from (4.3 c). The equation holds for any $x \in K$ of the form $Q(z)$ where $Q \in \mathfrak{J}$, since $v_{z;P}\, Q(z) = \delta_{P,Q}$ for all $P \in \mathfrak{J}$ and $v_{z;\infty}\, Q(z) = -\deg Q$. Using the homomorphic property of exponential valuations, we conclude that the equation holds for any non-zero $x \in K$. □

One should note that the "sum formula" in (4.4 d) is an analogue to the "product formula" proven in (1.16).

In the proof of (4.4) we have used the fact that $v_{z;\infty} = v_{z^{-1};X}$. More generally, for any $y \in K$ such that $K = K_o(y)$ there is a permutation π of $\mathfrak{J} \cup \{\infty\}$ such that $v_{z;P} = v_{y;\pi P}$ for all $P \in \mathfrak{J} \cup \{\infty\}$. From (4.4 c) we conclude that $\deg P = \deg \pi P$ for any $P \in \mathfrak{J} \cup \{\infty\}$.

The ideal theory in the ring A_v of a discrete exponential valuation v (which we may assume to be normalized) is particularly simple, as the following proposition shows. Its proof is trivial.

(4.5) The ring A_v of any normalized exponential valuation of K is a PID. For any $n \in \mathbb{Z}$ and any $t_n \in K$ such that $vt_n = n$ we have $\mathfrak{w}_n'(0) = \mathfrak{w}_{n-1}(0) = t_n \cdot A_v$, and $\mathfrak{w}_n'(0) = \mathfrak{M}_v^n$ for $n \in \mathbb{N}$.

Moreover, $n \mapsto \mathfrak{w}_n'(0)$ is an isomorphism from the additive monoid $\mathbb{N}$ (resp. group $\mathbb{Z}$) onto the multiplicative monoid (resp. group) of all non-zero ideals (resp. fractional ideals) of A_v.

Note that, on the contrary, for any non-discrete exponential valuation v we have $\mathfrak{M}_v^n = \mathfrak{M}_v$ for all $n > 0$, and all $\mathfrak{w}_\gamma(0)$ with $\gamma \geqq 0$ are non-principal ideals of A_v.

§5 Complete discretely valued fields

In this section, v denotes a normalized exponential valuation of K such that (K,v) is complete. For any $n \in \mathbb{Z}$ we choose $t_n \in K$ such that $vt_n = n$ (usually $t_n = t^n$ for some $t \in K$, $vt = 1$).

For any subset S of any ring we denote by $\times[S]$ (resp. $\times(S)$) the set of all families $(a_i)_{i\in\mathbb{Z}} \in S^{\mathbb{Z}}$ such that $a_i = 0$ for all (resp. almost all) $i < 0$. Obviously, $\times[S]$ can be identified with the set $S^{\mathbb{N}}$ of all sequences $(a_i)_{i\in\mathbb{N}}$ of elements of S. The proof of the following statement is also obvious.

(5.1) For any family $(a_i)_{i\in\mathbb{Z}} \in \times(A_v)$ the sequence $(\sum_{i=-\infty}^{n} a_i \cdot t_i)_{n\in\mathbb{N}}$ is v-convergent. Its v-limit is denoted by $\sum_{i=-\infty}^{\infty} a_i \cdot t_i$.

Let S be a set of representatives for (K,v) , i.e. S is a subset of A_v with $0 \in S$ and such that the restriction to S of the canonical homomorphism κ_v is a bijection $S \to \mathcal{K}_v$. We denote by $\sigma: \mathcal{K}_v \to S$ its inverse. Note that for all $\alpha,\beta \in \mathcal{K}_v$ we have $\alpha \neq \beta$ if and only if $v(\sigma\alpha - \sigma\beta) = 0$.

(5.2) THEOREM - The mapping $\times(\mathcal{K}_v) \to K$ defined by $(\alpha_i)_{i\in\mathbb{Z}} \mapsto$ $\mapsto \sum_{i=-\infty}^{\infty} \sigma\alpha_i \cdot t_i$ is a bijection. Moreover, for all $(\alpha_i)_{i\in\mathbb{Z}} \neq$ $\neq (0)_{i\in\mathbb{Z}}$ we have $v(\sum_{i=-\infty}^{\infty} \sigma\alpha_i \cdot t_i) = \min\{i | \alpha_i \neq 0\}$.

Proof: Let $(\alpha_i)_{i\in\mathbb{Z}} \neq (\beta_i)_{i\in\mathbb{Z}}$, $n = \min\{i|\alpha_i \neq \beta_i\}$. Then

$$v((\sigma\alpha_n - \sigma\beta_n)\cdot t_n) = n < v((\sigma\alpha_i - \sigma\beta_i)\cdot t_i) \quad \text{for all } i \neq n ,$$

hence $v(\sum_{i=-\infty}^{m} (\sigma\alpha_i - \sigma\beta_i)\cdot t_i) = n$ for all $m \geq n$, hence

$\sum_{i=-\infty}^{\infty} \sigma\alpha_i \cdot t_i - \sum_{i=-\infty}^{\infty} \sigma\beta_i \cdot t_i = \sum_{i=-\infty}^{\infty} (\sigma\alpha_i - \sigma\beta_i)\cdot t_i \neq 0$. Therefore the mapping is injective. To prove that it is surjective, let $x \in K$, $x \neq 0$, and assume (by induction) the existence of elements $\alpha_i \in \mathcal{K}_v$ for all $i < m$ such that $\alpha_i = 0$ for almost all $i < m$ and

$v(x - \sum_{i=-\infty}^{m-1} \sigma\alpha_i \cdot t_i) > m-1$; this assumption is, of course, true for $m \leq vx$. Let $\alpha_m = \kappa_v((x - \sum_{i=-\infty}^{m-1} \sigma\alpha_i \cdot t_i) \cdot t_m^{-1})$; then $v((x - \sum_{i=-\infty}^{m-1} \sigma\alpha_i \cdot t_i) \cdot t_m^{-1} - \sigma\alpha_m) > 0$, hence $v(x - \sum_{i=-\infty}^{m} \sigma\alpha_i \cdot t_i) > vt_m = m$. Therefore, $x = \sum_{i=-\infty}^{\infty} \sigma\alpha_i \cdot t_i$. Finally, for any $(\alpha_i)_{i\in\mathbb{Z}} \in X(\mathcal{K}_v)$ we have $v(\sum_{i=-\infty}^{\infty} \sigma\alpha_i \cdot t_i) = v(\sum_{i=-\infty}^{m} \sigma\alpha_i \cdot t_i)$ for sufficiently large $m \in \mathbb{Z}$, by (3.3), and this value is equal to $n = \min \{i | \alpha_i \neq 0\}$ as we have seen above. □

From (5.2) we conclude that any element of K can be written in a unique way as $\sum_{i=-\infty}^{\infty} a_i \cdot t_i$, where $a_i \in S$ for all $i \in \mathbb{Z}$ and $a_i = 0$ for almost all $i \in \mathbb{Z} \setminus \mathbb{N}$, and that

$A_v = \{\sum_{i=0}^{\infty} a_i \cdot t_i \mid a_i \in S$ for all $i \in \mathbb{N}\}$ and

$\mathfrak{M}_v = \{\sum_{i=1}^{\infty} a_i \cdot t_i \mid a_i \in S$ for all $i > 0\}$. Moreover, it follows that

(5.3) COROLLARY 1 - $\#K = \#A_v = (\#\mathcal{K}_v)^{\aleph_0} > \aleph_0$.

In particular, K cannot be a finitely generated extension of its prime field.

(5.4) COROLLARY 2 - If $\mathcal{K}_v$ is finite, then A_v is compact and K is locally compact.

Proof: Consider $\mathcal{K}_v$ with the discrete topology; then $\mathcal{K}_v^{\mathbb{N}}$, with the product topology, is compact. It suffices to show that the restriction $\mathcal{K}_v^{\mathbb{N}} \to A_v$ of the mapping in (5.2) is continuous. In fact, for any $(\alpha_i)_{i\in\mathbb{N}} \in \mathcal{K}_v^{\mathbb{N}}$ and $n \in \mathbb{N}$, the set $\{(\beta_i)_{i\in\mathbb{N}} \in \mathcal{K}_v^{\mathbb{N}} \mid \beta_0 = \alpha_0, \ldots, \beta_n = \alpha_n\}$ is a neighborhood of $(\alpha_i)_{i\in\mathbb{N}}$ in $\mathcal{K}_v^{\mathbb{N}}$, which is mapped into the neighborhood $\mathfrak{w}_n(x)$ of $x = \sum_{i=0}^{\infty} \sigma\alpha_i \cdot t_i$. □

We mention without proof that also the converse of corollary 2 holds (cf Bourbaki [5], Chap. 6,§5).

If (K,v) is the completion of (K_o,v_o) then one can choose $t_n \in K_o$, for all $n \in \mathbb{N}$, and $S \subseteq A_{v_o}$ (because of (3.12)). In particular, in the case $(K,v) = (\hat{\mathbb{Q}}_p,\hat{v}_p)$ one usually chooses $t_n = p^n$ and $S = \{0,1,\dots,p-1\}$. Thus any p-adic number has a "p-adic expansion" $\sum_{i=-\infty}^{\infty} a_i \cdot p^i$ with uniquely determined elements $a_i \in \{0,1,\dots,p-1\}$. In this form, the p-adic numbers were introduced by Hensel [17].

There is an analogy to the n-adic expansions of real numbers $x = \sum_{i=-\infty}^{\infty} a_i \cdot n^{-i}$, with $a_i \in \{0,1,\dots,n-1\}$ and $a_i = 0$ for almost all $i < 0$, where n is any natural number >1 (the most usual are the decimal expansions, where $n = 10$). Note that here the a_i are not always uniquely determined by x (e.g., $1 = 0.999\dots$). It is well known that $x \in \mathbb{Q}$ if and only if $(a_i)_{i\in\mathbb{Z}}$ is periodic. This is also true for the p-adic expansions in $\hat{\mathbb{Q}}_p$, as the following theorem shows.

(5.5) THEOREM - <u>Let</u> $\sum_{i=-\infty}^{\infty} a_i \cdot p^i$ (<u>with</u> $0 \leq a_i < p$) <u>be the p-adic expansion of</u> $\hat{x} \in \hat{\mathbb{Q}}_p$. <u>Then</u> $(a_i)_{i\in\mathbb{Z}}$ <u>is periodic if and only if</u> $\hat{x} \in \mathbb{Q}$.

<u>Proof</u>: We may assume $\hat{v}_p\hat{x} = 0$, hence $a_i = 0$ for $i < 0$. Assume that $(a_i)_{i\in\mathbb{Z}}$ is periodic, i.e., there exist $m,k \in \mathbb{N}$, $k \geq 1$, such that $a_{i+k} = a_i$ for all $i \geq m$. Let $s = \sum_{i=o}^{m-1} a_i \cdot p^i$ and $s' = \sum_{i=m}^{m+k-1} a_i \cdot p^i$; then $\hat{x} - s = \sum_{i=m}^{\infty} a_i \cdot p^i = s' + \sum_{i=m+k}^{\infty} a_i \cdot p^i =$ $= s' + p^k \cdot \sum_{i=m}^{\infty} a_i \cdot p^i = s' + p^k(\hat{x}-s)$, hence $(\hat{x}-s)\cdot(1-p^k) = s'$, $\hat{x} = s + s' \cdot (1-p^k)^{-1} \in \mathbb{Q}$. On the other hand, suppose that $x = \sum_{i=o}^{\infty} a_i \cdot p^i \in \mathbb{Q}$. We claim that there exist $m,k \in \mathbb{N}$, $k \geq 1$, and $t,u \in \mathbb{Z}$ such that $0 \leq t < p^m$, $0 \leq u < p^k$, and $x = t + u \cdot p^m(1-p^k)^{-1}$. In fact, we have $x = a \cdot d^{-1}$, where $a,d \in \mathbb{Z}$, $d > 0$ and $(p,d) = 1$, hence there exists a $k \geq 1$ such that $p^k \equiv 1 \mod d$, hence $x = b \cdot (p^k-1)^{-1}$ for some $b \in \mathbb{Z}$. Choose $m \in \mathbb{N}$ such that $-p^m \leq b <$

$< p^m$. Since $(p^m, p^k - 1) = 1$, there are $t, u \in \mathbb{Z}$ such that $b = t\cdot(p^k-1) - u\cdot p^m$, and u can be chosen such that $0 \leqq u < p^k - 1$ if $x > 0$ and $1 \leqq u < p^k$ if $x < 0$. In either case t and u have the desired properties. There exist $a_o,\dots,a_{m+k-1} \in \{0,\dots,p-1\}$ such that $t = \sum_{i=o}^{m-1} a_i\cdot p^i$ and $u = \sum_{i=o}^{k-1} a_{m+i}\cdot p^i$. Since $\sum_{i=o}^{\infty} p^{i\cdot k} = (1-p^k)^{-1}$, we have $x = \sum_{i=o}^{\infty} a_i\cdot p^i$, where $a_{i+k} = a_i$ for all $i \geqq m$; hence $(a_i)_{i\in\mathbb{N}}$ is periodic. □

The preceding proof yields a method of finding p-adic expansions of any $x \in \mathbb{Q}$.

Returning to the general case of a complete discretely valued field (K,v) , we wonder whether the set $\times(\mathcal{K}_v)$ can be endowed with a field structure, in a natural way, such that the bijection in (5.2) becomes an isomorphism [12]. Obviously we have to distinguish between the equi-characteristic case (Char $\mathcal{K}_v$ = Char K) and the case of distinct characteristics (Char $\mathcal{K}_v \neq$ Char K = 0). Considering first the equi-characteristic case we note:

(5.6) Let $\mathcal{K}$ be any field. Then the set $\times(\mathcal{K})$, endowed with the addition

$$(\alpha_i)_{i\in\mathbb{Z}} + (\beta_i)_{i\in\mathbb{Z}} = (\alpha_i+\beta_i)_{i\in\mathbb{Z}}$$

and the multiplication

$$(\alpha_i)_{i\in\mathbb{Z}} \cdot (\beta_i)_{i\in\mathbb{Z}} = (\sum_{i+j=k} \alpha_i\cdot\beta_j)_{k\in\mathbb{Z}}$$

is a field of the same characteristic as $\mathcal{K}$, with zero $0 = (0)_{i\in\mathbb{Z}}$ and unit element $1 = (\delta_{i0})_{i\in\mathbb{Z}}$.

The mapping $\alpha \mapsto (\alpha\cdot\delta_{i0})_{i\in\mathbb{Z}}$ is a monomorphism from $\mathcal{K}$

---

12 Of course, we can always transport the field structure from K onto the set $\times(\mathcal{K}_v)$ by means of the indicated bijection; but we don't consider this a natural procedure.

into $\mathcal{X}(\mathcal{K})$. By $u_L 0 = \infty$ and $u_{\mathcal{K}}(\alpha_i)_{i\in\mathbb{Z}} = \min\{i \mid \alpha_i \neq 0\}$, for all $(\alpha_i)_{i\in\mathbb{Z}} \neq 0$, a normalized exponential valuation $u_{\mathcal{K}}$ of $\mathcal{X}(\mathcal{K})$ is defined.

The proof is trivial. Note that $\mathcal{X}(\mathcal{K})$ with the above field structure is usually called the field of formal Laurent series over $\mathcal{K}$ and denoted by $\mathcal{K}((X))$, and $(\alpha_i)_{i\in\mathbb{Z}}$ is usually written as $\sum_{i=-\infty}^{\infty} \alpha_i \cdot X^i$. In (5.8) we shall see that $(\mathcal{K}((X)),u_{\mathcal{K}})$ is complete.

We are going to show that any complete discretely valued field (K,v) such that $\operatorname{Char} \mathcal{K}_v = \operatorname{Char} K$ is isomorphic to $(\mathcal{K}_v((X)),u_{\mathcal{K}_v})$. The proof is easy if we assume the existence of a field of representatives for (K,v) , that is, a subfield S of K which is a set of representatives of (K,v) . In fact, in this case $\kappa_v|S$ is an isomorphism from S onto $\mathcal{K}_v$ and so is its inverse; therefore we conclude from (5.2) and (5.6):

(5.7) THEOREM - Let (K,v) have a field of representatives S . Then for any $t \in K$ such that $vt = 1$, the mapping $\sum_{i=-\infty}^{\infty} \alpha_i \cdot X^i \mapsto \sum_{i=-\infty}^{\infty} (\sigma\alpha_i)\cdot t^i$ is an isomorphism from $(\mathcal{K}_v((X)),u_{\mathcal{K}_v})$ onto (K,v) .

Conversely, for any field $\mathcal{K}$, the field of formal Laurent series over $\mathcal{K}$ endowed with $u_{\mathcal{K}}$ is a complete discretely valued field; in fact:

(5.8) Let $\mathcal{K}(z)$ be a transcendental extension of the field $\mathcal{K}$ and let ν be the normalized exponential valuation of $\mathcal{K}(z)$ defined by $\nu 0 = \infty$ and $\nu(F(x)\cdot G(z)^{-1}) = \deg G - \deg F$ $(F,G \in \mathcal{K}[X] \setminus \{0\})$. Then the completion of $(\mathcal{K}(z),\nu)$ is isomorphic to $(\mathcal{K}((X)),u_{\mathcal{K}})$.

Proof: Let (K,v) be a completion of $(\mathcal{K}(z),\nu)$. By (3.12), v is a normalized exponential valuation of K and its residue field concides with that of ν which is turn is equal to $\kappa_\nu \mathcal{K}$, by (4.4 c);

therefore $\mathcal{K}$ is a field of representatives for (K,v) . By (5.8) we have $(K,v) \cong (\mathcal{K}_v((X)),u_{\mathcal{K}_v}) \cong (\mathcal{K}((X)),u_{\mathcal{K}})$. □

It is obvious that if there exists a field of representatives for (K,v) then its characteristic must be equal to both Char K and Char $\mathcal{K}_v$. Conversely, one can prove that in the equi-characteristic case there exists always a field of representatives. We shall give the proof only under certain hypotheses.

(5.9) THEOREM - Assume that

a) Char $\mathcal{K}_v$ = Char K = 0 or

b) Char $\mathcal{K}_v$ = Char K = $p \neq 0$ and $\mathcal{K}_v$ is finitely generated over $\kappa_v K_o$ for some perfect subfield K_o of K which is contained in A_v [13]. Then there exists a field of representatives for (K,v).

Proof: In either case there exists a transcendence basis $\mathfrak{T}$ of $\mathcal{K}_v$ over $\kappa_v K_o$ such that $\mathcal{K}_v$ is separable over $(\kappa_v K_o)(\mathfrak{T})$ (where K_o is the prime field of K in case a), cf (3.10)). Let T be a subset of A_v such that $\kappa_v|T: T \to \mathfrak{T}$ is a bijection. Then $K_o(T)$ is a subfield of K contained in A_v , and using Zorn's lemma one shows the existence of a maximal subfield K_1 of K such that $K_o(T) \subseteq K_1 \subseteq A_v$. We claim that $\kappa_v K_1 = \mathcal{K}_v$. In fact, for any $\alpha \in \mathcal{K}_v$ let P be a monic polynomial with coefficients in $A_v \cap K_1$ such that $\kappa_v P$ is the minimal polynomial of α over $\kappa_v K_1$. Since $\kappa_v P$ is separable and $X-\alpha$ is a factor of $\kappa_v P$ in $\mathcal{K}_v[X]$, we conclude from Hensel's lemma (which will be proven in §16) that there is an $a \in A_v$ such that $\kappa_v a = \alpha$ and $P(a) = 0$, hence $K_1 \subseteq K_1(a) = K_1[a] \subseteq A_v$. By the maximality of K_1 it follows that $a \in K_1$, hence $\alpha = \kappa_v a \in \kappa_v K_1$. Therefore K_1 is a field of representatives for (K,v) . □

---

13 The second part of assumption b) is satisfied whenever $\mathcal{K}_v$ is finitely generated over its prime field (cf (3.10)).

Note that this field of representatives depends on the choice of T and therefore, in general, is not unique. However, it is unique in the case in which $\mathcal{K}_v$ is a perfect field of characteristic $p \neq 0$, as will be shown in the following theorem, which is also fundamental in the case of distinct characteristics.

(5.10) THEOREM - Assume that $\mathcal{K}_v$ is a perfect field of characteristic $p \neq 0$ (which may be equal or not to the characteristic of K). Then there exists one and only one set of representatives S for (K,v) such that $S^p \subseteq S$, and S has the following properties:

a) $1 \in S$, $S \cdot S \subseteq S$, and the mapping $\sigma: \mathcal{K}_v \to S$ (the inverse of $\kappa_v|S: S \to \mathcal{K}_v$) is a multiplicative isomorphism.

b) The mapping $a \mapsto a^p$ $(a \in S)$ is a multiplicative automorphism of S.

c) If $\operatorname{Char} K = p$, then S is the only field of representatives for (K,v).

Proof: 1) For any $\alpha \in \mathcal{K}_v$ and $n \in \mathbb{N}$ choose $a_n \in A_v$ such that $\kappa_v a_n = \alpha^{p^{-n}}$; then $\kappa_v(a_{n+1}{}^p - a_n) = 0$, hence $v(a_{n+1}{}^p - a_n) \geq 1$ hence $v(a_{n+1}{}^{p^{n+1}} - a_n{}^{p^n}) \geq n+1$, by (3.9); therefore $(a_n{}^{p^n})_{n\in\mathbb{N}}$ is v-Cauchy, hence v-convergent. The v-limit $\sigma\alpha =$ $= \lim_v (a_n{}^{p^n})_{n\in\mathbb{N}}$ depends only on α, since if $\kappa_v b_n = \alpha^{p^{-n}}$ then $\kappa_v(a_n - b_n) = 0$, $v(a_n - b_n) \geq 1$, and $v(a_n{}^{p^n} - b_n{}^{p^n}) \geq n+1$, by (3.9). In particular, $\sigma 0 = 0$ and $\sigma 1 = 1$. Obviously, $\alpha \mapsto \sigma\alpha$ is a bijection from $\mathcal{K}_v$ onto a subset S of A_v, and $\kappa_v \circ \sigma = \iota_{\mathcal{K}_v}$; hence S is a set of representatives for (K,v).

2) We claim that σ is a multiplicative homomorphism. In fact, let $\alpha, \beta \in \mathcal{K}_v$, $\sigma\alpha = \lim_v (a_n{}^{p^n})_{n\in\mathbb{N}}$, $\beta = \lim_v (b_n{}^{p^n})_{n\in\mathbb{N}}$; then $\sigma\alpha \cdot \sigma\beta = \lim_v (a_n{}^{p^n} \cdot b_n{}^{p^n})_{n\in\mathbb{N}} = \lim_v ((a_n \cdot b_n)^{p^n})_{n\in\mathbb{N}} = \sigma(\alpha \cdot \beta)$, since $\kappa_v (a_n \cdot b_n) = \alpha^{p^{-n}} \cdot \beta^{p^{-n}} = (\alpha \cdot \beta)^{p^{-n}}$. Therefore $S \cdot S \subseteq S$ and in particular $S^p \subseteq S$. If $\operatorname{Char} K = p$, then a similar argument holds for addition; therefore in this case σ is also an additive homomorphism

and S a subfield of K .

3) The mapping in b) is composed of the multiplicative isomorphisms κ_A , $\alpha \mapsto \alpha^p$ $(\alpha \in \mathcal{K}_v)$, and σ ; hence b) is true.

4) Let T be any set of representatives for (K,v) such that $T^p \subseteq T$. We have to prove that $T \subseteq S$ (it follows that $T = S$). In fact, let $a \in T$ and $a_n \in T$ such that $\kappa_v a_n = (\kappa_v a)^{p^{-n}}$, for any $n \in \mathbb{N}$; then obviously $a_n{}^{p^n} = a$ for any $n \in \mathbb{N}$. On the other hand, $a = \lim_v (a_n{}^{p^n})_{n\in\mathbb{N}} = \sigma(\kappa_v(a)$ by 1), hence $a \in S$. Therefore, S is uniquely determined by $S^p \subseteq S$. In particular, S is the only set of representatives with $S \cdot S \subseteq S$ and, in case Char $K = p$, the only field of representatives for (K,v) . □

Now we are going to consider the case of distinct characteristics [14]. Fixing a prime number p , two sequences $(S_i)_{i\in\mathbb{N}}$, $(P_i)_{i\in\mathbb{N}}$ of polynomials S_i , $P_i \in \mathbb{Z}[X_o,\dots,X_i;Y_o,\dots,Y_i]$ (where $X_o,X_1,\dots;Y_o,Y_1,\dots$ are indeterminates) can be constructed by means of the following systems of equations:

$$\sum_{j=o}^{i} p^j \cdot S_j{}^{p^{i-j}} = \left(\sum_{j=o}^{i} p^j \cdot X_j{}^{p^{i-j}}\right) + \left(\sum_{j=o}^{i} p^j \cdot Y_j{}^{p^{i-j}}\right) \qquad (i \in \mathbb{N})$$

$$\sum_{j=o}^{i} p^j \cdot P_j{}^{p^{i-j}} = \left(\sum_{j=o}^{i} p^j \cdot X_j{}^{p^{i-j}}\right) \cdot \left(\sum_{j=o}^{i} p^j \cdot Y_j{}^{p^{i-j}}\right) \qquad (i \in \mathbb{N}) .$$

They will be used to endow the set $\mathsf{X}[\mathfrak{R}] = \mathfrak{R}^{\mathbb{N}}$ with a ring structure, where $\mathfrak{R}$ is any commutative ring with unit element 1. Let $\langle\alpha\rangle = (\alpha \cdot \delta_{o,i})_{i\in\mathbb{N}}$ for any $\alpha \in \mathfrak{R}$.

(5.11) <u>The set $\mathsf{X}[\mathfrak{R}]$, endowed with the addition and multiplication defined by</u>

$$(\alpha_i)_{i\in\mathbb{N}} + (\beta_i)_{i\in\mathbb{N}} = (S_i(\alpha_o,\dots,\alpha_i;\beta_o,\dots,\beta_i))_{i\in\mathbb{N}}$$

and

$$(\alpha_i)_{i\in\mathbb{N}} \cdot (\beta_i)_{i\in\mathbb{N}} = (P_i(\alpha_o,\dots,\alpha_i;\beta_o,\dots,\beta_i))_{i\in\mathbb{N}}$$

---

[14] The proofs of the following statements can be found for example in Hasse [16], Serre [33], or Endler [8].

is a commutative ring $W[\mathfrak{R}]$ with zero $\langle 0\rangle$ and unit element $\langle 1\rangle$, and the mapping $\alpha \mapsto \langle\alpha\rangle$ $(\alpha \in \mathfrak{R})$ is an injective multiplicative homomorphism from $\mathfrak{R}$ into $W[\mathfrak{R}]$.

If $\mathfrak{R}$ is an integral domain of characteristic p then $W[\mathfrak{R}]$ is an integral domain of characteristic 0, and for any $(\alpha_i)_{i\in\mathbb{N}}$ we have $p\cdot(\alpha_i)_{i\in\mathbb{N}} = (\alpha_{i-1}^p)_{i\in\mathbb{N}}$ (where $\alpha_{-1} = 0$).

The ring $W[\mathfrak{R}]$ is called the ring of Witt's vectors over $\mathfrak{R}$. We are particularly interested in the case in which $\mathfrak{R}$ is a perfect field $\mathcal{K}$ of characteristic p. In this case, $W[\mathcal{K}]$ is the valuation ring of a complete discretely valued field $(W(\mathcal{K}), u_{\mathcal{K}})$ of characteristic 0, and its residue field can be identified with $\mathcal{K}$, as the following theorem shows.

(5.12) THEOREM - Let $\mathcal{K}$ be a perfect field of characteristic p. Then:

a) The quotient field $W(\mathcal{K})$ of $W[\mathcal{K}]$ equals the ring of fractions of $W[\mathcal{K}]$ with respect to the multiplicative set $\{p^n\cdot\langle 1\rangle \mid n \in \mathbb{N}\}$ and can be identified with $\mathcal{X}(\mathcal{K})$ by means of the well-defined bijection

$$\frac{(\alpha_i)_{i\in\mathbb{N}}}{p^n\cdot\langle 1\rangle} \mapsto (\alpha_{i+n}^{p^{-n}})_{i\in\mathbb{Z}} \quad \text{(where } \alpha_i = 0 \text{ for all } i \in \mathbb{Z} \setminus \mathbb{N}\text{)}.$$

b) A normalized exponential valuation $u_{\mathcal{K}}$ of $W(\mathcal{K})$ is defined by $u_{\mathcal{K}}\langle 0\rangle = \infty$ and $u_{\mathcal{K}}(\alpha_i)_{i\in\mathbb{Z}} = \min\{i \mid \alpha_i \neq 0\}$ for all $(\alpha_i)_{i\in\mathbb{Z}} \neq \langle 0\rangle$. The ring of $u_{\mathcal{K}}$ is equal to $W[\mathcal{K}]$, the ideal of $u_{\mathcal{K}}$ is the principal ideal generated by $p\cdot\langle 1\rangle = (1\cdot\delta_{1,i})_{i\in\mathbb{Z}}$, and the mapping $(\alpha_i)_{i\in\mathbb{N}} \mapsto \alpha_0$ induces an isomorphism from the residue field of $u_{\mathcal{K}}$ onto $\mathcal{K}$.

c) $(W(\mathcal{K}), u_{\mathcal{K}})$ is complete, $S = \{\langle\alpha\rangle \mid \alpha \in \mathcal{K}\}$ is its unique set of representatives which satisfies $S^p \subseteq S$, and any $(\alpha_i)_{i\in\mathbb{Z}} \in W(\mathcal{K})$ can be written as $\sum_{i=-\infty}^{\infty} \langle\alpha_i^{p^{-i}}\rangle\cdot(p\cdot\langle 1\rangle)^i$.

On the other hand, any complete discretely valued field (K,v) of characteristic 0, with perfect residue field $\mathcal{K}$ of characteristic p, contains an isomorphic image of $(W(\mathcal{K}),u_{\mathcal{K}})$. More precisely:

(5.13) THEOREM - Let (K,v) have a perfect residue field $\mathcal{K}_v$ of characteristic p, whereas $\operatorname{Char} K = 0$ (hence we may assume $K \supseteq Q$). Let S be the unique set of representatives for (K,v) such that $S^p \subseteq S$ and σ the inverse of $\kappa_v|S: S \to \mathcal{K}_v$. Moreover, let $n = vp$ (>0). Then

$$\lambda: (\alpha_i)_{i\in\mathbb{Z}} \mapsto \sum_{i=-\infty}^{\infty} (\sigma\alpha_i{}^{p^{-i}})\cdot p^i \qquad ((\alpha_i)_{i\in\mathbb{Z}} \in W(\mathcal{K}_v))$$

is an isomorphism from $(W(\mathcal{K}_v),u_{\mathcal{K}_v})$ onto a valued subfield (K_1,v_1) of $(K,\frac{1}{n}v)$ such that $[K:K_1] = n$. For any $t \in K$ such that $vt = 1$ we have $K = K_1(t)$ and t is a root of some eisensteinian polynomial over (K_1,v_1) of degree n. [15]

In particular, in case $vp = 1$ we get an isomorphism λ from $(W(\mathcal{K}_v),u_{\mathcal{K}_v})$ onto (K,v), in analogy to theorem (5.7).

The valued field $(W(\mathcal{K}),u_{\mathcal{K}})$ can be considered as a generalization of the field of p-adic numbers. In fact, we get $(W(\mathbb{F}_p),u_{\mathbb{F}_p}) = (\hat{\mathbb{Q}}_p,\hat{v}_p)$ if we identify any element $(\alpha_i)_{i\in\mathbb{Z}} = \sum_{i=-\infty}^{\infty} \langle\alpha_i\rangle\cdot(p\cdot\langle 1\rangle)^i$ of $W(\mathbb{F}_p)$ with the corresponding p-adic number $\lambda(\alpha_i)_{i\in\mathbb{Z}} = \sum_{i=-\infty}^{\infty}(\sigma\alpha_i)\cdot p^i$. (Note that $\alpha^p = \alpha$ for any $\alpha \in \mathbb{F}_p$). The distinguished set of representatives for $(\hat{\mathbb{Q}}_p,\hat{v}_p)$ consists obviously of 0, $1,\hat{z},\hat{z}^2,\ldots,\hat{z}^{p-2}$, where $\hat{z} \in \hat{\mathbb{Q}}_p$ is a primitive $(p-1)$th root of unity. Note that $\hat{z} \notin \mathbb{Q}$ for $p \geqq 3$.

---

[15] A monic polynomial $X^n + a_1\cdot X^{n-1} + \ldots + a_n \in K[X]$ is called eisensteinian over (K,v) if $va_n = 1 \leqq \min\{va_1,\ldots,va_n\}$; it is irreducible in $K[X]$ (cf Exercise III- 13).

CHAPTER II

Valuation Rings

§6 Valuation rings

In §3 we have defined a valuation ring of a field K to be a subring A of K such that $x \in A$ or $x^{-1} \in A$ for any non-zero $x \in K$. Obviously K is the quotient field of A, and K itself is a valuation ring of K. If K is absolutely algebraic (i.e., algebraic over its prime field) and of prime characteristic, then K is the only valuation ring of K (since any subring of K is a field). We shall see later that all other fields K have valuation rings distinct from K.

In this section we give some elementary characterizations of valuation rings and study their ideals.

A binary operation $\leqq$ on a set S is called a quasi-ordering of S if $\leqq$ is reflexive (i.e., $s \leqq s$ for any $s \in S$) and transitive (i.e., $s_1 \leqq s_2$, $s_2 \leqq s_3$ imply $s_1 \leqq s_3$). It is called an ordering of S if $s_1 \leqq s_2$, $s_2 \leqq s_1$ imply $s_1 = s_2$, and it is called a total quasi-ordering if $s_1 \nleqq s_2$ implies $s_2 \leqq s_1$. In particular, the trivial quasi-ordering ($s \leqq s'$ for all $s, s' \in S$) is total.

A non-trivial quasi-ordering $\backslash$ of a field K is called a divisibility of K if for all $x,y,z \in K$, $x \backslash y$ implies $x \cdot z \backslash y \cdot z$, and $x \backslash y$, $x \backslash z$ imply $x \backslash (y-z)$. Note that $x \backslash 0$ and $0 \not\backslash x$ for any non-zero $x \in K$. The following proposition is obvious:

(6.1) *The divisibilities $\backslash$ of K are in 1-1 correspondence with the subrings R of K*[16] *by*

[16] We consider only unitary subrings of K, that is, those containing the unit element of K.

$$[x\backslash y \Leftrightarrow y\cdot x^{-1} \in R] \text{ and } R = \{x \in K \mid 1\backslash x\}.$$

<u>Moreover,</u> $U_R = \{x \in R \mid x\backslash 1\}$ <u>is the group of units of</u> R.

As a first characterization of valuation rings we prove:

(6.2) <u>A subring R of K is a valuation ring of K if and only if the corresponding divisibility is a total quasi-ordering.</u>

<u>Proof</u>: Let $\backslash$ be total. If $x \in K \setminus R$ then $1 \not\backslash x$, hence $x\backslash 1$, hence $1\backslash x^{-1}$, hence $x^{-1} \in R$, hence R is a valuation ring of K. On the other hand, let R be a valuation ring of K and let $x \not\backslash y$. If $x = 0$ then $y\backslash x$. If $x \neq 0$ then $1 \not\backslash (y\cdot x^{-1})$, hence $y\cdot x^{-1} \notin R$, hence $x\cdot y^{-1} \in R$, $1\backslash x\cdot y^{-1}$, $y\backslash x$. Therefore $\backslash$ is total. $\square$

A subset M of K is called R-<u>stable</u>, if $R\cdot M \subseteq M$, i.e., $a\cdot x \in M$ for all $a \in R$, $x \in M$. We show that in the case of a valuation ring R any R-stable non-empty subset of R (resp. K) is an ideal of R (resp. an R-submodule of K) and that this property characterizes valuation rings.

(6.3) THEOREM - <u>Let K be the quotient field of R and $\mathcal{I}$ (resp. $\mathcal{J}$) the set, ordered by inclusion, of all R-stable non-empty subsets of R (resp. K). Then the following conditions are equivalent:</u>

(i) R <u>is a valuation ring of</u> K.

(ii) $\mathcal{J}$ <u>is totally ordered.</u>

(iii) $\mathcal{I}$ <u>is totally ordered.</u>

(iv) <u>The subset of $\mathcal{I}$ consisting of all principal ideals of R is totally ordered.</u>

<u>In this case,</u> $\mathcal{I}$ (resp. $\mathcal{J}$) <u>is the set of all ideals of R (resp. R-submodules of K).</u>

<u>Proof</u>: (i) $\Rightarrow$ (ii): Suppose that $M,N \in \mathcal{J}$, $M \not\subseteq N$ and $N \not\subseteq M$. Let $x \in M \setminus N$ and $y \in N \setminus M$. Since $x = (x\cdot y^{-1})\cdot y \notin N$ we have

$x\cdot y^{-1} \notin R$, and since $y = (y\cdot x^{-1})\cdot x \notin M$, we have $y\cdot x^{-1} \notin R$; therefore R is not a valuation ring of K . (ii) ⇒ (iii) and (iii) ⇒ (iv) are trivial. (iv) ⇒ (i): Let $x = \frac{a}{b} \in K$, where $a,b \in R$, $a \neq 0 \neq b$. If $x \notin R$, then $R\cdot a \nsubseteq R\cdot b$, hence $R\cdot b \subseteq R\cdot a$, hence $x^{-1} = \frac{b}{a} \in R$. Hence R is a valuation ring of K .

For the last statement, it suffices to show that for any $M \in \mathcal{J}$ and $x,y \in M\setminus\{0\}$ we have $x - y \in M$. In fact, if $R\cdot x \subseteq R\cdot y$ then $x-y = (x\cdot y^{-1} - 1)\cdot y \in R\cdot y \subseteq M$; if $R\cdot x \nsubseteq R\cdot y$, then $R\cdot y \subseteq R\cdot x$ and $x-y = (1 - y\cdot x^{-1})\cdot x \in R\cdot x \subseteq M$. □

A ring R is called a <u>local ring</u> [17] , if it has only one maximal ideal $\mathfrak{M}_R$; obviously $\mathfrak{M}_R = R\setminus U_R$. For any valuation ring A the set $A\setminus U_A$ obviously is A-stable, hence by (6.3) we get:

(6.4) COROLLARY - <u>Any valuation ring is a local ring</u>.

Valuation rings are not noetherian, in general. However, we have

(6.5) COROLLARY - <u>Any finitely generated ideal of a valuation ring A is a principal ideal</u>.

<u>Proof</u>: Let $\mathfrak{A} = A\cdot a_1 + \dots + A\cdot a_m$. By (6.3), $\{A\cdot a_1,\dots,A\cdot a_m\}$ has a largest element, say $A\cdot a_1 \supseteq A\cdot a_i$ $(i=1,\dots,m)$. Then $\mathfrak{A} \subseteq A\cdot a_1 \subseteq \mathfrak{A}$. □

We shall use the fact that, for any subring R of K and any prime ideal $\mathfrak{P}_o$ of R , the prime ideals $\mathfrak{Q}$ of the ring of fractions $R_{\mathfrak{P}_o}$ are in 1-1 correspondence with those prime ideals $\mathfrak{P}$ of R which are contained in $\mathfrak{P}_o$, by $\mathfrak{Q} = \mathfrak{P}\cdot R_{\mathfrak{P}_o}$ and $\mathfrak{P} = \mathfrak{Q} \cap R$. In particular, $R_{\mathfrak{P}_o}$ is a local ring with the maximal ideal $\mathfrak{P}_o\cdot R_{\mathfrak{P}_o}$.

---

[17] Some authors prefer the name "quasi-local" and reserve "local" for noetherian local rings.

(6.6) THEOREM - *Let* A *be a valuation ring of* K , $\mathcal{P}$ *the set of all prime ideals* $\mathfrak{p}$ *of* A *and* $\mathcal{B}$ *the set of all subrings* B *of* K *which contain* A . *Then any* $B \in \mathcal{B}$ *is a valuation ring of* K *with* $\mathfrak{M}_B \subseteq A$, *and there is an inclusion inverting* 1-1 *correspondence* $\mathcal{B} \leftrightarrow \mathcal{P}$ *given by* $\mathfrak{p} = \mathfrak{M}_B$ *and* $B = A_{\mathfrak{p}}$.

Moreover, $\mathcal{B}$ *and* $\mathcal{P}$ *are totally ordered by inclusion.*

Proof: Let $B \in \mathcal{B}$. For any $x \in K$, $x \notin B$ implies $x \notin A$, hence $x^{-1} \in A \subseteq B$, whereas $x \in \mathfrak{M}_B$ implies $x^{-1} \notin B$, hence $x \in A$. For any $B \in \mathcal{B}$, $\mathfrak{M}_B \cap A = \mathfrak{M}_B$ is a prime ideal of A , and obviously $A_{\mathfrak{M}_B} \subseteq B$. Even $A_{\mathfrak{M}_B} = B$, since if $x \in B \setminus A$ then $x^{-1} \in A \subseteq B$, $x^{-1} \notin \mathfrak{M}_B$, hence $x \in A_{\mathfrak{M}_B}$. For any $\mathfrak{p} \in \mathcal{P}$ we have $A_{\mathfrak{p}} \in \mathcal{B}$ and $\mathfrak{M}_{A_{\mathfrak{p}}} = \mathfrak{M}_{A_{\mathfrak{p}}} \cap A = \mathfrak{p} \cdot A_{\mathfrak{p}} \cap A = \mathfrak{p}$. The correspondence $\mathcal{B} \leftrightarrow \mathcal{P}$ obviously is inclusion-inverting. By (6.3), $\mathcal{P}$ is totally ordered, hence so is $\mathcal{B}$. □

The order type of the set $\mathcal{B} \setminus \{K\}$ ordered by inclusion, is equal to the order type of the set $\mathcal{P} \setminus \{(0)\}$ ordered by the inverse inclusion, and is called the *rank* of A . In particular, the valuation ring A of K has finite rank n if and only if there is a non-refinable chain $A = B_1 \subset \dots \subset B_n \subset K$ of valuation rings B_i of K or, equivalently, a non-refinable chain $\mathfrak{M}_A = \mathfrak{p}_1 \supset \dots \supset \mathfrak{p}_n \supset \{0\}$ of prime ideals of A . K is the only valuation ring of K of rank 0 , and a valuation ring A of K has rank 1 if and only if $\mathfrak{M}_A$ is its only non-zero prime ideal. The following corollary is obvious:

(6.7) COROLLARY - *Let* $\mathcal{B}, \mathcal{P}$ (resp. $\mathcal{B}', \mathcal{P}'$) *be defined as in* (6.6) *with respect to the valuation ring* A (resp. A') *of* K . *If* $A' \subseteq A$, *then* $\mathcal{B} = \{B' \in \mathcal{B}' \mid A \subseteq B'\}$ *and* $\mathcal{P} = \{\mathfrak{p}' \in \mathcal{P}' \mid \mathfrak{p}' \subseteq \mathfrak{M}_A\}$, *and the* 1-1 *correspondence* $\mathcal{B} \leftrightarrow \mathcal{P}$ *is induced by* $\mathcal{B}' \leftrightarrow \mathcal{P}'$.

For any commutative ring R and any proper ideal $\mathfrak{A}$ of R , the set $\sqrt{\mathfrak{A}} = \{a \in R \mid a^n \in \mathfrak{A}$ for some $n \in \mathbb{N}\}$ is a proper ideal of R , called the <u>radical</u> of $\mathfrak{A}$, and $\mathfrak{A} \mapsto \sqrt{\mathfrak{A}}$ is a closure operation in the set of all proper ideals of R .

By means of the following proposition, we reduce the study of ideals of a valuation ring A to that of its prime ideals.

(6.8) <u>Let</u> $\mathfrak{A}$ <u>be a proper ideal of a valuation ring</u> A . <u>Then</u> $\sqrt{\mathfrak{A}}$ <u>is a prime ideal of</u> A .

<u>Proof</u>: Let $a,b \in A$ such that $a\cdot b \in \sqrt{\mathfrak{A}}$, say $(a\cdot b)^n \in \mathfrak{A}$. If $A\cdot b \not\subseteq A\cdot a$, then $b^{2n} \in A\cdot a^n\cdot b^n \subseteq \mathfrak{A}$, hence $b \in \sqrt{\mathfrak{A}}$. If $A\cdot b \subseteq A\cdot a$ then $A\cdot a \subseteq A\cdot b$, hence $a \in \sqrt{\mathfrak{A}}$. □

(6.9) <u>Let</u> A, A' <u>be valuation rings of</u> K <u>such that</u> $A' \subseteq A$. <u>Then any proper ideal</u> $\mathfrak{A}$ <u>of</u> A <u>is a proper ideal of</u> A'.

<u>Proof</u>: $\sqrt{\mathfrak{A}}$ is a prime ideal of A , hence a prime ideal of A' by (6.7), hence $\mathfrak{A} \subseteq \sqrt{\mathfrak{A}} \subseteq A'$. Therefore $\mathfrak{A} = \mathfrak{A} \cap A'$ is an ideal of A' and proper since $1 \notin \mathfrak{A}$. □

We consider now two arbitrary valuation rings A_1, A_2 of K and their product $A_1\cdot A_2$, i.e., the least subring of K containing both A_1 and A_2 . Obviously $A_1\cdot A_2$ consists of all finite sums $\sum_{i=1}^{r} x_i\cdot y_i$, where $x_1,\dots,x_r \in A_1$ and $y_1,\dots,y_r \in A_2$.

(6.10) THEOREM - <u>Let</u> A_1 , A_2 <u>be valuation rings of</u> K <u>and let</u> $A_3 = A_1\cdot A_2$. <u>Let</u> $\mathcal{I}_i$ (resp. $\mathcal{P}_i$) <u>be the set of all proper</u> (resp. prime) <u>ideals of</u> A_i (i=1,2,3) . <u>Then</u> $\mathcal{I}_3 = \mathcal{I}_1 \cap \mathcal{I}_2$, $\mathcal{P}_3 = \mathcal{P}_1 \cap \mathcal{P}_2 = \mathcal{I}_1 \cap \mathcal{P}_2 = \mathcal{P}_1 \cap \mathcal{I}_2$, <u>and</u> $\mathfrak{M} = \mathfrak{M}_{A_3}$ <u>is the largest element of both sets. Moreover</u>, $A_3 = (A_1)_{\mathfrak{M}} = (A_2)_{\mathfrak{M}}$.

<u>Proof</u>: $\mathcal{I}_3 \subseteq \mathcal{I}_1 \cap \mathcal{I}_2$ and $\mathcal{P}_3 \subseteq \mathcal{P}_1 \cap \mathcal{P}_2 \subseteq \mathcal{I}_i \cap \mathcal{P}_j$ $(i,j \in [1,2],\ i\neq j)$ follow from (6.9) and (6.7), respectively. Let $\mathfrak{A} \in \mathcal{I}_1 \cap \mathcal{I}_2$; then for any $a \in \mathfrak{A}$, $r \in \mathbb{N}$, and $x_i \in A_1$, $y_i \in A_2$ (i=1,...,r) , we have $(\sum_{i=1}^{r} x_i\cdot y_i)\cdot a = \sum_{i=1}^{r} x_i\cdot(y_i\cdot a) \in \mathfrak{A}$, hence $\mathfrak{A} \in \mathcal{I}_3$ and

therefore $\mathfrak{A} \subseteq \mathfrak{M}_{A_3}$. In particular if $\mathfrak{A} \in (\mathcal{J}_1 \cap \mathcal{P}_2) \cup (\mathcal{J}_2 \cap \mathcal{P}_1)$ then $\mathfrak{A} \in \mathcal{P}_3$ by (6.7). Therefore, we have $\mathcal{J}_3 = \mathcal{J}_1 \cap \mathcal{J}_2$, $\mathcal{P}_3 = \mathcal{P}_1 \cap \mathcal{P}_2 = \mathcal{J}_1 \cap \mathcal{P}_2 = \mathcal{P}_1 \cap \mathcal{J}_2$. Obviously $\mathfrak{M}_{A_3}$ is the largest element of both sets. The last assertion follows from (6.6). □

The following corollary is an immediate consequence of (6.10):

(6.11) COROLLARY - <u>Let</u> A_1, A_2 <u>be valuation rings of</u> K. <u>Then</u>:

$$A_1 \cdot A_2 = K \Leftrightarrow \mathfrak{M}_{A_1 \cdot A_2} = (0) \Leftrightarrow \mathcal{J}_1 \cap \mathcal{J}_2 = \{(0)\} \Leftrightarrow \mathcal{P}_1 \cap \mathcal{P}_2 = \{(0)\} .$$

In this case, A_1, A_2 are called <u>independent</u> (of each other).

(6.12) COROLLARY - <u>Let</u> A_1, A_2 <u>be valuation rings of</u> K. <u>Then</u>:

$$\mathfrak{M}_{A_1} \in \mathcal{J}_2 \Leftrightarrow A_2 \subseteq A_1 \Leftrightarrow \mathfrak{M}_{A_1} \in \mathcal{P}_2 .$$

<u>Proof</u>: Let $\mathfrak{M}_{A_1} \in \mathcal{J}_2$. Then $\mathfrak{M}_{A_1}$ is the largest element of $\mathcal{J}_1 \cap \mathcal{J}_2$, hence $\mathfrak{M}_{A_1} = \mathfrak{M}_{A_1 \cdot A_2}$ by (6.10), hence $A_2 \subseteq A_1 \cdot A_2 = A_1$ by (6.6). Let $A_2 \subseteq A_1$; then $\mathfrak{M}_{A_1} \in \mathcal{P}_2$ by (6.7). The rest follows from the inclusion $\mathcal{P}_2 \subseteq \mathcal{J}_2$. □

For a generalization of theorem (6.10) and Corollary (6.11) see Exercise II-2.

§7 <u>Krull valuations</u>

A <u>Krull valuation</u> of a field K is a surjective mapping $v: K \to \Gamma \cup \{\infty\}$ satisfying the conditions (E_1), (E_2), and (E_3) of §3, where Γ is an arbitrary totally ordered abelian group, written additively [18]; Γ is called the <u>value group</u> of v. This notion

---

[18] Of course the total ordering $\leqq$ of Γ is assumed to be compatible with the addition, i.e., $\gamma \leqq \gamma'$ implies $\gamma + \delta \leq \gamma' + \delta$

obviously generalizes the notion of exponential valuation inasmuch as the value group Γ need not be any longer a subgroup of R^+. Note that $v(x+y) = \min\{vx, vy\}$ whenever $vx \neq vy$.

Two Krull valuations v_1, v_2 of K with the value groups Γ_1, Γ_2, respectively, are called <u>equivalent</u> if there is an isomorphism (of ordered groups) $\iota: \Gamma_1 \to \Gamma_2$ such that $v_2 = \iota \circ v_1$ (with the convention $\iota\infty = \infty$). Note that any bijective order-preserving homomormophism $\iota: \Gamma_1 \to \Gamma_2$ is an isomorphism; in particular, its inverse is also order-preserving. Note also that this equivalence coincides for exponential valuations with the equivalence introduced in §3.

(7.1) <u>For any Krull valuation</u> v <u>of</u> K, <u>the set</u> $A_v = \{x \in K \mid vx \geqq 0\}$ <u>is a valuation ring of</u> K.

<u>The mapping</u> $v \mapsto A_v$ <u>induces a bijection from the set of all equivalence classes of Krull valuations of</u> K <u>onto the set of all valuation rings of</u> K.

<u>Proof</u>: A_v is a valuation ring of K, since $x \in K \setminus A_v$ implies $vx < 0$, $vx^{-1} > 0$, $x \in A_v$. For equivalent v_1, v_2 we have $v_1x \geqq 0$ if and only if $v_2x \geqq 0$ for all $x \in K$, hence $A_{v_1} = A_{v_2}$. For any valuation ring A of K we define a (the canonical) Krull valuation as follows: The divisibility of K corresponding to A is a total quasi-ordering of the multiplicative group K^* of K, by (6.2). The factor group $\Gamma_A = K^*/U_A$ is a totally ordered abelian group; we write it additively and denote its total ordering by $\leqq$. The canonical homomorphism $v_A: K^* \to \Gamma_A$, extended to K by setting $v_A 0 = \infty$, is a Krull valuation of K with value group Γ_A and

---

<u>18</u> for all $\gamma, \gamma', \delta \in \Gamma$. Note that $\leqq$ is uniquely determined by the subset $\Pi = \{\gamma \in \Gamma \mid \gamma \geqq 0\}$.

$A_{v_A} = A$. In fact, $v_A: K \to \Gamma_A \cup \{\infty\}$ is surjective and obviously satisfies (E_1) and (E_2), as well as $[x \in A \Leftrightarrow vx \geqq 0]$ for all $x \in K$. Let $x,y \in K$ such that $v_A x \leqq v_A y$; then $x^{-1}.y \in A$, hence $1 + x^{-1}\cdot y \in A$, hence $v_A(x+y) = v_A x + v_A(1 + x^{-1}\cdot y) \geqq v_A x = \min\{v_A x, v_A y\}$; therefore (E_3) is satisfied, too. We have still to show that if $A = A_v$ then v is equivalent to v_A. In fact, since v is surjective and has kernel U_A, it induces an order-preserving bijection ι from $\Gamma_A = K^*/U_A$ onto the value group Γ of v, and this is even an isomorphism $\iota: \Gamma_A \to \Gamma$ such that $v = \iota \circ v_A$. □

The preceding proof shows that the Krull valuations of K are essentially (up to equivalence) the canonical mappings $v_A: K^* \to \Gamma_A = K^*/U_A$ corresponding to valuation rings A of K. In fact, it would be possible (but sometimes inconvenient) to work only with these canonical Krull valuations.

. We mention without proof that any Krull valuation v of K defines in K a Hausdorff topology, compatible with its field structure and such that K is totally disconnected. The sets $\mathfrak{W}_\gamma(x) = \{y \in K \mid v(y-x) > \gamma\}$ $(\gamma \in \Gamma)$ form a fundamental system of neighborhoods of x. They are open and closed, and so are the sets $\mathfrak{W}'_\gamma(x) = \{y \in K \mid v(y-x) \geqq \gamma\}$. One can also construct the completion $(\hat{K},\hat{v})$ of (K,v), by means of Cauchy filters (see Bourbaki [5], Chap. 6, §5).

Some information on a Krull valuation can be obtained by studying its value group. For this purpose we define: An <u>isolated subgroup</u> of a totally ordered abelian group Γ is a subgroup Φ of Γ such that $\{\gamma \in \Gamma \mid 0 \leqq \gamma \leqq \varphi\} \subseteq \Phi$ for any $\varphi \in \Phi$. In particular, $\{0\}$ and Γ are isolated subgroups of Γ. It is easy to prove that the set $\mathcal{G}(\Gamma)$ of all isolated subgroups of Γ is totally ordered by inclusion. The order type of $\mathcal{G}(\Gamma)\setminus\{\Gamma\}$ is called the <u>rank of</u> Γ.

In particular, Γ has rank 0 if and only if $\Gamma = \{0\}$. We show:

(7.2) Γ has rank $\leqq 1$ if and only if Γ is archimedean (i.e., if for all positive $\alpha, \beta \in \Gamma$ there is some $n \in \mathbb{N}$ such that $n\alpha \geqq \beta$).

Proof: We can assume $\Gamma \neq \{0\}$. For any $\alpha \in \Gamma$, $\alpha > 0$, $\Phi_\alpha =$

$$= \bigcup_{n\in\mathbb{N}} \{\beta \in \Gamma \mid -n\alpha \leqq \beta \leqq n\alpha\}$$ is the smallest isolated subgroup of Γ containing α. We have rank $\Gamma = 1$ if and only if $\Phi_\alpha = \Gamma$ for all $\alpha > 0$, if and only if for all $\alpha > 0$ and $\beta > 0$ there is some $n \in N$ such that $\beta \leqq n\alpha$. □

We note without proof the well-known fact that a totally ordered abelian group is archimedean if and only if it is isomorphic to a subgroup or $\mathbb{R}^+$, endowed with its natural total ordering.

There exist totally ordered abelian groups of rank >1, for example products $\Gamma_1 \times \Gamma_2$ of two non-zero totally ordered groups with lexicographic ordering

$$(\gamma_1,\gamma_2) \leqq (\delta_1,\delta_2) \Leftrightarrow [\gamma_1 < \delta_1 \quad \text{or} \quad (\gamma_1 = \delta_1 \quad \text{and} \quad \gamma_2 \leqq \delta_2)].$$

In the theory of totally ordered abelian groups, the isolated subgroups play a role similar to that of the normal subgroups in the theory of arbitrary groups:

(7.3) a) For any isolated subgroup Φ of Γ, the quotient group $\bar{\Gamma} = \Gamma/\Phi$ is totally ordered by setting $\gamma + \Phi \geqq \bar{0}$ if and only if $\gamma \geqq \varphi$ for some $\varphi \in \Phi$. The canonical mapping $\Gamma \to \Gamma/\Phi$ is a homomorphism (of ordered groups).

b) If $f: \Gamma \to \Gamma'$ is a homomorphism from Γ onto a totally ordered group Γ', then the kernel Φ of f is an isolated subgroup of Γ, and f induces an isomorphism from the totally ordered group Γ/Φ onto Γ'.

Proof: a) The canonical group homomorphism $k: \Gamma \to \bar{\Gamma}$ maps $\Pi = \{\gamma \in \Gamma \mid \gamma \geqq 0\}$ onto a submonoid $\bar{\Pi}$ of $\bar{\Gamma}$ such that $\bar{\Pi} \cup -\bar{\Pi} = \bar{\Gamma}$. For any $\gamma_1, \gamma_2 \in \Pi$, $k\gamma_1 = -k\gamma_2$ implies $k(\gamma_1 + \gamma_2) = \bar{0}$, $\gamma_1 + \gamma_2 \in \Phi$, hence $\gamma_1 \in \Phi$, $k\gamma_1 = \bar{0}$; therefore $\bar{\Pi} \cap -\bar{\Pi} = \{\bar{0}\}$. Therefore $\bar{\Pi}$ defines a total ordering of $\bar{\Gamma}$, for which $\gamma + \Phi \geqq \bar{0} \Leftrightarrow$ $\Leftrightarrow k\gamma \in \bar{\Pi}$ $\Leftrightarrow$ $[k\gamma = k\pi$ for some $\pi \in \Pi] \Leftrightarrow [\gamma - \pi \in \Phi$ for some $\pi \in \Pi] \Leftrightarrow$ $\Leftrightarrow [\gamma \geqq \varphi$ for some $\varphi \in \Phi]$.

b) Let $\varphi \in \Phi$ (=kernel of f). For any $\gamma \in \Gamma$ such that $0 \leqq \gamma \leqq \varphi$ we have $0 \leqq f\gamma \leqq f\varphi = 0$, hence $\gamma \in \Phi$; therefore Φ is an isolated subgroup of Γ. Obviously, f induces a group-isomorphism $\bar{f}: \bar{\Gamma} \to \Gamma'$. If $\gamma + \Phi \geqq \bar{0}$, then $\gamma \geqq \varphi$ for some $\varphi \in \Phi$, hence $\bar{f}(\gamma + \Phi) = f\gamma \geqq f\varphi = 0$; therefore $\bar{f}$ is order-preserving and is hence an isomorphism (of ordered groups). □

Let v be a Krull valuation of K with valuation ring A and value group Γ, and let $\mathfrak{B}$ be the set of all subrings of K containing A. For any $B \in \mathfrak{B}$ we consider the canonical Krull valuation $v_B: K \to \Gamma_B \cup \{\infty\}$. Since the kernel U_A of v is contained in the kernel U_B of v_B, there exists one and only one group homomorphism $g_B: \Gamma \to \Gamma_B$ such that $v_B = g_B \circ v$; g_B is surjective, has kernel $\Phi_B = v(U_B)$, and is order-preserving.

(7.4) THEOREM - $B \mapsto \Phi_B$ <u>is a bijective mapping from</u> $\mathfrak{B}$ <u>onto</u> $\mathcal{G}(\Gamma)$. <u>It is inclusion-preserving, and so is its inverse</u>.

Proof: By (7.3), for any $B \in \mathfrak{B}$ we have $\Phi_B \in \mathcal{G}(\Gamma)$. For any $\Phi \in \mathcal{G}(\Gamma)$, Γ/Φ is a totally ordered abelian group, and the composition $w_\Phi = f \circ v$ of v with the canonical homomorphism $f: \Gamma \to \Gamma/\Phi$ is a Krull valuation of K with valuation ring $B_\Phi \supseteq A$. For any $B \in \mathfrak{B}$, $g_B: \Gamma \to \Gamma_B$ induces an isomorphism $\bar{g}_B: \Gamma/\Phi_B \to \Gamma_B$ such that $v_B = \bar{g}_B \circ w_{\Phi_B}$, hence $B_{\Phi_B} = B$, by (7.1). For any $\Phi \in \mathcal{G}(\Gamma)$ there is an isomorphism $\iota: \Gamma/\Phi \to \Gamma_{B_\Phi}$ such that $v_{B_\Phi} = \iota \circ w_\Phi$, hence

$g = \iota \circ f : \Gamma \to \Gamma_{B_\Phi}$ is a homomorphism with kernel Φ such that $v_{B_\Phi} = g \circ v$; hence $g = g_{B_\Phi}$, $\Phi = \Phi_{B_\Phi}$. Obviously $B \mapsto \Phi_B$ is inclusion-preserving, and so is its inverse, since $\mathfrak{B}$ is totally ordered. □

From (7.4) we conclude

(7.5) COROLLARY - The rank of the value group of any Krull valuation v of K coincides with the rank of the valuation ring corresponding to v.

The composition of the inclusion-inverting (resp. -preserving) 1-1 correspondeces $\mathfrak{P} \leftrightarrow \mathfrak{B}$ and $\mathfrak{B} \leftrightarrow \mathfrak{G}(\Gamma)$ yields an inclusion-inverting 1-1 correspondence $\mathfrak{P} \leftrightarrow \mathfrak{G}(\Gamma)$ (see Exercise II- 3).

Finally, (7.4), (7.2) and the remark following (7.2) give another characterization of rank 1 valuation rings:

(7.6) COROLLARY - Let A be a valuation ring of K and v a corresponding Krull valuation. The following conditions are equivalent:

(i) A has rank ≤ 1.

(ii) The value group of v is archimedean.

(iii) v is equivalent to some exponential valuation.

We are going to give some characterizations of discrete valuation rings (i.e., valuation rings of a discrete exponential valuation, see §4) within the class of local integral domains.

(7.7) THEOREM - Let R be a local integral domain distinct from its quotient field K. The following conditions are equivalent:

(i) R is a discrete valuation ring of K.

(ii) R is a noetherian valuation ring of K.

(iii) R is a principal ideal ring.

(iv) R is noetherian and $\mathfrak{M}_R$ is a principal ideal.

(v) $\bigcap_{n=0}^{\infty} \mathfrak{M}_R^n = (0)$ <u>and</u> $\mathfrak{M}_R$ <u>is a principal ideal.</u>

<u>Proof</u>: (i) ⇒ (ii) follows from (4.5). (ii) ⇒ (iii) follows from (6.5). (iii) ⇒ (iv) is trivial. (iv) ⇒ (v) follows from the well known "Intersection theorem" (see for example Zariski-Samuel [36] Chap. IV, §7, Th. 12). (v) ⇒ (i): We have $\mathfrak{M}_R = t\cdot R$ for some $t \in R$. From $\bigcap_{n=0}^{\infty} \mathfrak{M}_R^n = (0)$ it follows that for any $x \in R \setminus \{0\}$, $\{n \in \mathbb{N} \mid x \in t^n \cdot R\}$ has a maximum vx (say) in $\mathbb{N}$. Letting $v0 = \infty$, we obviously have $v(x+y) \geqq \min\{vx, vy\}$ for any $x,y \in R$. Let $x_i \in R \setminus \{0\}$; then $x_i = u_i \cdot t^{vx_i}$ for some $u_i \in U_R$ $(i = 1,2)$; since $x_1 \cdot x_2 = u_1 \cdot u_2 \cdot t^{vx_1+vx_2}$ and $u_1 \cdot u_2 \in U_R$ we have $v(x_1 \cdot x_2) = vx_1 + vx_2$, and the same equation is obvious whenever $x_1 = 0$ or $x_2 = 0$. The resulting map $v\colon R \to \mathbb{N} \cup \{\infty\}$ extends uniquely to a normalized discrete exponential valuation of the quotient field K of R. Obviously R is contained in the valuation ring A of v. On the other hand, if $x_i = u_i \cdot t^{vx_i} \in R \setminus \{0\}$ $(i=1,2)$ then $x_1 \cdot x_2^{-1} = u_1 \cdot u_2^{-1} \cdot t^{vx_1 - vx_2}$, $v(x_1 \cdot x_2^{-1}) = vx_1 - vx_2$; hence $v(x_1 \cdot x_2^{-1}) = n \geqq 0$ implies $x_1 \cdot x_2^{-1} = u_1 \cdot u_2^{-1} \cdot t^n \in R$. □

Other characterizations of discrete valuation rings, within the class of all commutative rings and the class of noetherian integral domains, can be found in Serre [33], Chap. I.

We conclude this section proving a lemma on "rational" valuation rings, which will be used in §12.

(7.8) <u>Let</u> A <u>be the ring of an exponential valuation</u> v <u>of</u> K <u>with value group</u> $\Gamma \subseteq \mathbb{Q}$, D <u>a subring of</u> K <u>not contained in</u> A, $R = D \cap A$, <u>and assume that</u> K <u>is the quotient field of</u> R. <u>Then</u> $\mathfrak{P} = \mathfrak{M}_A \cap R$ <u>is a minimal non-zero prime ideal of</u> R.

<u>Proof</u>: From $D \not\subseteq A$, $D \subseteq K$ we conclude $A \neq K$, $\mathfrak{M}_A \neq (0)$. Since K is the quotient field of R we have also $\mathfrak{P} \neq (0)$. Let

$\mathfrak{Q}$ be a prime ideal of R which is strictly contained in $\mathfrak{P}$. Let $y \in \mathfrak{P} \setminus \mathfrak{Q}$ and $z \in D \setminus A$. From $\Gamma \subseteq \mathbb{Q}$ we conclude that, for any non-zero $r \in R$, there exist $m, n, k \in \mathbb{N}$ such that $m \cdot vz = -n \cdot vr$ and $k \cdot vy \geqq vr$, hence $v(r^n . z^m) = 0$ and $v(r^{n-1} . z^m . y^k) \geqq 0$, hence $\{r^n . z^m , r^{n-1} . z^m . y^k\} \subseteq D \cap A = R$ and $r^n . z^m \notin \mathfrak{P}$. Suppose that $r \in \mathfrak{Q}$; then $r^n . z^m . y^k \in \mathfrak{Q}$, hence $r^n . z^m \in \mathfrak{Q}$ or $y \in \mathfrak{Q}$, a contradiction. Therefore $\mathfrak{Q} = (0)$. □

§8 Places

We have seen in §7 that the valuation rings A of K are in 1-1 correspondence with the equivalence classes of Krull valuations of K and that the Krull valuations corresponding to A are essentially the canonical homomorphism from the multiplicative group K^* of K onto the quotient group K^*/U_A, totally ordered by the divisibility corresponding to A.

Similarly we show now that the valuation rings A of K are in 1-1 correspondence with the equivalence classes of places of K and that the places corresponding to A are essentially the canonical homomorphism from A onto the residue field $A/\mathfrak{M}_A$.

For the definition of places, we have to extend fields to projective fields, adjoining an element ∞. More precisely, the <u>projective field</u> obtained from the field K is the set $\tilde{K} = K \cup \{\infty\}$ endowed with the addition and the multiplication of K extended to $\tilde{K}$ by

$$x + \infty = \infty + x = \infty \quad \text{for all } x \in K$$

$$x \cdot \infty = \infty \cdot x = \infty \quad \text{for all non-zero } x \in \tilde{K}.$$

Moreover, we set $0^{-1} = \infty$, $\infty^{-1} = 0$, and $-\infty = \infty$. Note that $\infty + \infty$, $0 \cdot \infty$, and $\infty \cdot 0$ are not defined.

A <u>place of</u> K <u>into</u> L is a mapping $\pi : \tilde{K} \to \tilde{L}$ satisfying the following conditions for all $x, y \in \tilde{K}$:

(P_1) If $x + y$ and $\pi x + \pi y$ are defined then $\pi(x+y) = \pi x + \pi y$.

(P_2) If $x \cdot y$ and $\pi x \cdot \pi y$ are defined then $\pi(x \cdot y) = \pi x \cdot \pi y$.

(P_3) There is some $z \in \tilde{K}$ such that $\pi z = 1$.

We state some elementary properties of places:

(8.1) a) $\pi 1 = 1$, $\pi 0 = 0$, $\pi\infty = \infty$.

b) <u>If</u> $\pi x + \pi y$ (<u>resp</u>. $\pi x \cdot \pi y$) <u>is defined then so is</u> $x + y$ (<u>resp</u>. $x \cdot y$).

c) $\pi(-x) = -\pi x$.

d) $\pi(x^{-1}) = (\pi x)^{-1}$.

e) $\pi^{-1}L$ <u>is a valuation ring</u> A_π <u>of</u> K , <u>and</u> $\pi | A_\pi : A_\pi \to L$ <u>is a (ring) homomorphism with kernel</u> $\mathfrak{M}_{A_\pi}$.

f) $\pi^{-1}L^* = U_{A_\pi}$, <u>and</u> $\pi | U_{A_\pi} : U_{A_\pi} \to L^*$ <u>is a multiplicative homomorphism with kernel</u> $1 + \mathfrak{M}_{A_\pi}$.

<u>Proof</u>: a) Let $z \in \tilde{K}$ such that $\pi z = 1$. Then $z \cdot 1$ and $\pi z \cdot \pi 1$ are defined, hence $1 = \pi z = \pi(z \cdot 1) = \pi z \cdot \pi 1 = 1 \cdot \pi 1 = \pi 1$. Since $1 + 0$ and $\pi 1 + \pi 0$ are defined, we have $1 = \pi 1 = \pi(1+0) = \pi 1 + \pi 0 = 1 + \pi 0$, hence $\pi 0 = 0$. Since $1 + \infty$ and $\pi 1 + \pi\infty$ are defined, we have $\pi\infty = \pi(1 + \infty) = \pi 1 + \pi\infty = 1 + \pi\infty$, hence $\pi\infty = \infty$.

b) If $\pi x + \pi y$ is defined then $(\pi x, \pi y) \neq (\infty, \infty)$, hence $(x,y) \neq (\infty,\infty)$ by a), hence $x + y$ is defined. If $\pi x \cdot \pi xy$ is defined then $(\pi x, \pi y) \notin \{(0,\infty),(\infty,0)\}$, hence $(x,y) \notin \{(0,\infty),(\infty,0)\}$ by a), hence $x \cdot y$ is defined.

c) If $\pi(-x) + \pi x$ is not defined then $\pi x = \pi(-x) = \infty$, hence $-\pi x = \infty$. If $\pi(-x) + \pi x$ is defined then so is $-x + x$, hence

$0 = \pi(-x + x) = \pi(-x) + x$, hence $\pi(-x) = -\pi x$.

d) If $\pi x^{-1}\cdot\pi x$ is not defined then $(\pi x^{-1}, \pi x) \in \{(0,\infty),(\infty,0)\}$, hence $\pi(x^{-1}) = (\pi x)^{-1}$. If $\pi x^{-1}\cdot\pi x$ is defined, then so is $x^{-1}\cdot x$, hence $1 = \pi(x^{-1}\cdot x) = \pi x^{-1}\cdot\pi x$, hence $\pi x^{-1} = (\pi x)^{-1}$.

e) We have $\pi^{-1}L \subseteq K$ since $\pi\infty = \infty$. Let $x,y \in \pi^{-1}L$; then $\pi x + \pi y$ and $\pi x\cdot\pi y$ are defined and so are $x + y$ and $x\cdot y$ by b), hence $\pi(x+y) = \pi x + \pi y \in L$ and $\pi(x\cdot y) = \pi x\cdot\pi y \in L$, hence $x+y$, $x\cdot y \in \pi^{-1}L$. Moreover $-x \in \pi^{-1}L$, by c). Therefore $\pi^{-1}L$ is a subring A_π of K . If $x \in L \setminus A_\pi$, then $\pi x = \infty$, hence $\pi x^{-1} = 0$, by d), hence $x^{-1} \in A_\pi$; therefore A_π is a valuation ring of K . Obviously $\pi|A_\pi : A_\pi \to L$ is a ring homomorphism with kernel $\mathfrak{N}_\pi \subseteq \mathfrak{M}_{A_\pi}$. We have even $\mathfrak{N}_\pi = \mathfrak{M}_{A_\pi}$, since $x \in \mathfrak{M}_{A_\pi}$ implies $x^{-1} \notin A_\pi$, $(\pi x)^{-1} = \pi x^{-1} = \infty$, $\pi x = 0$.

f) $\pi^{-1}L^* = \{x \in A_\pi \mid \pi x \neq 0\} = A_\pi \setminus \mathfrak{M}_{A_\pi}$, so $\pi|U_{A_\pi} : U_{A_\pi} \to L^*$ is a multiplicative homomorphism. Its kernel is $1 + \mathfrak{M}_{A_\pi}$, since $\pi x = 1$ if and only if $\pi(x-1) = 0$, if and only if $x - 1 \in \mathfrak{M}_{A_\pi}$. □

By (8.1) e), any place of K into L induces a homorphism $\lambda : A \to L$ from a valuation ring A of K into L, with kernel $\mathfrak{M}_A$. The converse is also true:

(8.2) <u>Let</u> A <u>be a valuation ring of</u> K <u>and</u> $\lambda : A \to L$ <u>a homomorphism into a field</u> L , <u>with kernel</u> $\mathfrak{M}_A$. <u>Then the mapping</u> $\pi : \tilde{K} \to \tilde{L}$, <u>defined by</u> $\pi x = \lambda x$ <u>for all</u> $x \in A$ <u>and</u> $\pi x = \infty$ <u>for all</u> $\tilde{K} \setminus A$, <u>is a place of</u> K <u>into</u> L <u>with</u> $A_\pi = A$.

<u>Proof</u>: It suffices to verify (P_1) and (P_2). $\pi x + \pi y$ is defined only if $(\pi x, \pi y) \neq (\infty,\infty)$. If $\pi x \neq \infty = \pi y$, then $x \in A$, $y \in \tilde{K} \setminus A$, hence $\pi(x+y) = \infty = \pi x + \pi y$. If $\pi x \neq \infty \neq \pi y$, then $x,y \in A$, hence $\pi(x+y) = \lambda(x+y) = \lambda x + \lambda y = \pi x + \pi y$. If $(\pi x, \pi y) \in \{(0,\infty),(\infty,0)\}$ then $\pi x\cdot\pi y$ is not defined. Let $x,y \in \tilde{K}$ such that $\pi x = \infty$, $\pi y \neq 0$. If $x = \infty$ or $y = \infty$ then $\pi(x\cdot y) = \infty = \pi x\cdot\pi y$; if

$x \neq \infty \neq y$ then $x \in K \setminus A$, $y \in U_A$, hence $x \cdot y \in K \setminus A$, $\pi(x \cdot y) = \infty = \pi x \cdot \pi y$. If $\pi x \neq \infty \neq \pi y$ then $x, y \in A$, hence $\pi(x \cdot y) = \lambda(x \cdot y) = \lambda x \cdot \lambda y = \pi x \cdot \pi y$. □

A place $\pi: \widetilde{K} \to \widetilde{L}$ is called <u>trivial</u> if $A_\pi = K$ or, equivalently, $\mathfrak{M}_{A_\pi} = (0)$. The trivial places of K into L are exactly the monomorphisms $\mu: K \to L$ extended by $\mu\infty = \infty$.

Let K_o be any subfield of K. It is obvious that $\widetilde{K}_o$ is a projective subfield of $\widetilde{K}$ (i.e., addition and multiplication in $\widetilde{K}_o$ are induced by those in $\widetilde{K}$) and that, for any place $\pi: K \to L$, the restriction $\pi|K_o: \widetilde{K}_o \to \widetilde{L}$ is a place of K_o into L. In particular, it is checked easily that the restriction of π to the prime field of K is non-trivial if and only if Char L $\neq$ Char K, and in this case Char K = 0. (Note that this statement generalizes (i) $\Leftrightarrow$ (ii) of (3.10). As to condition (iii) of (3.10), cf. Exercise II-4).

For any place $\pi: \widetilde{K} \to \widetilde{L}$, the image πA_π of A_π is a subfield of L, called the <u>residue field</u> of π, and π can also be considered as a place of K into πA_π. We have $L = \pi A_\pi$ if and only if $\pi: \widetilde{K} \to \widetilde{L}$ is surjective; in this case, π is called a place of K <u>onto</u> L, or a <u>surjective place</u>. In particular, for any valuation ring A of K, the canonical homomorphism $\varkappa_A: A \to A/\mathfrak{M}_A$ extends to a place π_A of K onto $A/\mathfrak{M}_A$, by (8.2); π_A is called the <u>canonical place</u> corresponding to A.

Places can be composed similarly as homomorphisms:

(8.3) <u>Let</u> $\pi: \widetilde{K} \to \widetilde{L}$ <u>and</u> $\zeta: \widetilde{L} \to \widetilde{M}$ <u>be places and let</u> A_π (<u>resp.</u> B_ζ) <u>be the valuation ring of</u> K (<u>resp.</u> L) <u>corresponding to</u> π (<u>resp.</u> ζ). <u>Then</u> $\zeta \circ \pi: \widetilde{K} \to \widetilde{M}$ <u>is a place and</u> $A_{\zeta \circ \pi} = \pi^{-1} B_\zeta \subseteq A_\pi$. <u>If</u> π <u>is a place of</u> K <u>onto</u> L, <u>then</u> $\pi A_{\zeta \circ \pi} = B_\zeta$.

Proof: $\zeta\circ\pi$ satisfies (P_1), since if $(\zeta\circ\pi)x + (\zeta\circ\pi)y$ is defined then, by (8.1 b), $\pi x + \pi y$ and $x + y$ are defined, hence $(\zeta\circ\pi)x + (\zeta\circ\pi)y = \zeta(\pi x + \pi y) = (\zeta\circ\pi)(x+y)$. Similarly (P_2) is verified. (P_3) follows from $\zeta(\pi 1) = \zeta 1 = 1$. For any $x \in \widetilde{K}$ we have $x \in A_{\zeta\circ\pi}$ if and only if $\zeta(\pi x) \neq \infty$, if and only if $\pi x \in B_\zeta$, if and only if $x \in \pi^{-1}B_\zeta \subseteq \pi^{-1}L = A_\pi$. If $\pi: \widetilde{K} \to \widetilde{L}$ is surjective, then $A_{\zeta\circ\pi} = \pi^{-1}B_\zeta$ implies $\pi A_{\zeta\circ\pi} = B_\zeta$. □

We use the composition of places for defining a quasi-ordering on the class of all surjective places of a fixed field K. Let π_o, π_1 be surjective places of K ; we write $\pi_1 < \pi_o$ if the equivalent conditions of the following proposition hold:

(8.4) Let π_i be a place of K onto L_i and $A_i = \pi_i^{-1}L_i$ (i=0,1). The following conditions are equivalent:

(i) $A_1 \subseteq A_o$.

(ii) There exists a mapping $\zeta: \widetilde{L}_o \to \widetilde{L}_1$ such that $\pi_1 = \zeta\circ\pi_o$.

In this case ζ is a uniquely determined place of L_o onto L_1.

Proof: (i) ⇒ (ii): It suffices to prove that $\zeta: \pi_o x \mapsto \pi_1 x$ $(x \in \widetilde{K})$ is well-defined. Let $x,y \in \widetilde{K}$ such that $\pi_o x = \pi_o y$.
a) If $x \notin A_o$, then $\pi_o x = \pi_o y = \infty$, $y \notin A_o$, hence $x,y \notin A_1$, hence $\pi_1 x = \pi_1 y = \infty$. b) If $x \in A_o$ then $\pi_o x \neq \infty$, hence $\pi_o x + \pi_o(-y)$ is defined and $\pi_o(x-y) = \pi_o x + \pi_o(-y) = \pi_o x - \pi_o y = 0$, hence $x-y \in \mathfrak{M}_{A_o} \subseteq \mathfrak{M}_{A_1} \subseteq A_1$ by (6.6). Let $x \notin A_1$; then $y \notin A_1$, $\pi_1 x = \infty = \pi_1 y$. Let $x \in A_1$; then $\pi_1 x \neq \infty$, hence $\pi_1 x + \pi_1(-y)$ is defined and $\pi_1 x - \pi_1 y = \pi_1 x + \pi_1(-y) = \pi_1(x-y) = 0$, hence $\pi_1 x = \pi_1 y$.

(ii) ⇒ (i): To prove that ζ is a place of L_o, it suffices to verify (P_1) and (P_2) for ζ. Let $\bar{x},\bar{y} \in \widetilde{L}_o$ such that $\bar{x} + \bar{y}$ and $\zeta\bar{x} + \zeta\bar{y}$ are defined, and let $x,y \in \widetilde{K}$ such that $\bar{x} = \pi_o x$, $\bar{y} =$

$= \pi_o y$; then $\pi_1 x = \zeta\bar{x}$, $\pi_1 y = \zeta\bar{y}$. Since $\pi_o x + \pi_o y$ (resp. $\pi_1 x + \pi_1 y$) is defined, we have $\pi_o(x + y) = \pi_o x + \pi_o y$ (resp. $\pi_1(x + y) = \pi_1 x + \pi_1 y$), by (8.1 b); hence $\zeta(\bar{x} + \bar{y}) = \zeta(\pi_o(x+y)) = \pi_1(x + y) = \pi_1 x + \pi_1 y = \zeta\bar{x} + \zeta\bar{y}$. (P_2) is proven similarly. Hence ζ is a place of L_o , and (i) follows from (8.3). Since π_o and π_1 are surjective, ζ is uniquely determined and is a place of L_o onto L_1 . □

In particular, for any surjective place π of K we have $\pi < \iota_K$ (the trivial place determined by the identity of K), and $\iota_K < \pi$ if and only if π is trivial.

Two surjective places π_o, π_1 of K are called equivalent, if $\pi_o < \pi_1$ and $\pi_1 < \pi_o$. We conclude from (8.4):

(8.5) With the notations of (8.4) the following conditions are equivalent:

(i) $A_1 = A_o$.

(ii) $\pi_1 = \zeta\circ\pi_o$ for some bijective mapping $\zeta: \tilde{L}_o \to \tilde{L}_1$.

(iii) $\pi_1 = \zeta\circ\pi_o$ for some trivial place ζ of L_o .

(iv) π_1 is equivalent to π_o .

Moreover, (8.2) and (8.5) yield the following statement, similar to (7.1):

(8.6) The mapping $\pi \mapsto A_\pi$ induces a bijection from the set of all equivalence classes of surjective places of K onto the set of all valuation rings of K .

By means of the composition of places, one gets a survey on the set of all valuation rings of K contained in some given valuation ring A_o of K . In fact, these valuation rings are in 1-1 correspondence with the valuation rings of the residue field

$A_o/\mathfrak{M}_{A_o}$, as the following theorem shows.

(8.7) THEOREM - Let A_o be a valuation ring of K and π_o a place of K onto L_o with $A_{\pi_o} = A_o$. Then there is an inclusion preserving 1-1 correspondence between the set $\mathfrak{A}_o$ of all valuation rings A of K contained in A_o and the set $\mathfrak{B}$ of all valuation rings B of L_o , given by $B = \pi_o A$ and $A = \pi_o^{-1} B$.

Moreover, if B corresponds to A then

$$\text{rank }(A) = \text{rank }(B) + \text{rank }(A_o)$$ [19]

Proof: For any $A \in \mathfrak{A}_o$ we have $\pi_A = \zeta \circ \pi_o$ for some place ζ of L_o onto $A/\mathfrak{M}_A$, by (8.4), and $\pi_o A = B_\zeta \in \mathfrak{B}$ by (8.3). Since $A = \pi_o^{-1} B_\zeta$ by (8.3), the mapping $\mathfrak{A}_o \to \mathfrak{B}$ defined by $A \mapsto \pi_o A$ is injective. It is also surjective, since if $B \in \mathfrak{B}$ then $B = B_\zeta$ for some place ζ of L_o , and $\pi = \zeta \circ \pi_o$ is a place of K with $A_\pi \in \mathfrak{A}_o$ and $\pi_o A_o = B$, by (8.3). Moreover, this mapping obviously is inclusion-preserving and so is its inverse. For any $A \in \mathfrak{A}_o$ the rank of A is equal to $r_1 + r_2$ where r_2 is the rank of A_o and r_1 the order type of the totally ordered set $\{A' \in \mathfrak{A}_o \mid A \subseteq A' \subset A_o\}$, which coincides with the order type of $\{B' \mid \pi_o A \subseteq B' \subset L_o\}$, i.e., the rank of $\pi_o A$. □

Note that the sets $\mathfrak{A}_o$ and $\mathfrak{B}$ are generally not totally ordered by inclusion.

Theorem (8.7) permits the "decomposition" of any valuation ring A of K into a valuation ring A_o of K containing A and a valuation ring B of the residue field $A_o/\mathfrak{M}_{A_o}$ (namely $B = A/\mathfrak{M}_{A_o}$)

---

19 Let r_i be the order type of the totally ordered set S_i $(i=1,2)$ and assume $S_1 \cap S_2 = \emptyset$. Then r_1+r_2 denotes the order type of $S_1 \cup S_2$ with respect to the total ordering which induces those of S_1 and S_2 and is such that $s_1 < s_2$ for all $s_1 \in S_1$, $s_2 \in S_2$. Note that in general $r_1 + r_2 \neq r_2 + r_1$.

such that $\operatorname{rank}(A) = \operatorname{rank}(B) + \operatorname{rank}(A_o)$. Also any Krull valuation v of K corresponding to A, with value group Γ, is "decomposed" into a Krull valuation v_o of K corresponding to A_o, with value group Γ/Φ_{A_o}, and a Krull valuation w of $A_o/\mathfrak{M}_{A_o}$ corresponding to B, with value group Φ_{A_o} (see (7.4) and Exercise II-6). In particular, by induction one can decompose a given valuation ring A of K of rank n into n valuation rings of rank 1, namely B_1 of $L_o = K$, B_2 of $L_1 = B_1/\mathfrak{M}_{B_1}, \ldots, B_n$ of $L_{n-1} = B_{n-1}/\mathfrak{M}_{B_{n-1}}$.

On the other hand, this theorem serves to construct valuation rings of higher rank. In fact, if A_o is a valuation ring of K and B a valuation of $A_o/\mathfrak{M}_{A_o}$, then $\varkappa_{A_o}^{-1}B$ is a valuation ring A of K contained in A_o and such that $\operatorname{rank}(A) = \operatorname{rank}(B) + \operatorname{rank}(A_o)$. For example, if $K = \mathbb{Q}(z)$, z transcendental, let A_o be the valuation ring of K corresponding to $v_{z;X}$ (see (4.4)) and B the valuation ring $\mathbb{Z}_{p \cdot \mathbb{Z}}$ of the p-adic valuation of $\mathbb{Q}$; then $\varkappa_{A_o}^{-1}B = \{\frac{F(z)}{G(z)} \mid F,G \in \mathbb{Z}[X],\ G(0) \neq 0,\ \frac{F(0)}{G(0)} \in \mathbb{Z}_{p\cdot\mathbb{Z}}\}$ is a valuation ring of K of rank 2, strictly contained in A_o.

Note however that the construction of a valuation ring $A \subset A_o$ is possible if and only if $A_o/\mathfrak{M}_{A_o}$ has a non-trivial valuation ring. In particular, $\mathbb{Q}$ has no valuation ring of rank > 1, since the residue field of the p-adic valuation is finite, for any prime number p.

§9 The extension theorem

Roughly speaking, the extension theorem states that any homomorphism from a subring of K into an algebraically closed field Ω is the restriction of some place of K into Ω. This theorem, or rather the first statement of its corollary (9.7), plays a most important role in valuation theory.

We need some preliminary propositions.

(9.1) Let R be a subring [16] of K and $\mathfrak{A}$ a proper ideal of R. Then for any non-zero $x \in K$, $\mathfrak{A}\cdot R[x]$ is a proper ideal of $R[x]$ or $\mathfrak{A}\cdot R[x^{-1}]$ is a proper ideal of $R[x^{-1}]$.

Proof: In fact, otherwise there exist $m,n \in \mathbb{N}$ and $a_o,\dots,a_m$, $b_1,\dots,b_n \in \mathfrak{A}$ such that $1 = \sum_{i=o}^{m} a_i\cdot x^i = \sum_{j=o}^{n} b_j\cdot x^{-j}$, and m,n may be chosen minimal. If $m \geqq n$, then $1-(b_o+a_o-a_o\cdot b_o) = (1-b_o)\cdot(1-a_o) = (1-b_o)\cdot\sum_{i=1}^{m-1} a_i\cdot x^i + a_m\cdot x^m\cdot\sum_{j=1}^{n} b_j\cdot x^{-j}$, hence $1 = \sum_{i=o}^{m-1} c_i\cdot x^i$ with $c_o,\dots,c_{m-1} \in \mathfrak{A}$, contrary to the minimal choice of m. If $n \geqq m$, the reasoning is similar. □

Let $\mathfrak{H}(K,\Omega)$ be the set of pairs (R,μ) such that R is a subring of K and $\mu:R \to \Omega$ a homomorphism. In $\mathfrak{H}(K,\Omega)$ an ordering is defined by

$$(R,\mu) \leqq (S,\nu) \Leftrightarrow [R \subseteq S \text{ and } \mu = \nu|R].$$

(9.2) For any $(R,\mu) \in \mathfrak{H}(K,\Omega)$ and any non-zero $x \in K$ there is a pair $(S,\nu) \in \mathfrak{H}(K,\Omega)$ such that $(R,\mu) \leqq (S,\nu)$ and $x \in S$ or $x^{-1} \in S$.

Proof: The kernel of μ being a prime ideal $\mathfrak{P}$ of R, μ extends to a homomorphism $\mu_{\mathfrak{P}}: R_{\mathfrak{P}} \to \Omega$ by $\mu_{\mathfrak{P}}(a\cdot b^{-1}) = \mu a\cdot(\mu b)^{-1}$, with kernel $\mathfrak{P}\cdot R_{\mathfrak{P}} = \mathfrak{M}_{R_{\mathfrak{P}}}$. We therefore may assume that R is a local ring and $\mathfrak{P} = \mathfrak{M}_R$, hence $\Lambda = \mu R$ is a subfield of Ω. Let μ_X be the homomorphism from $R[X]$ onto $\Lambda[X]$ defined by $\mu_X(\sum_{i=o}^{m} a_i\cdot X^i) = \sum_{i=o}^{n} \mu a_i\cdot X^i$, where $a_o,\dots,a_m \in R$. Assume that $\mathfrak{M}_R\cdot R[x]$ is a proper ideal of $R[x]$ (otherwise $\mathfrak{M}_R\cdot R[x^{-1}]$ is a proper ideal of $R[x^{-1}]$, by (9.1), and x has to be replaced by x^{-1} in the following considerations). Let $\mathfrak{N}$ be the kernel of the R-homomorphism

$R[X] \to R[x]$ determined by $X \mapsto x$; then $\mu_X \mathfrak{N}$ is an ideal of $\Lambda[X]$, and since $\Lambda[X]$ is a PID, we have $\mu_X\mathfrak{N} = \mu_X Q\cdot\Lambda[X]$ for some $Q \in \mathfrak{N}$. We claim that $\mu_X Q$ has a zero $\xi \in \Omega$. In fact, otherwise $\mu_X Q$ would be a non-zero element of Λ , hence $Q = \sum_{i=o}^{m} a_i \cdot X^i$ for appropriate $a_1,\ldots,a_m \in \mathfrak{M}_R$ and $a_o \in R\setminus\mathfrak{M}_R = U_R$, hence $a_o = a_o - Q(x) \in$ $\in \mathfrak{M}_R\cdot R[x]$, contradicting the above assumption. The mapping $F(x) \mapsto (\mu_X F)(\xi)$ (where $F \in R[X]$) is well-defined; in fact, if $F(x) = 0$, then $F \in \mathfrak{N}$, $\mu_X F \in \mu_X\mathfrak{N}$, hence $\mu_X Q$ divides $\mu_X F$ in $\Lambda[X]$, hence $(\mu_X F)(\xi) = 0$. Obviously this mapping is a homomorphism $\nu\colon R[x] \to \Omega$ such that $\nu|R = \mu$, hence $(R[x],\nu)$ has the desired property. □

The following theorem characterizes the places of K into Ω as being (essentially) those homomorphisms $R \to \Omega$ which are not extendable within K .

(9.3) THEOREM - $\pi \mapsto (A_\pi,\pi|A_\pi)$ <u>is a bijective mapping from the set of all places of</u> K <u>into</u> Ω <u>onto the set of all maximal elements of</u> $\mathfrak{H}(K,\Omega)$.

<u>Proof</u>: Obviously it is an injective mapping into $\mathfrak{H}(K,\Omega)$. We claim that, for any place π of K into Ω, $(A_\pi, \pi|A_\pi)$ is maximal in $\mathfrak{H}(K,\Omega)$. In fact, otherwise we have $(A_\pi, \pi|A_\pi) \leqq (S,\nu)$ for some $(S,\nu) \in \mathfrak{H}(K,\Omega)$ with $S \neq A_\pi$; choosing $x \in S \setminus A_\pi$ we have $x^{-1} \in \mathfrak{M}_\pi \subseteq A_\pi \subseteq S$, $\nu x^{-1} = 0$, $0 = \nu x^{-1}\cdot\nu x = \nu(x^{-1}\cdot x) = \nu 1 = 1$, a contradiction. On the other hand, let (R,μ) be a maximal element of $\mathfrak{H}(K,\Omega)$. By (9.2), for every non-zero $x \in K$ we have $x \in R$ or $x^{-1} \in R$, hence R is a valuation ring of K . Since μ extends to a homomorphism $\mu_\mathfrak{P}\colon R_\mathfrak{P} \to \Omega$, where $\mathfrak{P}$ is the kernel of μ , it follows from the maximality of (R,μ) that $R = R_\mathfrak{P}$, hence $\mathfrak{P} = \mathfrak{M}_R$, by (6.6). By (8.2), μ is the restriction of a place π of K into Ω such that $A_\pi = R$. □

Any homomorphism $\mu: R \to \Omega$ of any subring R of K can be extended to a homomorphism $\nu: S \to \Omega$ which is no longer extendable within K. In other words:

(9.4) For any $(R,\mu) \in \mathfrak{H}(K,\Omega)$ there is a maximal element $(S,\nu) \in \mathfrak{H}(K,\Omega)$ such that $(R,\mu) \leq (S,\nu)$.

Proof: The set of all $(S',\nu') \in \mathfrak{H}(K,\Omega)$ such that $(R,\mu) \leq (S',\nu')$ is non-empty and inductively ordered, as is checked easily. Hence by Zorn's lemma it has a maximal element, and this is a maximal element of $\mathfrak{H}(K,\Omega)$. □

Combining (9.3) and (9.4) we get immediately:

(9.5) EXTENSION THEOREM - For any $(R,\mu) \in \mathfrak{H}(K,\Omega)$ there is a place π of K into Ω such that $(R,\mu) \leq (A_\pi, \pi|A_\pi)$.

It is easy to strengthen the extension theorem in the following way:

(9.6) COROLLARY - Let $(R,\mu) \in \mathfrak{H}(K,\Omega)$, T a transcendence basis of K over the quotient field of R, and τ a mapping from T into Ω. Then there exists a place π of K into Ω such that $\pi|R = \mu$ and $\pi|T = \tau$.

Proof: It suffices to apply (9.5) on $(R[T],\nu) \in \mathfrak{H}(K,\Omega)$, where ν is the uniquely determined homomorphism from $R[T]$ into Ω which extends μ and τ. □

This corollary yields the following statement on the existence of valuation rings.

(9.7) COROLLARY - Let R be a subring of K and $\mathfrak{P}$ a prime ideal of R. Then there exists a valuation ring A of K such that $R \subseteq A$ and $\mathfrak{P} = \mathfrak{M}_A \cap R$.

Moreover, for any transcendence basis T of K over the

quotient field K_o of R and any disjoint subsets T_1, T_2 with $T = T_1 \cup T_2$, A can be chosen such that $R[T] \subseteq A$, $T_1 \subseteq \mathfrak{M}_A$, and $\varkappa_A T_2$ is a transcendence basis of $A/\mathfrak{M}_A$ over the quotient field of $\varkappa_A R$.

Proof: Let Λ_o be the quotient field of $R/\mathfrak{P}$. There exists a bijective mapping τ_2 from T_2 onto a transcendence basis of some algebraically closed field extension Ω of Λ_o. Let τ be the extension of τ_2 to T defined by $\tau T_1 = \{0\}$. Let π be a place as in (9.6), where $\mu: R \to R/\mathfrak{P} \subseteq \Omega$ is the canonical homomorphism, and let $A = A_\pi$, $\Lambda = \pi A$. Obviously $R \subseteq R[T] \subseteq A$, $\mathfrak{P} = \mathfrak{M}_A \cap R$ and there is an isomorphism $\iota: \Lambda \to A/\mathfrak{M}_A$ such that $\iota \circ \pi | A = \varkappa_A$. Obviously $\varkappa_A T_2 = \iota(\tau T_2)$ is a transcendence basis of $A/\mathfrak{M}_A$ over $\iota \Lambda_o$ and $\varkappa_A T_1 = \iota(\tau T_1) = \{0\}$, hence $T_1 \subseteq \mathfrak{M}_A$. □

In particular, for any cardinal number α less than or equal to the transcendence degree δ of K over K_o, there is a valuation ring A of K such that $R \subseteq A$, $\mathfrak{P} = \mathfrak{M}_A \cap R$, and $A/\mathfrak{M}_A$ has transcendence degree α over Λ_o. Note that $A \neq K$ whenever $\alpha < \delta$ or $\mathfrak{P} \neq (0)$. The ring R has no non-zero prime ideal if and only if it is a field; in this case (9.7) yields:

(9.8) COROLLARY - Let K_o be a subfield of K. Then the following conditions are equivalent:

(i) K is not algebraic over K_o.

(ii) There is a valuation ring A of K such that $K_o \subseteq A \neq K$.

In this case A can be chosen such that the transcendence degree of $A/\mathfrak{M}_A$ over $\varkappa_A K_o$ is any preassigned cardinal number less than the transcendence degree of K over K_o.

Substituting for K_o the prime field of K, we conclude that there is a valuation ring A of K such that $A \neq K$, $\mathrm{Char}\, A/\mathfrak{M}_A = \mathrm{Char}\, K$ if and only if K is not absolutely algebraic.

Note that the characteristic condition is irrelevant whenever K has prime characteristic.

In the rest of this section, only the first statement of (9.7) will be used. We use it first for a characterization of rank 1 valuation rings of K .

(9.9) THEOREM - For any subring R of K the following conditions are equivalent:

(i) R is a valuation ring of K of rank 1 .

(ii) R is a maximal element of the set of those subrings of K which are not fields.

Proof: (i) ⇒ (ii) follows from (6.6). (ii) ⇒ (i): R is not a field, hence it has a non-zero prime ideal $\mathfrak{P}$. By (9.7) there exists a valuation ring A of K such that $R \subseteq A$ and $\mathfrak{P} = \mathfrak{M}_A \cap R$, hence A is not a field. If R is not a valuation ring, then $R \neq A$, contradicting (ii). If R is a valuation ring with rank (R) > 1, one can choose $\mathfrak{P} \neq \mathfrak{M}_R$; then $R \neq R_{\mathfrak{P}} = A$ by (6.6), contradicting (ii). Hence R is a valuation ring of rank 1. □

Next we characterize valuation rings of K within the set $\mathcal{L}(K)$ of all local [17] subrings R of K (we don't assume that K is the quotient field of R). For this purpose we introduce an ordering ≤ in $\mathcal{L}(K)$: Let $S, R \in \mathcal{L}(K)$; we write $R \leq S$ and say that S dominates R , if $R \subseteq S$ and the equivalent conditions of the following proposition are satisfied:

(9.10) For any $R, S \in \mathcal{L}(K)$ such that $R \subseteq S$ the following conditions are equivalent:

(i) $\mathfrak{M}_R \subseteq \mathfrak{M}_S$.

(ii) $\mathfrak{M}_R = \mathfrak{M}_S \cap R$.

(iii) $\mathfrak{M}_R \cdot S$ is a proper ideal of S .

<u>Proof</u>: (i) ⇒ (ii): We have $\mathfrak{M}_R \subseteq \mathfrak{M}_S \cap R$. The equality holds since $\mathfrak{M}_R$ is a maximal and $\mathfrak{M}_S \cap R$ a proper ideal of R .
(ii) ⇒ (iii) follows from $\mathfrak{M}_R \cdot S = (\mathfrak{M}_S \cap R) \cdot S \subseteq \mathfrak{M}_S$. (iii) ⇒ (i): Since $\mathfrak{M}_S$ is the largest proper ideal of S , we have $\mathfrak{M}_R \subseteq \mathfrak{M}_R \cdot S \subseteq \subseteq \mathfrak{M}_S$. □

(9.11) THEOREM - <u>The valuation rings of</u> K <u>are the maximal elements of</u> $\mathfrak{L}(K)$ <u>with respect to the ordering</u> $\leq$. <u>For any</u> $R \in \mathfrak{L}(K)$ <u>there is a valuation ring</u> A <u>of</u> K <u>such that</u> $R \leq A$.

<u>Proof</u>: From (9.7) we conclude that for any $R \in \mathfrak{L}(K)$ there is a valuation ring A of K such that $R \subseteq A$ and $\mathfrak{M}_R = \mathfrak{M}_A \cap R$, hence $R \leq A$. In particular, if R is maximal in $\mathfrak{L}(K)$ with respect to $\leq$, then $R = A$. On the other hand, let A be a valuation ring of K . We claim that, for any $S \in \mathfrak{L}(K)$, $A \leq S$ implies $A = S$. In fact, we have $A \subseteq S$, and for any non-zero $x \in S$ we have $x^{-1} \cdot x = 1 \notin \mathfrak{M}_A \cdot S$, hence $x^{-1} \notin \mathfrak{M}_A$, hence $x \in A$. □

It should be mentioned that the first statement of theorem (9.11) can be proven without using the extension theorem or its corollaries; in fact, only theorem (9.3) is needed.

In §13, we shall apply theorem (9.11) to prove that for any valuation ring A of K and any field extension $L|K$ there is a valuation ring B of L such that $B \cap K = A$. In fact, this equality will turn out to be equivalent to $A \leq B$.

§10 Integrally closed domains

Let R and S be integral domains such that $R \subseteq S$; we consider S as an R-module. An element $x \in S$ is called *integral over* (or *integrally dependent on*) R , if $F(x) = 0$ for some monic polynomial $F \in R[X]$. In this case, x is also integral over any ring between R and S .

(10.1) *For any* $x \in S$ *the following conditions are equivalent:*

(i) x *is integral over* R .

(ii) $R[x]$ *is a finite R-submodule of* S .

(iii) *There is a non-zero finite R-submodule* M *of* S *such that* $x \cdot M \subseteq M$.

Proof: (i) ⇒ (ii): Let $F \in R[X]$ be monic and such that $F(x) = 0$ and let $n = \deg F$. By induction one shows that $x^n, x^{n+1}, \ldots \in R + R\cdot x + \ldots + R\cdot x^{n-1}$, hence $R[x] = R + R\cdot x + \ldots + R\cdot x^{n-1}$.

(ii) ⇒ (iii): Take $M = R[x]$.

(iii) ⇒ (i): Let $M = \sum_{i=1}^{m} R\cdot y_i \neq 0$, and let $a_{ij} \in R$ such that $x\cdot y_i = \sum_{j=1}^{m} a_{ij}\cdot y_j$ $(i,j = 1,\ldots,m)$. Then the determinant $\det(\delta_{ij}\cdot x - a_{ij})_{i,j=1,\ldots,m}$ is zero, hence $F(x) = 0$ for the monic polynomial $F = \det(\delta_{ij}\cdot X - a_{ij})_{i,j=1,\ldots,m} \in R[X]$. □

Conditions (ii) and (iii) are used in the following corollaries.

(10.2) *Let* $x_1,\ldots,x_k \in S$ *be integral over* R . *Then* $R[x_1,\ldots,x_k]$ *is a finite R-submodule of* S .

Proof: Assume that $R_i = R[x_1,\ldots,x_i]$ is a finite R-submodule of S (this is trivial for $i = 0$). Since x_{i+1} is integral over R, it is integral over R_i , hence $R_{i+1} = R_i[x_{i+1}]$ is a finite R_i-submodule of S and therefore also a finite R-submodule of S . □

The set $I_S(R)$ of those elements $x \in S$ which are integral over R is called the <u>integral closure</u> of R <u>in</u> S . We show:

(10.3) a) $I_S(R)$ <u>is a subring of</u> S <u>which contains</u> R .
b) $I_S(R) = I_S(I_{R'}(R)) \subseteq I_S(R')$ <u>whenever</u> $R \subseteq R' \subseteq S$.

<u>Proof</u>: a) Any $a \in R$ is a root of the monic polynomial $X-a \in R[X]$; therefore $R \subseteq I_S(R)$. Let $x,y \in I_S(R)$; then $R[x,y]$ is a finite non-zero R-submodule of S, by (10.2). Since $(x-y)\cdot R[x,y] \subseteq$ $\subseteq R[x,y]$ and $(x\cdot y)\cdot R[x,y] \subseteq R[x,y]$, the elements $x-y$ and $x\cdot y$ are integral over R . Therefore $I_S(R)$ is a subring of S .

b) Since $R \subseteq I_{R'}(R) \subseteq R'$ by a), we have $I_S(R) \subseteq I_S(I_{R'}(R)) \subseteq I_S(R')$. Let $z \in I_S(I_{R'}(R))$; then z is a root of some monic polynomial $X^m + y_1\cdot X^{m-1} + \ldots + y_m \in I_{R'}(R)[X]$. Therefore z is integral over $R'' = R[y_1,\ldots,y_m]$, hence $R''[z]$ is a finite R''-submodule of S , and even a finite R-submodule of S . Moreover, $z\cdot R''[z] \subseteq R''[z] \neq$ $\neq \{0\}$, hence z is integral over R , i.e., $z \in I_S(R)$. □

Let $\mathfrak{R}(S)$ be the set of all subrings of S . We conclude from (10.3) that the mapping $R \mapsto I_S(R)$ is a closure operation in $\mathfrak{R}(S)$; this justifies the name "integral closure".

We say that S is <u>integral over</u> the subring R if any $x \in S$ is integral over R or, equivalently, $I_S(R) = S$. It is obvious that if $R \subseteq R' \subseteq S$ and S is integral over R , then S is integral over R' and R' is integral over R . On the other hand, (10.3 b) yields the following transitivity.

(10.4) <u>If</u> S <u>is integral over</u> R' <u>and</u> R' <u>is integral over</u> R <u>then</u> S <u>is integral over</u> R .

The following proposition shows how the units and the maximal ideals of R and S are related when S is integral over R.

(10.5) Let S be integral over R. Then:

a) $U_R = U_S \cap R$.

b) S is a field if and only if R is a field.

c) Let $\mathfrak{P}$ be a prime ideal of S; then $\mathfrak{P}$ is a maximal ideal of S if and only if $\mathfrak{P} \cap R$ is a maximal ideal of R.

Proof: a) $U_R \subseteq U_S \cap R$ is obvious. Let $x \in U_S \cap R$ and let $F \in R[X]$ be monic and such that $F(x^{-1}) = 0$. Multiplication by x^{n-1}, where $n = \deg F$, yields $x^{-1} \in R[x] \subseteq R$, hence $x \in U_R$.

b) If S is a field then $U_S = S^*$, hence $U_R = S^* \cap R = R \setminus \{0\}$, hence R is a field. Let R be a field. Then any non-zero $x \in S$ is algebraic over R, hence $x^{-1} \in R(x) = R[x] \subseteq S$; therefore S is a field.

c) Let $\kappa: S \to S/\mathfrak{P}$ be the canonical homomorphism. Obviously $S/\mathfrak{P}$ is integral over κR, and κR is isomorphic to $R/(\mathfrak{P} \cap R)$. By b), $S/\mathfrak{P}$ is a field if and only if $R/(\mathfrak{P} \cap R)$ is a field, hence $\mathfrak{P}$ is maximal if and only if $\mathfrak{P} \cap R$ is maximal. □

For more information about the relations between prime ideals of R and prime ideals of S see Zariski & Samuel [36], Vol. I, Chap. V, §2-3.

In the following we consider only integral closures $I_L(R)$ such that L is a field. We have $R \subseteq I_L(R) \subseteq L$ and, by (10.5 b), $I_L(R) = L$ if and only if R is a field and L is algebraic over R. We say that R is integrally closed in L if $R = I_L(R)$. In particular, we say that R is integrally closed if it is integrally closed in its quotient field. It is well known, and easy to prove, that any UFD is integrally closed.

(10.6) THEOREM - Any valuation ring is integrally closed.

Proof: Let A be a valuation ring of K ; hence K is its quotient field. Suppose there is an $x \in I_K(A)$ such that $x \notin A$; then $x^{-1} \in A \subseteq I_K(A)$, hence $x^{-1} \in U_{I_K(A)} \cap A$, hence $x^{-1} \in U_A$ by (10.5 a); this contradicts $x \notin A$. □

This theorem, together with (7.6) and (7.7), yields the implication (i) ⇒ (ii) of the following corollary, which characterizes discrete valuation rings within the class of integral domains. For the proof of the implication (ii) ⇒ (i), see for example Serre [33], Chap. I, Prop. 3).

(10.7) COROLLARY - For any integral domain R the following conditions are equivalent:

(i) R is a discrete valuation ring.

(ii) R is noetherian, integrally closed, and has exactly one non-zero prime ideal.

We are going to characterize arbitrary integrally closed integral domains as intersections of valuation rings. More generally, we prove for an arbitrary subring R of any field L :

(10.8) THEOREM - Given $R \subseteq L$, let $\mathfrak{B}$ be the set of all valuation rings of L containing R and let $\mathfrak{B}'$ be the set of all those $B \in \mathfrak{B}$ for which $\mathfrak{M}_B \cap R$ is a maximal ideal of R . Then

$$I_L(R) = \bigcap_{B \in \mathfrak{B}} B = \bigcap_{B \in \mathfrak{B}'} B .$$

Proof: Let $x \in I_L(R)$. For any $B \in \mathfrak{B}$, x is integral over B , hence $x \in B$, by (10.6). Therefore $I_L(R) \subseteq \bigcap_{B \in \mathfrak{B}'} B \subseteq \bigcap_{B \in \mathfrak{B}'} B$.
To prove the equalities, it suffices to show that for any $x \in L \setminus I_L(R)$ there is a $B \in \mathfrak{B}'$ such that $x \notin B$. Let $x \in L \setminus I_L(R)$; then $x \notin R[x^{-1}]$, since $-x = a_o + a_1 \cdot x^{-1} + \dots + a_m \cdot x^{-m}$ with

$a_o,\ldots,a_m \in R$ would imply $F(x) = 0$ for $F = X^{m+1} + a_o \cdot X^m + \ldots + a_m$. Hence x^{-1} is a non-unit of $R[x^{-1}]$ and therefore $x^{-1} \in \mathfrak{N}$ for some maximal ideal $\mathfrak{N}$ of $R[x^{-1}]$. Since the homomorphism $R \to R[x^{-1}]/\mathfrak{N}$ defined by $a \mapsto a + \mathfrak{N}$ is surjective, its kernel $\mathfrak{N} \cap R$ is a maximal ideal of R. By (9.7) there is a valuation ring B of L such that $R[x^{-1}] \subseteq B$ and $\mathfrak{N} = \mathfrak{M}_B \cap R[x^{-1}]$, and $B \in \mathfrak{B}'$ since $\mathfrak{M}_B \cap R = \mathfrak{N} \cap R$. Moreover, $x \notin B$ since $x^{-1} \in \mathfrak{N} \subseteq \mathfrak{M}_B$. □

In the preceding proof we used only the first statement of the corollary (9.7) of the extension theorem. By using its second statement, one can find other subsets $\mathfrak{B}''$ of $\mathfrak{B}$ such that $I_L(R) = \bigcap_{B \in \mathfrak{B}''} B$ (for example the set of all $B \in \mathfrak{B}'$ such that the residue field of B is algebraic over $\varkappa_B R$).

Note that if R is a local ring then $\mathfrak{B}'$ is the set of those valuation rings of L which dominate R.

Theorem (10.8) yields the following characterization of subrings R of L which are integrally closed in L:

(10.9) COROLLARY - Any subring R of L is integrally closed in L if and only if it is the intersection of some set of valuation rings of L.

Proof: If R is integrally closed in L, then $R = I_L(R) = \bigcap_{B \in \mathfrak{B}} B$, by (10.8). On the other hand, if $R = \bigcap_{B \in \mathfrak{B}_o} B$ for some set $\mathfrak{B}_o$ of valuation rings of L, then $\mathfrak{B}_o \subseteq \mathfrak{B}$, hence $R \subseteq I_L(R) = \bigcap_{B \in \mathfrak{B}} B \subseteq \bigcap_{B \in \mathfrak{B}_o} B = R$, hence R is integrally closed in L. □

We terminate this section stating some properties of domains R which are integrally closed.

(10.10) Let R be integrally closed (in its quotient field K).

a) <u>If</u> F , $G \in K[X]$ <u>are monic and</u> $F \cdot G \in R[X]$, <u>then</u> F , $G \in$ $\in R[X]$.

b) <u>If</u> $x \in L \supseteq K$ <u>is integral over</u> R <u>then</u> x <u>is algebraic over</u> K <u>and its minimal polynomial</u> $P_{x|K}$ <u>over</u> K <u>is in</u> $R[X]$.

<u>Proof</u>: a) There exist a field extension L of K and y_i , $z_j \in L$ such that $F = \prod_{i=1}^{r}(X-y_i)$ and $G = \prod_{j=1}^{s}(X-z_j)$. Then $y_1,\dots,y_r,z_1,\dots,z_s \in I_L(R)$ since they are roots of the monic polynomial $F \cdot G \in R[X]$. Therefore F, $G \in (I_L(R))[X]$ and even $\in R[X]$ since $I_L(R) \cap K = I_K(R) = R$.

b) Let $F \in R[X]$ be monic and such that $F(x) = 0$. Then x is algebraic over K and $F = P_{x|K} \cdot H$ for some $H \in K[X]$. From a) it follows that $P_{x|K} \in R[X]$. □

The following statement will be used in §11.

(10.11) <u>Let</u> R <u>and</u> K <u>be as in (10.10)</u>, $\mathfrak{P}$ <u>a prime ideal of</u> R <u>and</u> $x \in K$, $x \neq 0$. <u>If</u> $F(x) = 0$ <u>for some (not necessarily monic) polynomial</u> $F \in R[X] \setminus \mathfrak{P}[X]$ <u>then</u> $x \in R_{\mathfrak{P}}$ <u>or</u> $x^{-1} \in R_{\mathfrak{P}}$.

<u>Proof</u>: By (9.7) there is a valuation ring A_o of K such that $R \subseteq A_o$ and $\mathfrak{P} = \mathfrak{M}_{A_o} \cap R$. We may assume $x \in A_o$ (otherwise in what follows x must be replaced by x^{-1}) . Let $F = \sum_{i=o}^{n} a_i \cdot X^{n-i}$ and $k = \min\{i \in \mathbb{N} \mid a_i \notin \mathfrak{P}\}$; obviously $0 \leq k \leq n$. We have even $k < n$, since otherwise $a_n = -a_o \cdot x^n - \dots - a_{n-1} \cdot x \in \mathfrak{P} \cdot A_o \cap R \subseteq$ $\subseteq \mathfrak{M}_{A_o} \cap R = \mathfrak{P}$, contradicting $F \notin \mathfrak{P}[X]$. Let $y = a_o \cdot x^k + \dots + a_k$ and $z = a_{k+1} + a_{k+2} \cdot x^{-1} + \dots + a_n \cdot x^{-n+k+1}$; then $y \notin \mathfrak{M}_{A_o}$ (since otherwise $a_k \in \mathfrak{M}_{A_o} \cap R = \mathfrak{P}$ contrary to the choice of k) and $y + z \cdot x^{-1} = x^{-n+k} \cdot F(x) = 0$, hence $x = -\frac{z}{y}$. In order to prove that $x \in R_{\mathfrak{P}}$ it suffices to show that z , $y \in R$; by (10.8) it suffices to prove that z, $y \in A$ for any valuation ring A of K containing R . If $x \in A$ then $y \in A$, $z = -x \cdot y \in A$. If $x \notin A$ then $x^{-1} \in A$, hence

$z \in A$ and $y = -\frac{z}{x} \in A$. □

§11 Prüfer rings. Approximation theorems

Let R be a subring of a field K . A valuation ring A of K is said to be essential for R if A is a ring of fractions of R , i.e. $A = R_M$ for some multiplicative subset M of R . Let $\mathfrak{B}(R)$ (resp. $\mathfrak{E}(R)$) be the set of all valuation rings of K which contain (resp. are essential for) R . Obviously $\mathfrak{E}(R) \subseteq \mathfrak{B}(R)$, and $\mathfrak{E}(R) \neq \emptyset$ if and only if K is the quotient field of R ; in this case $K \in \mathfrak{E}(R)$. We denote by $\mathfrak{P}(R)$ (resp. $\mathfrak{m}(R)$) the set of all prime (resp. maximal) ideals of R and we set $\mathfrak{P}(R;\mathfrak{p}) =$
$= \{\mathfrak{Q} \in \mathfrak{P}(R) \mid \mathfrak{Q} \subseteq \mathfrak{p}\}$.

(11.1) Let $A \in \mathfrak{B}(R)$ and $\mathfrak{p} = \mathfrak{M}_A \cap R$ [20]. Then A dominates $R_{\mathfrak{p}}$, and the following conditions are equivalent:

(i) $A = R_{\mathfrak{p}}$.

(ii) $R_{\mathfrak{p}}$ is a valuation ring of K .

(iii) $A \in \mathfrak{E}(R)$.

In this case we have $\mathfrak{B}(A) \subseteq \mathfrak{E}(R)$, and an inclusion-inverting 1-1 correspondence $\mathfrak{B}(A) \leftrightarrow \mathfrak{P}(R;\mathfrak{p})$ is given by

$$B \mapsto \mathfrak{M}_B \cap R \quad (B \in \mathfrak{B}(A)) \quad \text{and} \quad \mathfrak{Q} \mapsto R_{\mathfrak{Q}} \quad (\mathfrak{Q} \in \mathfrak{P}(R;\mathfrak{p})) .$$

Proof: A dominates $R_{\mathfrak{p}}$ since $R_{\mathfrak{p}} \subseteq A$ and $\mathfrak{p} \cdot R_{\mathfrak{p}} \subseteq \mathfrak{M}_A$. The implications (i) ⇒ (ii) and (i) ⇒ (iii) are trivial. (ii) ⇒ (i) follows from (9.11). (iii) ⇒ (i): Let $A = R_M$, where M is a multiplicative subset of R . Obviously $M \cap \mathfrak{M}_A = \emptyset$, hence $M \subseteq R \setminus \mathfrak{p}$, hence $A = R_M \subseteq R_{\mathfrak{p}} \subseteq A$.

---

20 The prime ideal $\mathfrak{M}_A \cap R$ is called the center of A on R .

Let $B \in \mathfrak{B}(A)$. By (6.6) we have $\mathfrak{M}_A \supseteq \mathfrak{M}_B$, hence $\mathfrak{M}_B \cap R \in \in \mathcal{P}(R;\mathfrak{P})$, $A = R_{\mathfrak{P}} \subseteq R_{\mathfrak{M}_B \cap R}$, and $R_{\mathfrak{M}_B \cap R}$ is a valuation ring of K; therefore $B = R_{\mathfrak{M}_B \cap R} \in \mathcal{E}(R)$. Let $\mathfrak{Q} \in \mathcal{P}(R;\mathfrak{P})$. We have $A = R_{\mathfrak{P}} \subseteq \subseteq R_{\mathfrak{Q}}$, hence $R_{\mathfrak{Q}}$ is a valuation ring of K; therefore $R_{\mathfrak{Q}} \in \mathfrak{B}(A)$ and $\mathfrak{M}_{R_{\mathfrak{Q}}} \cap R = \mathfrak{Q}\cdot R_{\mathfrak{Q}} \cap R = \mathfrak{Q}$. Thus the last statements of (11.1) hold. □

Note that for any valuation ring A of K, $\mathfrak{B}(A)$ is the set of all rings between R and K and $\mathfrak{B}(A) = \mathcal{E}(A)$.

(11.2) <u>For any multiplicative subset</u> M <u>of</u> R <u>we have</u> $\mathcal{E}(R_M) = \mathcal{E}(R) \cap \mathfrak{B}(R_M)$.

<u>Proof</u>: The inclusion $\subseteq$ holds since any ring of fractions of R_M is a ring of fractions of R. Let $A \in \mathcal{E}(R) \cap \mathfrak{B}(R_M)$; then A is a ring of fractions of R_M, hence $A \in \mathcal{E}(R_M)$. □

For any $A \in \mathfrak{B}(R)$ let $\varkappa_A : A \to A/\mathfrak{M}_A$ be the canonical homomorphism. Obviously $\varkappa_A R \subseteq A/\mathfrak{M}_A$, and if $\varkappa_A R = A/\mathfrak{M}_A$ then $\mathfrak{M}_A \cap R$ is a maximal ideal of R. The converse statement is true if we assume that A is essential for R; in fact:

(11.3) <u>Let</u> $A \in \mathcal{E}(R)$. <u>We have</u> $\varkappa_A R = A/\mathfrak{M}_A$ <u>if and only if</u> $\mathfrak{M}_A \cap R \in \mathfrak{m}(R)$.

<u>Proof</u>: Let $\mathfrak{P} = \mathfrak{M}_A \cap R$. By hypothesis we have $R_{\mathfrak{P}} = A$, $\mathfrak{P}\cdot R_{\mathfrak{P}} = \mathfrak{M}_A$ and therefore $R_{\mathfrak{P}}/\mathfrak{P}\cdot R_{\mathfrak{P}} = A/\mathfrak{M}_A$. If $\mathfrak{P} \in \mathfrak{m}(R)$, then the restriction to R of the canonical homomorphism $R_{\mathfrak{P}} \to R_{\mathfrak{P}}/\mathfrak{P}\cdot R_{\mathfrak{P}}$ is surjective, hence $\varkappa_A R = A/\mathfrak{M}_A$. □

A subring R of K is called a <u>Prüfer ring</u> of K if $\mathcal{E}(R) = \mathfrak{B}(R)$. Prüfer rings can be characterized in different ways:

(11.4) THEOREM - <u>For any subring</u> R <u>of</u> K <u>the following conditions</u>

are equivalent:

(i) R is a Prüfer ring of K .

(ii) $R_{\mathfrak{M}}$ is a valuation ring for any maximal ideal $\mathfrak{M}$ of R .

(iii) $R_{\mathfrak{P}}$ is a valuation ring for any prime ideal $\mathfrak{P}$ of R .

In this case, K is the quotient field of R , R is integrally closed, and an inclusion-inverting 1-1 correspondence $\mathcal{B}(R) \leftrightarrow \mathcal{P}(R)$ is given by $\mathfrak{P} = \mathfrak{M}_A \cap R$, $A = R_{\mathfrak{P}}$.

Proof: (i) ⇒ (ii): For any $\mathfrak{M} \in \mathfrak{m}(R)$ there is an $A \in \mathcal{B}(R) = \mathcal{C}(R)$ such that $\mathfrak{M}_A \cap R = \mathfrak{M}$, by (9.7). By (11.1), $R_{\mathfrak{M}}$ is a valuation ring of K (actually, $R_{\mathfrak{M}} = A$). (ii) ⇒ (iii): Any $\mathfrak{P} \in \mathcal{P}(R)$ is contained in some $\mathfrak{M} \in \mathfrak{m}(R)$, hence $R_{\mathfrak{M}} \subseteq R_{\mathfrak{P}}$. Since $R_{\mathfrak{M}}$ is a valuation ring of K , so is $R_{\mathfrak{P}}$. (iii) ⇒ (i): Let $A \in \mathcal{B}(R)$ and $\mathfrak{P} = \mathfrak{M}_A \cap R$. Since $R_{\mathfrak{P}}$ is a valuation ring of K , we have $A \in \mathcal{C}(R)$, by (11.1).

Since $K \in \mathcal{C}(R)$, K is the quotient field of R , and $R = \bigcap_{\mathfrak{M} \in \mathfrak{m}(R)} R_{\mathfrak{M}}$ is integrally closed, by (10.8). Since $A = R_{\mathfrak{M}_A \cap R}$ for all $A \in \mathcal{C}(R)$ and $\mathfrak{P} = \mathfrak{P} \cdot R_{\mathfrak{P}} \cap R$ for all $\mathfrak{P} \in \mathcal{P}(R)$, the last statement of the theorem holds. □

In particular, any PID is a Prüfer ring. More precisely, theorem (4.3) yields:

(11.5) Let R be a PID and K its quotient field. Then R is a Prüfer ring of K and any $A \in \mathcal{B}(R)$, distinct from K , is a discrete valuation ring of K (namely $A = R_{p \cdot R}$, where $\mathfrak{M}_A = p \cdot A$, p irreducible in R).

Theorem (11.4) yields the following characterization of valuation rings:

(11.6) COROLLARY - For any subring R of K the following conditions are equivalent:

(i) R *is a valuation ring of* K .

(ii) R *is a Prüfer ring of* K *and* R *is a local ring*.

Note that for any Prüfer ring R the sets $\mathfrak{B}(R)$ and $\mathfrak{P}(R)$ are not totally ordered by inclusion unless R is a valuation ring. However, we conclude from (11.4) that their orderings are of the following simple type:

(11.7) COROLLARY - *Let* R *be a Prüfer ring of* K . *Then we have*:

a) *For any* $\mathfrak{P} \in \mathfrak{P}(R)$, *the set* $\mathfrak{P}(R;\mathfrak{P})$ *is totally ordered by inclusion.*

b) *Any* $A \in \mathfrak{B}(R)$ *contains a minimal element of* $\mathfrak{B}(R)$.

From (11.4) and (11.3) we conclude:

(11.8) COROLLARY - *Let* R *be a Prüfer ring of* K *and let* $A \in \mathfrak{B}(R)$. *We have* $\varkappa_A R = A/\mathfrak{M}_A$ *if and only if* A *is a minimal element of* $\mathfrak{B}(R)$.

The following statement generalizes well known facts on valuation rings (see (6.6) and (6.7)). For a more detailed generalization see Gilmer [10], (22.1).

(11.9) *Let* R *and* S *be subrings of* K *such that* $R \subseteq S$. *If* R *is a Prüfer ring of* K *then so is* S , *and for any prime ideal* $\mathfrak{P}$ *of* S *we have* $S_{\mathfrak{P}} = R_{\mathfrak{P} \cap R}$.

Proof: Let $\mathfrak{P} \in \mathfrak{P}(S)$ and $\mathfrak{Q} = \mathfrak{P} \cap R$. Obviously $R_{\mathfrak{Q}} \subseteq S_{\mathfrak{P}}$ and $\mathfrak{Q} \cdot R_{\mathfrak{Q}} \subseteq \mathfrak{P} \cdot S_{\mathfrak{P}}$, hence $S_{\mathfrak{P}}$ dominates $R_{\mathfrak{Q}}$. Since $R_{\mathfrak{Q}}$ is a valuation ring of K we have $R_{\mathfrak{Q}} = S_{\mathfrak{P}}$, by (9.11). By (11.4), S is a Prüfer ring of K . □

The following theorem characterizes the Prüfer rings of K within the set of all integrally closed subrings of K .

(11.10) THEOREM - Let R be an integrally closed subring of K. The following conditions are equivalent:

(i) R is a Prüfer ring of K.

(ii) Any ring S between R and K is integrally closed.

(iii) For any $x \in K$ we have $R[x] = \bigcap_{n \in \mathbb{N}} R[x^n]$.

(iv) For any $x \in K$ there is some $m \geq 1$ such that $x^m \in \sum_{i \in \mathbb{N} \setminus \{m\}} R \cdot x^i$.

(v) For any $x \in K$ and any maximal ideal $\mathfrak{M}$ of R, x is a root of some polynomial $F_{x,\mathfrak{M}} \in R[x] \setminus \mathfrak{M}[X]$.

In this case the polynomials $F_{x,\mathfrak{M}}$ can be chosen such that $\deg F_{x,\mathfrak{M}} = 1$.

Proof: (i) ⇒ (ii) follows from (11.4) and (11.9).

(ii) ⇒ (iii): For any $x \in K$ and $n \in \mathbb{N}$, x is integral over $R[x^n]$. Since $R[x^n]$ is integrally closed, we have $x \in R[x^n]$. Therefore $x \in \bigcap_{n \in N} R[x^n]$. Since this intersection is a ring, it is equal to $R[x]$.

(iii) ⇒ (iv): Obviously $x \in R[x^2] \subseteq \sum_{i \in \mathbb{N} \setminus \{1\}} R \cdot x^i$ for any $x \in K$.

(iv) ⇒ (v): For any $x \in K$ there is some $m \geq 1$ such that $x^m \in \sum_{i \in \mathbb{N} \setminus \{m\}} R \cdot x^i$, hence x is a root of some polynomial $F_x \in R[X]$ whose m-th coefficient is equal to 1. In particular, we have $F_x \notin \mathfrak{M}[X]$ for any $\mathfrak{M} \in \mathfrak{m}(R)$.

(v) ⇒ (i): Let $\mathfrak{M} \in \mathfrak{m}(R)$. For any non-zero $x \in K$ we have $x \in R_{\mathfrak{M}}$ or $x^{-1} \in R_{\mathfrak{M}}$, by (10.11); hence $R_{\mathfrak{M}}$ is a valuation ring of K. By (11.4), R is a Prüfer ring of K.

Assume that R is a Prüfer ring of K. By (11.4), $R_{\mathfrak{M}}$ is a valuation ring of K for any $\mathfrak{M} \in \mathfrak{m}(R)$, hence for any $x \in K$ there exist $r \in R$, $s \in R \setminus \mathfrak{M}$ such that $x = \frac{r}{s}$ or $x = \frac{s}{r}$. If $x = \frac{r}{s}$ (resp. $= \frac{s}{r}$) we may choose $F_{x,\mathfrak{M}} = s \cdot X - r$ (resp. $= r \cdot X - s$). Therefore the last statement of (11.10) holds. □

We note that theorem (11.10) yields a characterization of valuation rings of K within the set of all integrally closed local subrings of K. In fact, if we assume that R is a local ring, theorem (11.10) remains true if one replaces (i) by

(i') R _is a valuation ring of_ K.

Moreover, under this assumption, (v) can be replaced by

(v') _Any_ $x \in K$ _is a root of some polynomial_ $F_x \in R[X] \setminus \mathfrak{M}_R[X]$.

For more characterizations of Prüfer rings see Gilmer [10], Chapter IV.

We are going to prove that any finite intersection of valuation rings of K is a Prüfer ring of K. We denote by $\mathfrak{F}$ the set of all monic polynomials F of degree $\geqq 1$ with coefficients in the prime ring of K and constant coefficient $F(0) = 1$ and prove:

(11.11) _Let_ $A_1, \ldots, A_n$ _be valuation rings of_ K. _For any non-zero_ $x \in A_1 \cap \ldots \cap A_n$ _there exists an_ $F \in \mathfrak{F}$ _with the following properties:_

a) $F(x) \in U_{A_1} \cap \ldots \cap U_{A_n}$.

b) _For any valuation ring_ A _of_ K _such that_ $x \notin A$ _we have_ $\frac{x}{F(x)} \in A$.

Proof: For each $i \in \{1, \ldots, n\}$ let π_i be a place corresponding to A_i. We choose $F_i \in \mathfrak{F}$ such that $\pi_i(F_i(x)) = 0$ whenever such a polynomial exists, otherwise we set $F_i = 1$; then $F = 1 + X \cdot F_1 \cdot \ldots \cdot F_n \in \mathfrak{F}$. a) Obviously $F(x) \in A_i$. Suppose $F(x) \notin U_{A_i}$; then $\pi_i(F(x)) = 0$, hence $\pi_i(F_i(x)) = 0$, hence $\pi_i(F(x)) = \pi_i(1 + x \cdot F_1(x) \cdot \ldots \cdot F_n(x)) = \pi_i 1 \neq 0$, a contradiction.

b) For any valuation ring A of K we have $F \in A[X]$. If $x \notin A$ then $F(x) \cdot x^{-\deg F} - 1 \in \mathfrak{M}_A$, hence $F(x) \cdot x^{-\deg F} \in U_A$ and $\frac{x}{F(x)} = (F(x) \cdot x^{-\deg F})^{-1} \cdot x^{1-\deg F} \in A$. □

Let $R = A_1 \cap \ldots \cap A_k$ where $A_1,\ldots,A_k$ are valuation rings of K ; then R is equal to the intersection of the minimal elements of the set $\{A_1,\ldots,A_k\}$ (ordered by inclusion). Therefore, considering finite intersections of valuation rings of K we may assume without loss of generality that these valuation rings are incomparable (with respect to inclusion).

(11.12) THEOREM - _Let_ $A_1,\ldots,A_k$ _be pairwise incomparable valuation rings of_ K . _Then_ $R = A_1 \cap \ldots \cap A_k$ _is a Prüfer ring of_ K _with exactly_ k _maximal ideals, namely_ $\mathfrak{M}_{A_1} \cap R,\ldots,\mathfrak{M}_{A_k} \cap R$.

Proof: Let $i \in \{1,\ldots,k\}$ and $\mathfrak{P}_i = \mathfrak{M}_{A_i} \cap R$; obviously $R_{\mathfrak{P}_i} \subseteq A_i$.

Let $x \in A_i$, $x \neq 0$, and let $I_x = \{j \mid 1 \leq j \leq k,\ x \in A_j\}$. By (11.11) there exists an $F \in \mathfrak{F}$ such that $F(x) \in U_{A_j}$ for all $j \in I_x$ and $\frac{x}{F(x)} \in A_j$ for all $j \in \{1,\ldots,k\} \setminus I_x$. Obviously $\frac{1}{F(x)}, \frac{x}{F(x)} \in A_j$ for all $j \in \{1,\ldots,k\}$, hence both elements are in R . Since $\frac{1}{F(x)} \notin \mathfrak{P}_i$ we have $x \in R_{\mathfrak{P}_i}$. Therefore $R_{\mathfrak{P}_i} = A_i$ for $i = 1,\ldots,k$, and since $A_1,\ldots,A_k$ are pairwise incomparable, so are $\mathfrak{P}_1,\ldots,\mathfrak{P}_k$. Obviously $R \setminus (\mathfrak{P}_1 \cup \ldots \cup \mathfrak{P}_k)$ is the set of all units of R ; hence any proper ideal of R is contained in $\mathfrak{P}_1 \cup \ldots \cup \mathfrak{P}_k$ and therefore in one of the prime ideals $\mathfrak{P}_1,\ldots,\mathfrak{P}_k$ (see Northcott [25] 1.4). Hence $\{\mathfrak{P}_1,\ldots,\mathfrak{P}_k\}$ is the set of all maximal ideals of R , and by (11.4) R is a Prüfer ring of K . □

We conclude from (11.4) that the valuation rings $A \in \mathfrak{B}(R) = \mathfrak{E}(R)$ are in 1-1 correspondence with the prime ideals $\mathfrak{P}$ of R , and that $A_1,\ldots,A_k$ are the minimal elements of $\mathfrak{B}(R)$. From (11.7) and (11.8) we conclude:

(11.13) COROLLARY - _Let_ $A_1,\ldots,A_k$ _and_ R _be as in_ (11.12). _For any_ $A \in \mathfrak{B}(R)$ _we have_

a) A _is essential for_ R _and_ $A \supseteq A_i$ _for some_ $i \in \{1,\ldots,k\}$.

b) $\varkappa_A R = A/\mathfrak{M}_A$ _if and only if_ $A \in \{A_1,\ldots,A_k\}$.

An integral domain R is said to satisfy the Chinese remainder theorem if for any finite number of ideals $\mathfrak{A}_1,\dots,\mathfrak{A}_k$ of R and any elements $x_1,\dots,x_k \in R$ such that $x_i - x_j \in \mathfrak{A}_i + \mathfrak{A}_j$ $(i,j=1,\dots,k)$ there is an $x \in R$ such that $x-x_1 \in \mathfrak{A}_1,\dots,x-x_k \in \mathfrak{A}_k$. We mention without proof that R has this property if and only if $\mathfrak{A} \cap (\mathfrak{B} + \mathfrak{C}) = (\mathfrak{A} \cap \mathfrak{B}) + (\mathfrak{A} \cap \mathfrak{C})$ for all ideals $\mathfrak{A}$, $\mathfrak{B}$, $\mathfrak{C}$ of R, if and only if R is a Prüfer ring. (See for example Gilmer [10], §21.)

On the other hand, it is well known that for any commutative ring R the following weak Chinese remainder theorem holds: For any finite number of pairwise comaximal ideals $\mathfrak{A}_1,\dots,\mathfrak{A}_k$ of R (i.e. $\mathfrak{A}_i + \mathfrak{A}_j = R$ for $i \neq j$) and any elements $x_1,\dots,x_k \in R$ there is an $x \in R$ such that $x-x_1 \in \mathfrak{A}_1,\dots,x-x_k \in \mathfrak{A}_k$ (see for example S. Lang [24], Chap. II, §2). In other words, for any finite number of pairwise comaximal ideals $\mathfrak{A}_1,\dots,\mathfrak{A}_k$ of R, the homomorphism $R \to R/\mathfrak{A}_1 \times\dots\times R/\mathfrak{A}_k$ defined by $x \mapsto (x+\mathfrak{A}_1,\dots,x+\mathfrak{A}_k)$ is surjective. From this fact and (11.13) we get the following weak form of the approximation theorem, which is valid for pairwise incomparable (not necessarily independent) valuation rings of K.

(11.14) THEOREM - Let $A_1,\dots,A_k$ be pairwise incomparable valuation rings of K and $R = A_1 \cap\dots\cap A_k$. Then for any $(a_1,\dots,a_k) \in A_1 \times\dots\times A_k$ there is a $d \in R$ such that $(d-a_1,\dots,d-a_k) \in \mathfrak{M}_{A_1} \times\dots\times \mathfrak{M}_{A_k}$.

Proof: By (11.13) there exist elements $d_i \in R$ such that $d_i - a_i \in \mathfrak{M}_{A_i}$ $(i=1,\dots,k)$, and $\mathfrak{M}_{A_1} \cap R,\dots,\mathfrak{M}_{A_k} \cap R$ are distinct maximal (hence pairwise comaximal) ideals of R. By the weak Chinese remainder theorem there is a $d \in R$ such that $d-d_i \in \mathfrak{M}_{A_i} \cap R$ and therefore $d-a_i \in \mathfrak{M}_{A_i}$ for all $i \in \{1,\dots,k\}$. □

For proving the usual approximation theorem we need the fact that for any proper non-zero ideals $\mathfrak{A}_i$ of A_i $(i=1,\dots,k)$ the ideals $\mathfrak{A}_1 \cap R,\dots,\mathfrak{A}_k \cap R$ are pairwise comaximal. This is not always true for incomparable valuation rings. In fact, let $k=2$ and $A_1 \cdot A_2 \neq K$; then $\mathfrak{A}_1 = \mathfrak{M}_{A_1 \cdot A_2} = \mathfrak{A}_2$ is a proper non-zero ideal of both A_1 and A_2 by (6.10), but $\mathfrak{A}_1 \cap R$, $\mathfrak{A}_2 \cap R$ are not comaximal. On the other hand, it <u>is</u> true in the case of pairwise independent valuation rings, as we prove now:

(11.15) <u>Let</u> $A_1,\dots,A_k$ <u>be pairwise independent valuation rings of</u> K , $R = A_1 \cap\dots\cap A_k$, <u>and let</u> $\mathfrak{A}_i$ <u>be a proper non-zero ideal of</u> A_i $(i=1,\dots,k)$. <u>Then</u> $\mathfrak{A}_1 \cap R,\dots,\mathfrak{A}_k \cap R$ <u>are pairwise comaximal ideals of</u> R.

<u>Proof</u>: We may assume $k \geq 2$. Let $\mathfrak{P}_i = \mathfrak{M}_{A_i} \cap R$ and $\mathfrak{C}_i = \mathfrak{A}_i \cap R$; then $\mathfrak{C}_i \subseteq \mathfrak{P}_i$. Since $A_i = R_{\mathfrak{P}_i}$ by (11.12), we have $\mathfrak{A}_i = (\mathfrak{A}_i \cap R)\cdot A_i$, hence $\mathfrak{C}_i \neq (0)$. Suppose $\mathfrak{C}_1,\dots,\mathfrak{C}_k$ are not comaximal; then there exist $i,j \in \{1,\dots,k\}$, $i \neq j$, such that $\mathfrak{C}_i + \mathfrak{C}_j$ is contained in some maximal ideal $\mathfrak{N}$ of R . By (11.12) $\mathfrak{N} = \mathfrak{P}_h$ for some $h \in \{1,\dots,k\}$. Therefore we may assume that $\mathfrak{C}_1 \subseteq \mathfrak{P}_2$. By (6.8) $\sqrt{\mathfrak{A}_1}$ is a prime ideal of A_1 , hence $\sqrt{\mathfrak{C}_1} = \sqrt{\mathfrak{A}_1} \cap R$ is a prime ideal $\mathfrak{P}$ (say) of R , and $\mathfrak{C}_1 \subseteq \mathfrak{P}_j$ implies $\mathfrak{P} = \sqrt{\mathfrak{C}_j} \subseteq \sqrt{\mathfrak{P}_j} = \mathfrak{P}_j$ $(j=1,2)$. Therefore $A_j = R_{\mathfrak{P}_j} \subseteq R_{\mathfrak{P}}$ $(j=1,2)$, hence $R_{\mathfrak{P}} = K$ by hypothesis, hence $\mathfrak{P}\cdot R_{\mathfrak{P}} = (0)$, hence $\mathfrak{P} = (0)$, $\mathfrak{C}_1 = (0)$, a contradiction. □

(11.16) APPROXIMATION THEOREM - <u>Let</u> $A_1,\dots,A_k$ <u>be pairwise independent valuation rings of</u> K <u>and let</u> v_i <u>be a valuation corresponding to</u> A_i <u>with value group</u> Γ_i $(i=1,\dots,k)$. <u>Then for any</u> $(x_1,\dots,x_k) \in K \times\dots\times K$ <u>and any</u> $(\gamma_1,\dots,\gamma_k) \in \Gamma_1 \times\dots\times \Gamma_k$ <u>there exist infinitely many</u> $x \in K$ <u>such that</u> $v_i(x-x_i) = \gamma_i$ $(i=1,\dots,k)$.

<u>Proof</u>: We may assume that $K \notin \{A_1,\dots,A_k\}$. Instead of the desired equalities we prove first the existence of infinitely many $x \in K$ satisfying $v_i(x-x_i) > \gamma_i$ $(i=1,\dots,k)$. By (11.12), K is the quotient field of $R = A_1 \cap\dots\cap A_k$, hence $x_i = d_i \cdot c^{-1}$ for

appropriate $c, d_1, \ldots, d_k \in R$, $c \neq 0$. Let $\mathfrak{A}_i =$ $= \{a \in \mathfrak{M}_{A_i} \mid v_i\, a > v_i\, c + \gamma_i\}$ and $\mathfrak{C}_i = \mathfrak{A}_i \cap R$. From (11.15) and the weak Chinese remainder theorem it follows that $(d-d_1, \ldots, d-d_k) \in$ $\in \mathfrak{C}_1 \times \ldots \times \mathfrak{C}_k$ for some $d \in R$, hence $v_i(d \cdot c^{-1} - d_i \cdot c^{-1}) =$ $= v_i(d-d_i) - v_i c > \gamma_i$ $(i=1,\ldots,k)$, hence $x = \frac{d}{c}$ has the desired property. Obviously d may be replaced by any element of the infinite set $d + \mathfrak{C}_1 \cdot \ldots \cdot \mathfrak{C}_k$. To prove the original statement of the theorem, let $y_i \in K$ such that $v_i y_i = \gamma_i$ $(i=1,\ldots,k)$ and choose $y \in K$ such that $v_i(y-y_i) > \gamma_i$ $(i=1,\ldots,k)$. Since $v_i y_i =$ $= \gamma_i < v_i(y-y_i)$ we have $v_i y = v_i((y-y_i) + y_i) = \gamma_i < v(x-x_i)$, hence $v_i(y+x-x_i) = \gamma_i$ $(i=1,\ldots,k)$. Hence $y + x$ has the desired property. □

The special case $(x_1,\ldots,x_k) = (0,\ldots,0)$ is often referred to as the "Independence theorem" or the "Weak Approximation theorem":

(11.17) COROLLARY - Let A_i, v_i, Γ_i as in (11.16). For any $(\gamma_1,\ldots,\gamma_k) \in \Gamma_1 \times \ldots \times \Gamma_k$ there exist infinitely many $x \in K$ such that $v_1 x = \gamma_1, \ldots, v_k x = \gamma_k$.

With the weaker hypothesis that $A_1,\ldots,A_k$ are pairwise incomparable (but not necessarily independent), the independence theorem does not hold for arbitrary k-tuples $(\gamma_1,\ldots,\gamma_k)$ (see Exercise II-18); it holds however for those k-tuples which are compatible in the following sense. Let w_{ij} be the canonical valuation corresponding to the valuation ring $A_i \cdot A_j$ of K and Γ_{ij} its value group, and let $g_{ij}\colon \Gamma_i \to \Gamma_{ij}$ be the homomorphism determined by $w_{ij} = g_{ij} \circ w_i$ ($i \neq j$; see §7). The k-tupel $(\gamma_1,\ldots,\gamma_k) \in$ $\in \Gamma_1 \times \ldots \times \Gamma_k$ is called compatible, if $g_{ij}\, \gamma_i = g_{ji}\, \gamma_j$ for all $i,j \in \{1,\ldots,k\}$, $i \neq j$. Obviously $A_1,\ldots,A_k$ are pairwise independent if and only if $\Gamma_{ij} = \{0\}$ for all $i, j, i \neq j$, if and only if all k-tuples $(\gamma_1,\ldots,\gamma_k) \in \Gamma_1 \times \ldots \times \Gamma_k$ are compatible. For the proof of this generalized independence theorem see Ribenboim [30]. There a generalization of the approximation theorem can also be found.

Note that valuation rings $A_1,\dots,A_k$ of K of rank 1 are pairwise independent if and only if they are pairwise incomparable, if and only if they are pairwise distinct. Therefore (11.16) (resp. (11.17)) is a generalization of (3.13) (resp. (3.14)).

§12 Krull rings and Dedekind rings

In §11 we discussed Prüfer rings as a generalization of finite intersections of arbitrary valuation rings. In this section, we discuss another generalization of finite intersections of discrete (resp. rank 1) valuation rings, namely Krull (resp. generalized Krull) rings. Those Krull (resp. generalized Krull) rings which are also Prüfer rings are called Dedekind (resp. generalized Dedekind) rings.

Fixing a field K we consider sets $\mathfrak{A}$ of arbitrary valuation rings of K. As in §11, we denote by $\mathfrak{B}(R)$ (resp. $\mathfrak{E}(R)$) the set of all valuation rings of K which contain (resp. are essential for) a given subring R of K. Note that $R \subseteq \bigcap_{A\in\mathfrak{A}} A$ if and only if $\mathfrak{A} \subseteq \mathfrak{B}(R)$. We say that R is <u>defined by</u> $\mathfrak{A}$ if $R = \bigcap_{A\in\mathfrak{A}} A$. We recall that R is defined by some set $\mathfrak{A}$ if and only if R is integrally closed in K; in this case, R is defined by $\mathfrak{B}(R)$ and certain subsets of it (cf §10). If R is defined by $\mathfrak{A}$ then it is also defined by $\mathfrak{A}^* = \mathfrak{A}\setminus\{K\}$. In particular, K is defined only by $\mathfrak{A} = \{K\}$ and by $\mathfrak{A} = \emptyset$.

We say that $\mathfrak{A}$ is <u>of finite character</u> if any non-zero $x \in K$ is a unit in almost all $A \in \mathfrak{A}$ (i.e., if $\{A \in \mathfrak{A} \mid x \notin U_A\}$ is finite for all $x \in K$, $x \neq 0$). This occurs in particular whenever $\mathfrak{A}$ is finite. We say that $\mathfrak{A}$ is <u>of discrete</u> (resp. <u>real</u>) <u>character</u> if any $A \in \mathfrak{A}$ is a discrete (resp. rank 1) valuation ring of K, i.e. the ring of a discrete (resp. non trivial) exponential valuation of K. We say that $\mathfrak{A}$ is <u>of rational character</u>, if any $A \in \mathfrak{A}$ is the ring of a non-trivial exponential valuation of K with value group $\Gamma_A \subseteq \mathbb{Q}$.

(12.1) <u>Let R be defined by a set $\mathfrak{A}$ of finite real (resp. discrete) character. For any multiplicative subset M of R, the ring of fractions R_M is defined by the set $\mathfrak{A} \cap \mathfrak{B}(R_M)$, which is also of finite real (resp. discrete) character. Moreover, if $\mathfrak{A} \subseteq \mathcal{E}(R)$ then $\mathfrak{A} \cap \mathfrak{B}(R_M) \subseteq \mathcal{E}(R_M)$.</u>

<u>Proof</u>: Let $\mathfrak{A}_M = \mathfrak{A} \cap \mathfrak{B}(R_M)$. Since $\mathfrak{A}$ is of finite real (resp. discrete) character, so is $\mathfrak{A}_M$ and, for any $x \in K$, the set $\mathfrak{A}_x = \{A \in \mathfrak{A} \mid x \notin A\}$ is finite. Obviously $R_M \subseteq \bigcap_{A \in \mathfrak{A}_M} A$. To prove the equality, let $x \in \bigcap_{A \in \mathfrak{A}_M} A$; then $\mathfrak{A}_x \subseteq \mathfrak{A} \setminus \mathfrak{A}_M$, say $\mathfrak{A}_x = \{A_1, \ldots, A_m\}$. For any $i \in \{1, \ldots, m\}$ we have $R_M \nsubseteq A_i$, hence $M \nsubseteq U_{A_i}$; therefore there is a non-zero $y_i \in M \cap \mathfrak{M}_{A_i} \subseteq R$. Since A_i has rank 1, we conclude from (7.6) that $x \cdot y_i^{k_i} \in A_i$ for sufficiently large $k_i \in \mathbb{N}$. Setting $k = \max\{k_1, \ldots, k_m\}$, $y = (y_1 \cdot \ldots \cdot y_m)^k$, and $z = x \cdot y$, we have $z \in A$ for all $A \in \mathfrak{A} \setminus \mathfrak{A}_x$ and $z = (x \cdot y_i^k) \cdot (y_1 \cdot \ldots \cdot y_{i-1} \cdot y_{i+1} \cdot \ldots \cdot y_m)^k \in A_i$ for $i = 1, \ldots, m$; hence $z \in \bigcap_{A \in \mathfrak{A}} A = R$, $x = z \cdot y^{-1} \in R_M$. Therefore R_M is defined by $\mathfrak{A}_M$. If $\mathfrak{A} \subseteq \mathcal{E}(R)$ then $\mathfrak{A} \cap \mathfrak{B}(R_M) \subseteq \mathcal{E}(R) \cap \mathfrak{B}(R_M) = \mathcal{E}(R_M)$, by (11.2). □

We denote by $\mathfrak{h}(R)$ the set of all minimal non-zero prime ideals of R (i.e., minimal elements of the set $\mathcal{P}(R) \setminus \{(0)\}$, ordered by inclusion) and prove:

(12.2) THEOREM - <u>Let R be defined by some set $\mathfrak{A}$ of finite real</u>

(resp. discrete) character. Then:

a) $\mathcal{C}(R)^*$ is of finite real (resp. discrete) character and $\mathcal{C}(R)^* \subseteq \mathfrak{a}$.

b) For any $A \in \mathcal{C}(R)^*$ we have $A = R_{\mathfrak{P}}$ where $\mathfrak{P} = \mathfrak{M}_A \cap R \in \mathfrak{h}(R)$.

c) If K is the quotient field of R then a 1-1 correspondence $\mathcal{C}(R)^* \leftrightarrow \mathfrak{h}(R)$ is given by $\mathfrak{P} = \mathfrak{M}_A \cap R$ and $A = R_{\mathfrak{P}}$.

Proof: a) By (12.1), any $A \in \mathcal{C}(R)$ is defined by some subset $\mathfrak{a}'$ of $\mathfrak{a} \cap \mathfrak{B}(R)$. Since $\mathfrak{B}(A)$ is totally ordered, by (6.6), and $\mathfrak{a}$ is of real character, we have $\# \mathfrak{a}' \leqq 1$, hence $A = K$ or $\mathfrak{a}' = \{A\}$. Therefore $\mathcal{C}(R)^* \subseteq \mathfrak{a}$. Since $\mathfrak{a}$ is of finite real (resp. discrete) character, so is $\mathcal{C}(R)^*$.

b) Let $A \in \mathcal{C}(R)^*$ and $\mathfrak{P} = \mathfrak{M}_A \cap R$. By (11.1), we have $A = R_{\mathfrak{P}}$, and $\mathfrak{P} \in \mathfrak{h}(R)$ since $\mathfrak{B}(A) = \{A,K\}$.

c) Let $\mathfrak{P} \in \mathfrak{h}(R)$. By (12.1), $R_{\mathfrak{P}}$ is defined by some subset $\mathfrak{a}''$ of $\mathfrak{a}$. We assume that K is the quotient field of R , hence also of $R_{\mathfrak{P}}$. Then the prime ideal $\mathfrak{M}_{A''} \cap R_{\mathfrak{P}}$ is distinct from the zero ideal, for any $A'' \in \mathfrak{a}''$, and therefore coincides with the unique non-zero prime ideal $\mathfrak{P} \cdot R_{\mathfrak{P}}$ of $R_{\mathfrak{P}}$. Let $x \in \mathfrak{P}$, $x \neq 0$; since $x \in \mathfrak{M}_{A''}$ for all $A'' \in \mathfrak{a}''$ and $\mathfrak{a}''$ is of finite character, $\mathfrak{a}''$ is even finite. Therefore $R_{\mathfrak{P}}$ is a Prüfer ring of K , by (11.12), and even a valuation ring of K , by (11.6); hence $R_{\mathfrak{P}} \in \mathcal{C}(R)^*$ and $\mathfrak{M}_{R_{\mathfrak{P}}} \cap R = \mathfrak{P}$. We conclude that the mappings $A \mapsto \mathfrak{M}_A \cap R$ $(A \in \mathcal{C}(R)^*)$ and $\mathfrak{P} \mapsto R_{\mathfrak{P}}$ $(\mathfrak{P} \in \mathfrak{h}(R))$ are inverse to each other. □

A subring R of K is called a Krull ring (resp. generalized Krull ring) of K if it is defined by some set $\mathfrak{a}_0$ of finite discrete (resp. real) character contained in $\mathcal{C}(R)^*$. For example, it follows immediately from (4.3) that

(12.3) Any UFD is a Krull ring of its quotient field.

We are going to show that we have always $\mathcal{A}_o = \mathcal{C}(R)^*$. More precisely:

(12.4) THEOREM - Let R be a generalized Krull ring of K . Then:

a) K is the quotient field of R .

b) For any set $\mathcal{A}$ of finite real character we have $\mathcal{C}(R)^* \subseteq \mathcal{A} \subseteq \mathcal{B}(R)^*$ if and only if R is defined by $\mathcal{A}$.

Proof: By hypothesis, R is defined by some set $\mathcal{A}_o$ of finite real character contained in $\mathcal{C}(R)^*$. a) Any $A \in \mathcal{A}_o$ is a ring of fractions of R , and K is the quotient field of A ; therefore K is the quotient field of R whenever $\mathcal{A}_o \neq \emptyset$. If $\mathcal{A}_o = \emptyset$ then $R = K$. b) If R is defined by $\mathcal{A}$, a set of finite real character, then $\mathcal{A} \subseteq \mathcal{B}(R)^*$ and $\mathcal{C}(R)^* \subseteq \mathcal{A}$, by (12.2). On the other hand, if $\mathcal{C}(R)^* \subseteq \mathcal{A} \subseteq \mathcal{B}(R)^*$ then $R = \bigcap_{A \in \mathcal{A}_o} A \supseteq \bigcap_{A \in \mathcal{A}} A \supseteq \bigcap_{A \in \mathcal{B}(R)^*} A \supseteq R$, hence R is defined by $\mathcal{A}$. □

Note that there may exist sets $\mathcal{A}$ of infinite discrete character which define a Krull ring R but do not contain $\mathcal{C}(R)^*$ (cf. Krull [21], §5, Nº 37). There are also examples of subrings R of K , with quotient field K , which are defined by some set $\mathcal{A}$ of finite real character but not by $\mathcal{C}(R)^*$ (cf Ohm [26] or Griffin [13]); in this case, $\mathcal{C}(R)^*$ is strictly contained in $\mathcal{A}$ and R is strictly contained in the ring R' defined by $\mathcal{C}(R)^*$. In the general case, we have the following relationship between the rings R and R' .

(12.5) COROLLARY - Let R be defined by some set $\mathcal{A}$ of finite real character. Then the ring R' defined by $\mathcal{C}(R)^*$ is a generalized Krull ring of K such that $R \subseteq R'$ and $\mathcal{C}(R)^* = \mathcal{C}(R')^* \subseteq \mathcal{A}$.

Proof: By (12.2), $\mathcal{C}(R)^*$ is of finite real character and contained in $\mathcal{A}$, hence $R' \supseteq R$. Any $A \in \mathcal{C}(R)^*$ contains R' , and since A is a ring of fractions of R it is also a ring of fractions of R' ; therefore $\mathcal{C}(R)^* \subseteq \mathcal{C}(R')^*$. We conclude that R' is a generalized Krull ring of K , and by (12.4) we have $\mathcal{C}(R')^* \subseteq \mathcal{C}(R)^*$.

□

We are going to prove that $R' = R$ whenever K is the quotient field of R and $\mathfrak{A}$ is of finite rational character. Given any set $\mathfrak{A}$ of valuation rings of K , we say that $A_o \in \mathfrak{A}$ is <u>irredundant in</u> $\mathfrak{A}$ if $\bigcap_{A\in\mathfrak{A}\setminus\{A_o\}} A \neq \bigcap_{A\in\mathfrak{A}} A$. We denote by $\mathfrak{A}^I$ the set of all $A \in \mathfrak{A}$ which are irredundant in $\mathfrak{A}$ and prove:

(12.6) <u>Let</u> R <u>be defined by some set</u> $\mathfrak{A}$ <u>of finite real character.</u> <u>Then:</u>

a) $\mathcal{C}(R)^* \subseteq \mathfrak{A}^I$.

b) <u>If</u> $\mathfrak{A}$ <u>is of finite rational character and</u> K <u>is the quotient field of</u> R <u>then</u> $\mathcal{C}(R)^* = \mathfrak{A}^I$.

<u>Proof</u>: a) Let $A \in \mathfrak{A} \setminus \mathfrak{A}^I$; then $\mathfrak{A}\setminus\{A\}$ is a set of finite real character which defines R , hence $\mathcal{C}(R)^* \subseteq \mathfrak{A}\setminus\{A\}$, by (12.2). Therefore $A \in \mathfrak{A} \setminus \mathcal{C}(R)^*$.

b) It suffices to show that $\mathfrak{A}^I \subseteq \mathcal{C}(R)^*$. Let $A_o \in \mathfrak{A}^I$ and $D = \bigcap_{A\in\mathfrak{A}\setminus\{A_o\}} A$. Then $R = D \cap A_o \neq D$, hence $D \not\subseteq A_o$ and, by (7.8), $\mathfrak{P}_o = \mathfrak{M}_{A_o} \cap R$ is a minimal prime ideal of R . By (12.2 c) we have $A_o = R_{\mathfrak{P}_o} \in \mathcal{C}(R)^*$. □

(12.7) THEOREM - <u>Let</u> R <u>be defined by some set</u> $\mathfrak{A}$ <u>of finite rational character and let</u> K <u>be the quotient field of</u> R . <u>Then</u> R <u>is a generalized Krull ring of</u> K .

<u>Proof</u>: By (12.2) we have $\mathcal{C}(R)^* \subseteq \mathfrak{A}$, hence $R \subseteq \bigcap_{A\in\mathcal{C}(R)^*} A$. To prove the equality, let $x \in \bigcap_{A\in\mathcal{C}(R)^*} A$. Since $\mathfrak{A}$ is of finite character, there are finitely many valuation rings $A_1,\dots,A_m \in \mathfrak{A} \setminus \mathcal{C}(R)^*$ such that $x \in A$ for all $A \in \mathfrak{A} \setminus \{A_1,\dots,A_m\}$. We assume

by induction that R is defined by $\mathfrak{a}_k = \mathfrak{a} \setminus \{A_1,\dots,A_{k-1}\}$ for some $k \in \{1,\dots,m\}$; this is trivial for $k = 1$. Since $\mathfrak{a}_k$ is of finite rational character and $A_k \notin \mathcal{E}(R)^*$, we have $A_k \in \mathfrak{a}_k \setminus \mathfrak{a}_k^I$ by (12.6 b), hence $\mathfrak{a}_{k+1} = \mathfrak{a}_k \setminus \{A_k\}$ also defines R . In particular, R is defined by $\mathfrak{a} \setminus \{A_1,\dots,A_m\}$; therefore $x \in R$. □

In particular, theorem (12.7) yields:

(12.8) COROLLARY - Let R be defined by some set of finite discrete character and let K be the quotient field of R. Then R is a Krull ring of K .

For noetherian subrings R of K , with quotient field K, the property of being integrally closed is not only necessary but also sufficient for R to be a Krull ring of K . In fact:

(12.9) Let R be noetherian and integrally closed and K its quotient field. Then R is a Krull ring of K and $\mathcal{E}(R)^* = \{R_{\mathfrak{p}} \mid \mathfrak{p} \in \mathfrak{h}(R)\}$.

A proof of this statement can be found in Zariski & Samuel [36], Chap. VI, §10, or in Gilmer [10], theorem (35.4). It is based on (10.7) and the representation of non-zero principal ideals as finite intersection of symbolic powers of minimal prime ideals. For a different proof, see Bourbaki [5], Chap. 7, §1.

An integral domain R is said to have dimension ≤ 1 if its non-zero prime ideals are pairwise incomparable with respect to inclusion; this occurs if and only if $\mathcal{P}(R)\setminus\{(0)\} \subseteq \mathfrak{m}(R)$, if and only if $\mathcal{P}(R)\setminus\{(0)\} \subseteq \mathfrak{h}(R)$. A subring R of K will be called a Dedekind ring (resp. generalized Dedekind ring) of K if it safisfies the equivalent conditions of the following theorem.

(12.10) THEOREM - For any subring R of K the following conditions are equivalent:

(i) R <u>is a Prüfer ring of</u> K <u>and a Krull (resp. generalized Krull) ring of</u> K .

(ii) R <u>is a Prüfer ring of</u> K <u>and is defined by some set of finite discrete (resp. real) character</u>.

(iii) K <u>is the field of quotients of</u> R , R <u>has dimension</u> ≤ 1 <u>and is defined by some set of finite discrete (resp. real) character</u>.

(iv) R <u>is defined by some set</u> $\mathfrak{a}$ <u>of finite character, and</u> $\{R_{\mathfrak{p}} \mid \mathfrak{p} \in \mathfrak{m}(R)\}^*$ <u>is a set of discrete (resp. real) character</u>.

<u>In this case,</u> $\mathfrak{B}(R)^* = \mathfrak{C}(R)^*$ <u>is of finite discrete (resp. real) character, and</u> R <u>is defined only by</u> $\mathfrak{B}(R)^*$ <u>and by</u> $\mathfrak{B}(R)$.

<u>Proof</u>: (i) ⇒ (ii) is trivial.

(ii) ⇒ (iii): The first and the last statements of (iii) are obvious. For any non-zero $\mathfrak{p} \in \mathfrak{P}(R)$ we have $R_{\mathfrak{p}} \in \mathfrak{C}(R)^*$ by (11.4), hence $\mathfrak{p} = \mathfrak{p}\cdot R_{\mathfrak{p}} \cap R \in \mathfrak{h}(R)$ by (12.2 b); therefore R has dimension ≤ 1 .

(iii) ⇒ (iv): The first statement is obvious. Since $\mathfrak{m}(R) \subseteq \mathfrak{h}(R) \cup \cup \{(0)\}$, it follows from (12.2 c) that $R_{\mathfrak{p}} \in \mathfrak{C}(R)$ for any $\mathfrak{p} \in \mathfrak{m}(R)$. Since $\mathfrak{C}(R)^*$ is of finite discrete (resp. real) character, by (12.2 a), so is $\{R_{\mathfrak{p}} \mid \mathfrak{p} \in \mathfrak{m}(R)\}^*$.

(iv) ⇒ (i): By (11.4), R is a Prüfer ring of K , hence $\mathfrak{a} \subseteq \mathfrak{B}(R) = \mathfrak{C}(R)$. For any $A \in \mathfrak{a}$ there is some $\mathfrak{p} \in \mathfrak{m}(R)$ such that $\mathfrak{M}_A \cap R \subseteq \subseteq \mathfrak{p}$, hence $R_{\mathfrak{p}} \subseteq A \subseteq K$ by (11.1) and (6.6); since $R_{\mathfrak{p}}$ has rank 1 we conclude that $A \in \{R_{\mathfrak{p}}, K\}$. Therefore $\mathfrak{a}^*$ is of finite discrete (resp. real) character and contained in $\mathfrak{C}(R)^*$, hence R is a Krull (resp. generalized Krull) ring of K .

The last statement of the theorem follows from (12.4). □

For example, we get as an immediate consequence of (11.12), (11.5), and (12.3):

(12.11) a) <u>Any finite intersection of discrete (resp. rank 1) valuation rings of</u> K <u>is a Dedekind (resp. generalized Dedekind) ring of</u> K .

b) <u>Any PID is a Dedekind ring of its quotient field.</u>

Moreover, one can show that an integral domain is a PID if and only if it is a UFD and a Prüfer ring of its quotient field (cf Gilmer [10], Propos. 31.6).

Next we show that any Dedekind ring is noetherian. This fact yields some more characterizations of Dedekind rings. Some of them are used as definitions by other authors.

(12.12) THEOREM - <u>For any subring</u> R <u>of</u> K <u>the following conditions are equivalent:</u>

(i) R <u>is a Dedekind ring of</u> K .

(ii) R <u>is a noetherian generalized Dedekind ring of</u> K .

(iii) R <u>is a noetherian Prüfer ring of</u> K .

(iv) R <u>is a noetherian integrally closed domain of dimension</u> ≤ 1 <u>and</u> K <u>is its quotient field.</u>

<u>Proof:</u> (ii) ⇒ (iii) and (ii) ⇒ (iv) are trivial. (iv) ⇒ (i) and (iii) ⇒ (i) follows from (12.9). (i) ⇒ (ii): It suffices to show that any Dedekind ring R of K is noetherian. We may assume $R \neq K$; then $\mathfrak{n}(R) = \mathfrak{m}(R)$. By (12.2), the sets $\mathfrak{m}(R)$, $\mathcal{C}(R)^*$ are in 1-1 correspondence by $A = R_{\mathfrak{P}}$, $\mathfrak{P} = \mathfrak{M}_A \cap R$, and any $A \in \mathcal{C}(R)^*$ is a discrete valuation ring, hence a PID (cf (7.7)). Let $\mathfrak{A}$ be a non-zero ideal of R and choose a non-zero $x \in \mathfrak{A}$. Since $\mathcal{C}(R)^*$ is of finite character, we have $x \in U_A$ for almost all $A \in \mathcal{C}(R)^*$, hence there are finitely many $\mathfrak{P}_1,\dots,\mathfrak{P}_r \in \mathfrak{m}(R)$ such that $x \notin \mathfrak{P}$ for all $\mathfrak{P} \in \mathfrak{m}(R) \setminus \{\mathfrak{P}_1,\dots,\mathfrak{P}_r\}$. For any $i \in \{1,\dots,r\}$ let $A_i = R_{\mathfrak{P}_i}$; then $\mathfrak{A} \cdot R_i = t_i \cdot R_i$ for some $t_i \in \mathfrak{A}$. We consider the ideal $\mathfrak{C} = x \cdot R + t_1 \cdot R + \dots + t_r \cdot R$ of R . Obviously $\mathfrak{C} \subseteq \mathfrak{A}$. Since $\mathfrak{C} \nsubseteq \mathfrak{P}$ for all

$\mathfrak{p} \in \mathfrak{m}(R)\setminus\{\mathfrak{p}_1,\ldots,\mathfrak{p}_r\}$, we have $\mathfrak{C} = \bigcap_{i=1}^{r} (\mathfrak{C}\cdot R_i \cap R) \supseteq \bigcap_{i=1}^{r} (t_i\cdot R_i \cap R) \supseteq$ $\supseteq \mathfrak{A}$ (cf Gilmer [10], theorem (3.10)). Therefore any ideal of R is finitely generated, i.e., R is noetherian. □

For other characterizations of Krull rings and Dedekind rings see for example Zariski & Samuel [36], Chap. V-VI, Bourbaki [5], Chap. 7, or Gilmer [10], Chap. VI. In Borevich & Shafarevich [3], Chap. 3, Krull rings are characterized by the existence of a divisor theory. Rings which are defined by sets of finite character consisting of valuation rings of K of arbitrary rank, have been studied by Griffin [13], [14], [15] .

Let $\mathfrak{a}$ be a set of real character. For any $A \in \mathfrak{a}$ we choose an exponential valuation v_A of K corresponding to A and denote by Γ_A its value group. We say that $\mathfrak{a}$ satisfies the <u>strong</u> (resp. <u>weak</u>) <u>approximation condition</u> if for any finitely many pairwise distinct $A_1,\ldots,A_n \in \mathfrak{a}$, for any $\gamma_1 \in \Gamma_{A_1},\ldots,\gamma_n \in \Gamma_{A_n}$, and for any $x_1,\ldots,x_n \in K$ (resp. for $x_1 = \ldots = x_n = 0$) there is an $x \in K$ such that $v_{A_i}(x-x_i) = \gamma_i$ $(i=1,\ldots,n)$ and $v_A x \geqq 0$ for all $A \in \mathfrak{a}\setminus\{A_1,\ldots,A_n\}$. These conditions are obviously independent of the choice of the exponential valuations v_A . If $A \in \mathfrak{a}$ is a discrete valuation ring, it is convenient to choose for v_A the unique normalized exponential valuation corresponding to A (so $\Gamma_A = \mathbb{Z}$).

(12.13) THEOREM - <u>Let</u> R <u>be defined by some set of finite real character and assume that</u> K <u>is the quotient field of</u> R . <u>Then</u> $\mathcal{C}(R)^*$ <u>satisfies the weak approximation condition.</u>

<u>Proof</u>: We first show that for any finitely many $A_1,\ldots,A_m \in \mathcal{C}(R)^*$ and any non-negative $\delta_1 \in \Gamma_{A_1},\ldots,\delta_m \in \Gamma_{A_m}$ there is an element $c \in R$ such that $v_{A_i}\, c = \delta_i$ $(i=1,\ldots,m)$. In fact, for any $i \in \{1,\ldots,m\}$ let $a_i \in A_i$ such that $v_{A_i}\, a_i = \delta_i$. By (11.1) there exist $d_i \in R$ and $c_i \in R\setminus\mathfrak{M}_{A_i}$ such that $a_i = d_i\cdot c_i^{-1}$, hence

$v_{A_i} d_i = \delta_i$. For all $i,j \in \{1,\dots,m\}$, $i \neq j$, we have $\mathfrak{M}_{A_j} \cap R \not\subseteq \not\subseteq \mathfrak{M}_{A_i} \cap R$ by (12.2 c); hence there exists an element $b_{i,j} \in \mathfrak{M}_{A_j} \cap \cap R \cap U_{A_i}$. Let $b_i = b_{i,1} \cdot \dots \cdot b_{i,i-1} \cdot b_{i,i+1} \cdot b_{i,i+1} \cdot \dots \cdot b_{i,m}$ $(i=1,\dots,m)$. For sufficiently large $k \in \mathbb{N}$ we have $v_{A_j}(d_i \cdot b_i{}^k) > \delta_j$ for all $j \neq i$, whereas $v_{A_i}(d_i \cdot b_i{}^k) = \delta_i$ $(i=1,\dots,m)$. Setting $c = d_1 \cdot b_1^k + \dots + d_m \cdot b_m^k$, we have $c \in R$ and $v_{A_i} c = \delta_i$ for all $i \in \{1,\dots,m\}$.

To prove the weak approximation condition for $\mathcal{C}(R)^*$, let $A_1,\dots,A_n \in \mathcal{C}(R)^*$ and $\gamma_1 \in \Gamma_{A_1},\dots,\gamma_n \in \Gamma_{A_n}$. By the first part of the proof, there is an element $c \in R$ such that $v_{A_i} c = \max\{0,-\gamma_i\}$ for all $i \in \{1,\dots,m\}$. Since $\mathcal{C}(R)^*$ is of finite character, by (12.2), there is a finite subset $\mathfrak{A}'$ of $\mathcal{C}(R)^*$ containing $\{A_1,\dots,A_n\}$ such that $v_A c = 0$ for all $A \in \mathcal{C}(R)^* \setminus \mathfrak{A}'$. Again by the first part of the proof, there is an element $d \in R$ such that $v_{A_i} d = \max\{0,\gamma_i\}$ for all $i \in \{1,\dots,n\}$ and $v_A d = v_A c$ for all $A \in \mathfrak{A}' \setminus \{A_1,\dots,A_n\}$. Let $x = d \cdot c^{-1}$; then $v_{A_i} x = v_{A_i} d - v_{A_i} c = = \gamma_i$ $(i=1,\dots,n)$ and $v_A x = v_A d - v_A c \geq 0$ for all $A \in \mathcal{C}(R)^* \setminus \setminus \{A_1,\dots,A_n\}$; therefore x has the desired property. □

Theorem (12.13) applies in particular to any generalized Krull ring of K . Conversely, we show that any set $\mathfrak{A}$ of finite real character which satisfies the weak approximation condition is equal to $\mathcal{C}(R)^*$, where R is the generalized Krull ring defined by $\mathfrak{A}$.

(12.14) COROLLARY - Let R be defined by a set $\mathfrak{A}$ of finite real character. The following conditions are equivalent:

(i) $\mathfrak{A}$ satisfies the weak approximation condition.

(ii) For any distinct $A_1, A_2 \in \mathfrak{A}$ we have $\mathfrak{M}_{A_1} \cap R \not\subseteq \mathfrak{M}_{A_2} \cap R$.

(iii) For any distinct $A, A' \in \mathfrak{A}$ we have $R_{\mathfrak{M}_A \cap R} \not\subseteq A'$.

(iv) $\mathfrak{A} \subseteq \mathcal{C}(R)^*$.

(v) R is a generalized Krull ring of K and $\mathfrak{A} = \mathcal{C}(R)^*$.

<u>Proof</u>: (i) ⇒ (ii): There is an $x \in K$ such that $v_{A_1} x > 0$, $v_{A_2} x = 0$, and $v_A x \geqq 0$ for all $A \in \mathfrak{G} \setminus \{A_1, A_2\}$. Therefore $x \in \bigcap_{A \in \mathfrak{G}} A = R$, $x \in \mathfrak{M}_{A_1}$, and $x \notin \mathfrak{M}_{A_2}$. (ii) ⇒ (iii): Let $x \in (\mathfrak{M}_{A'} \cap R) \setminus (\mathfrak{M}_A \cap R)$. Then $x^{-1} \in R_{\mathfrak{M}_A \cap R} \setminus A'$. (iii) ⇒ (iv): For any $A \in \mathfrak{G}$ we have $\mathfrak{G} \cap \mathfrak{B}(R_{\mathfrak{M}_A \cap R}) = \{A\}$, hence $R_{\mathfrak{M}_A \cap R} = A$, by (12.1). Therefore $\mathfrak{G} \subseteq \mathcal{C}(R)^*$. (iv) ⇒ (v): Follows from (12.4 b). (v) ⇒ (i): Follows from (12.13). □

It is obvious that this corollary remains valid if one replaces "real" by "discrete" and "generalized Krull ring" by "Krull ring". Its principal assertion is the characterization of Krull (resp. generalized) Krull rings as those subrings of K which are defined by some set $\mathfrak{G}$ of finite discrete (resp. real) character satisfying the weak approximation condition. Similarly, Dedekind (resp. generalized Dedekind) rings can be characterized as those subrings of K which are defined by some set $\mathfrak{G}$ of finite discrete (resp. real) character satisfying the strong approximation condition (cf Ribenboim [30]). We prove here only the following analogue of theorem (12.13).

(12.15) THEOREM - <u>For any Dedekind ring R of K the set $\mathcal{C}(R)^*$ satisfies the strong approximation condition.</u>

<u>Proof</u>: We may assume that $\Gamma_A = \mathbb{Z}$ for any $A \in \mathcal{C}(R)^*$. We first claim that, for any $A \in \mathcal{C}(R)^*$, any $u \in R \setminus \mathfrak{M}_A$, and any positive integer k, there is an $a \in R$ such that $v(u \cdot a - 1) \geqq k$. This is true for $k = 1$; in fact, since $\varkappa_A R$ is a field, by (11.3), the element $\varkappa_A u \neq 0$ has an inverse in $\varkappa_A R$, i.e., we have $u \cdot a' - 1 \in \mathfrak{M}_A$ for some $a' \in R$. Assume, by induction, that the claim is true for some $k \geqq 1$. Since $\mathcal{C}(R)^*$ satisfies the weak approximation condition, by (12.13), and is of finite character, there exist $t, u_1 \in R$ such that $v_A t = k$, $v_A u_1 = 0$, and $v_{A'} u_1 = v_{A'} t$ for all those (finite-

ly many) $A' \in \mathcal{C}(R)^*$ for which $t \notin U_{A'}$. Let $b = u_1 \cdot (u \cdot a - 1) \cdot t^{-1}$; obviously $b \in R$ and, since $u_1 \cdot u \in R \setminus \mathfrak{M}_A$, there is a $c \in R$ such that $v_A(u_1 \cdot u \cdot c - b) \geqq 1$. Let $d = a - t \cdot c$; obviously $d \in R$, and $v_A(u \cdot d - 1) \geqq k+1$ since $u_1 \cdot (u \cdot d - 1) = u_1 \cdot (u \cdot a - 1) - u_1 \cdot u \cdot t \cdot c =$ $= t \cdot (b - u_1 \cdot u \cdot c)$.

Next we claim that, for any finitely many distinct $A_1, \ldots, A_m \in \mathcal{C}(R)^*$, any $x_1, \ldots, x_m \in K$, and any $k \in \mathbb{N}$, there is an $x \in K$ such that $v_{A_i}(x - x_i) \geqq k$ $(i=1,\ldots,m)$ and $v_A x \geqq 0$ for all $A \in \mathcal{C}(R)^* \setminus \{A_1, \ldots, A_m\}$. We may write $x_i = b_i \cdot c^{-1}$ for appropriate b_i , $c \in R$ $(i=1,\ldots,m)$. Since $\mathcal{C}(R)^*$ is of finite character, there are finitely many $A_{m+1}, \ldots, A_n \in \mathcal{C}(R)^* \setminus \{A_1, \ldots, A_m\}$ such that $c \in U_A$ for all $A \in \mathcal{C}(R)^* \setminus \{A_1, \ldots, A_m, \ldots, A_n\}$. We set $b_{m+1} = \ldots = b_n = 0$, $k' = \max \{k + v_{A_i} c \mid i=1,\ldots,n\}$. Using again the weak approximation condition, we choose elements $u_1, \ldots, u_n \in R$ such that $v_{A_i} u_i = 0$ and $v_{A_j} u_i \geqq k'$ for all $i,j \in \{1,\ldots,n\}$, $i \neq j$. By the first part of the proof, there is an $a_i \in R$ such that $v_{A_i}(u_i \cdot a_i - b_i) \geqq k'$ $(i=1,\ldots,n)$. Let $b = u_1 \cdot a_1 + \ldots + u_n \cdot a_n$; obviously $b \in R$ and $v_{A_i}(b - b_i) \geqq \min \{v_{A_i}(u_j \cdot a_j - \delta_{i,j} \cdot b_j) \mid j=1,\ldots,n\} \geqq k'$ for all $i \in \{1,\ldots,n\}$. Therefore $x = b \cdot c^{-1}$ has the desired property.

To prove the strong approximation condition, let $A_1, \ldots, A_n \in \mathcal{C}(R)^*$, $k_1, \ldots, k_n \in \mathbb{Z}$, and $x_1, \ldots, x_n \in K$. We choose elements $y_1, \ldots, y_n \in K$ such that $v_{A_i} y_i = k_i$ $(k=1,\ldots,n)$ and choose an integer $k > \max \{0, k_1, \ldots, k_n\}$. By the second part of the proof there exist $x, y \in K$ such that $v_{A_i}(x - x_i) \geqq k$, $v_{A_i}(y - y_i) \geqq k$, and $v_A x \geqq 0$, $v_A y \geqq 0$ for all $A \in \mathcal{C}(R)^* \setminus \{A_1, \ldots, A_n\}$. For any $i \in \{1,\ldots n\}$ we have $v_{A_i}((x - x_i) + (y - y_i)) \geqq k > k_i = v_{A_i} y_i$, hence $v_{A_i}(x + y - x_i) = k_i$; moreover $v_A(x+y) \geqq 0$ for all $A \in \mathcal{C}(R)^* \setminus$ $\setminus \{A_1, \ldots, A_n\}$. Therefore $x + y$ has the desired property. □

CHAPTER III

Extension of Valuation Rings

§13 The case of an algebraic field extension

We first consider an arbitrary (not necessarily algebraic) field extension $L|K$. Let A (resp. B) be a valuation ring of K (resp. L). We say that B lies over A if $A = B \cap K$; in this case B is also called an extension of A. Considering A and B as local subrings of L we get:

(13.1) B lies over A if and only if B dominates A.

Proof: If B lies over A then $A = B \cap K \subseteq B$ and, for any non-zero $x \in \mathfrak{M}_A$, we have $x^{-1} \in K \setminus A \subseteq L \setminus B$, hence $x \in \mathfrak{M}_B$; therefore B dominates A. On the other hand, if B dominates A then $A \subseteq B \cap K$ and, for any non-zero $x \in B \cap K$, we have $x^{-1} \in K \setminus \mathfrak{M}_B \subseteq K \setminus \mathfrak{M}_A$, hence $x \in A$; therefore B lies over A. □

Obviously any valuation ring B of L lies over exactly one valuation ring of K, namely $B \cap K$. On the other hand, we conclude from (9.11) and (13.1):

(13.2) THEOREM - For any valuation ring A of K there is at least one valuation ring B of L which lies over A.

Moreover, we can conclude from (9.7) that B may be chosen such that the transcendence degree of $B/\mathfrak{M}_B$ over $\varkappa_B A$ $(\cong A/\mathfrak{M}_A)$ is any preassigned cardinal number less than or equal to the transcendence degree of $L|K$.

Given a valuation ring A of K and a field extension $L|K$, we want to get a survey on the set $\mathfrak{B}$ of all those valuation rings B of L which lie over (or, equivalently, dominate) A. We denote by D the integral closure $I_L(A)$ of A in L , by $\mathcal{C}$ the set of all valuation rings of L which contain D , and by $\mathfrak{P}$ (resp. $\mathfrak{M}$) the set of all prime (resp. maximal) ideals of D.

(13.3) a) $\mathcal{B} = \{C \in \mathcal{C} \mid \mathfrak{M}_C \cap D \in \mathfrak{M}\}$.

b) $D = \bigcap_{C \in \mathcal{C}} C = \bigcap_{B \in \mathcal{B}} B$.

c) $C \mapsto \mathfrak{M}_C \cap D$ $(C \in \mathcal{C})$ <u>is a mapping from</u> $\mathcal{C}$ <u>onto</u> $\mathfrak{P}$ <u>and induces a mapping from</u> $\mathcal{B}$ <u>onto</u> $\mathfrak{M}$.

d) <u>The quotient field of</u> D <u>equals the relative algebraic closure of</u> K <u>in</u> L .

<u>Proof</u>: a) For any $B \in \mathcal{B}$ we have $B = I_L(B) \supseteq I_L(A) = D$, by (10.3 b) and (10.6), hence $B \in \mathcal{C}$. Since $(\mathfrak{M}_B \cap D) \cap A = \mathfrak{M}_A$, we have $\mathfrak{M}_B \cap D \in \mathfrak{M}$ by (10.5 c). On the other hand, if $C \in \mathcal{C}$ such that $\mathfrak{M}_C \cap D \in \mathfrak{M}$ then $C \supseteq D \supseteq A$ and $\mathfrak{M}_C \cap A = (\mathfrak{M}_C \cap D) \cap A = \mathfrak{M}_A$ by (10.5 c); hence C dominates A , i.e., $C \in \mathcal{B}$.

b) follows from (10.8) and a).

c) is a consequence of (9.7) and a).

d) For any $x \in L$ which is algebraic over K there is a non-zero $c \in \mathfrak{M}_A$ such that $c \cdot x \in D$, as is checked easily; therefore $x = \frac{c \cdot x}{c}$ is an element of the quotient field of D . On the other hand, since any $y \in D$ is algebraic over K , so is any quotient $\frac{y}{z}$ of elements $y, z \in D$, $z \neq 0$. □

We are particularly interested in the case in which $L|K$ is an algebraic field extension. In this case D is a Prüfer ring of L , the mappings indicated in (13.3 c) are bijective, and the valuation rings $B \in \mathcal{B}$ are pairwise incomparable. These are the main statements of the following theorem.

(13.4) THEOREM - <u>The following conditions are equivalent</u>:

(i) $L|K$ <u>is an algebraic extension.</u>

(ii) D <u>is a Prüfer ring of</u> L .

(iii) <u>There is a</u> $\mathfrak{P} \in \mathcal{P}$ <u>such that</u> $\# \{C \in \mathcal{C} \mid \mathfrak{M}_C \cap D = \mathfrak{P}\} = 1$.

<u>In this case, the following statements hold:</u>

a) <u>There are inclusion-inverting 1-1 correspondences</u> $\mathcal{C} \leftrightarrow \mathcal{P}$ <u>and</u> $\mathcal{B} \leftrightarrow \mathfrak{m}$ <u>given by</u> $\mathfrak{P} = \mathfrak{M}_C \cap D$ <u>and</u> $C = D_{\mathfrak{P}}$.

b) $\mathcal{B}$ <u>is the set of all minimal elements of</u> $\mathcal{C}$, <u>and the elements of</u> $\mathcal{B}$ <u>are pairwise incomparable (with respect to inclusion.)</u>

c) <u>For any</u> $B \in \mathcal{B}$ <u>we have</u> $\varkappa_B D = B/\mathfrak{M}_B$.

<u>Proof</u>: (i) ⇒ (ii): For any $x \in L$ there is some $a \in K$ such that $a \cdot P_{x|K} \in A[X] \setminus \mathfrak{M}_A[X]$. For any $\mathfrak{N} \in \mathfrak{m}$ we have $\mathfrak{N} \cap A = \mathfrak{M}_A$ by (10.5 c), hence $a \cdot P_{x|K} \in D[X] \setminus \mathfrak{N}[X]$. Therefore D is a Prüfer ring of L , by (11.10).

(ii) ⇒ (iii) follows from (11.4).

(iii) ⇒ (i): If L is transcendental over K then L is also transcendental over the quotient field of D , by (13.3d). Therefore, for any $\mathfrak{P} \in \mathcal{P}$ there exists more than one element $C \in \mathcal{C}$ such that $\mathfrak{M}_C \cap D = \mathfrak{P}$, by (9.7).

Statements a) and b) follow from (11.4) and (13.3). Statement c) follows from b) and (11.8). □

We get as an immediate consequence of (13.3) and (13.4):

(13.5) COROLLARY - <u>The following conditions are equivalent:</u>

(i) $L|K$ <u>is an algebraic extension and</u> D <u>is a local ring.</u>

(ii) $\# \mathcal{B} = 1$.

(iii) D <u>is a valuation ring of</u> L .

<u>In this case,</u> $\mathcal{B} = \{D\}$, <u>i.e.,</u> D <u>is the unique valuation ring of</u> L <u>which lies over</u> A .

We say that A is <u>indecomposed</u> in L if the equivalent conditions of (13.5) are satisfied.

(13.6) a) Let L' be any field between K and L. The valuation ring A of K is indecomposed in L if and only if A is indecomposed in L' and $I_{L'}(A)$ is indecomposed in L.

b) A is indecomposed in L if and only if A is indecomposed in any finite subextension L' of $L|K$.

Proof: a) is an easy consequence of (13.2).

b) If A is not indecomposed in L then there exist at least two different valuation rings B_1, B_2 of L which lie over A; we may assume that $B_1 \not\subseteq B_2$. Let $y \in B_1 \setminus B_2$ and $L' = K(y)$. Then $B_1 \cap L'$, $B_2 \cap L'$ are distinct valuation rings of L' which lie over A; therefore A is not indecomposed in the finite subextension L' of $L|K$. The converse follows from a). □

Valuation rings A of K which are indecomposed in a given normal extension N of K (in particular, in an algebraic closure of K) will be studied in §16.

Given an algebraic extension $L|K$ there may exist infinitely many valuation rings of L lying over A. However, their number is finite whenever the separability degree $[L:K]_{sep}$ ($= [L_{sep}:K]$, where L_{sep} is the maximal separable subextension of $L|K$) is finite. In fact:

(13.7) THEOREM - Let $L|K$ be an algebraic extension such that $[L:K]_{sep} < \infty$. Then $\# \mathfrak{B} \leq [L:K]_{sep}$.

Proof: Let $B_1,\dots,B_k$ be distinct valuation rings of L which lie over A. By (13.4 b) they are pairwise incomparable. By (11.14) there exist $y_i \in B_1 \cap \dots \cap B_k$ such that $y_i - \delta_{ij} \in \mathfrak{M}_{B_j}$ $(i,j=1,\dots,k)$, and there is an $n \in \mathbb{N}$ such that $y_1^{p^n},\dots,y_k^{p^n}$ are separable over K, where p is the characteristic exponent of K

i.e., $p = \max\{1, \text{Char } K\}$. We claim that $y_1^{p^n},\dots,y_k^{p^n}$ are linearly independent over K. In fact, otherwise there are $a_1,\dots,a_k \in K$, not all zero, such that $\sum_{i=1}^{k} a_i \cdot y_i^{p^n} = 0$. Because of (6.3) we may assume that $a_1 \cdot A \supseteq a_i \cdot A$ for all $i \in \{1,\dots,k\}$. Then $a_1 \neq 0$ and $y_1^{p^n} = -\sum_{i=2}^{k} a_1^{-1} \cdot a_i \cdot y_i^{p^n} \in \mathfrak{M}_{B_1}$, hence $y_1 \in \mathfrak{M}_{B_1}$ in contradiction to $y_1 - 1 \in \mathfrak{M}_{B_1}$. □

In particular, (13.7) yields

(13.8) COROLLARY - A <u>is indecomposed in any purely inseparable extension of</u> K.

Let $L|K$ be again an arbitrary field extension and let v (resp. w) be a Krull valuation of K (resp. L). We say that w <u>lies over</u> (or: <u>extends</u>) v if $v = w|K$; in this case the valuation ring B of L corresponding to w lies over the valuation ring A of K which corresponds to v. Obviously, any Krull valuation w of L lies over exactly one Krull valuation of K, namely over $w|K$. On the other hand, given a Krull valuation v of K and a valuation ring B of L which lies over the valuation ring A corresponding to v, the restriction to K of any Krull valuation w of L corresponding to B is only equivalent to v. However it is easy to show that w can be chosen such that $w|K = v$, and in this case the value group Δ of w contains the value group Γ of v. It is obvious that the pair (Δ,Γ) is determined by B and A up to an isomorphism (of ordered groups), and therefore the index $(\Delta:\Gamma)$ depends only on B and K. This index is denoted by $e_{B|K}$ and is called the <u>ramification index</u> of $B|K$ [21].

Analogous considerations hold for places π (resp. ζ) of K (resp. L). Assume that ζ <u>lies over</u> π (i.e., $\pi = \zeta|K$) and let

[21] $e_{B|K}$ and $f_{B|K}$ may be finite (≥ 1) or infinite ($=\infty$). We do not distinguish among infinite cardinalities.

$\mathcal{K}$ (resp. $\mathcal{L}$) be the residue field of π (resp. ζ). The pair $(\mathcal{L},\mathcal{K})$ is determined, up to an isomorphism (of fields), by the valuation rings B and $A = B \cap K$ corresponding to ζ and π, respectively, and therefore the degree $[\mathcal{L}:\mathcal{K}]$ depends only on B and K. This degree is denoted by $f_{B|K}$ and is called the residue degree of $B|K$ [21].

We are going to show that if $L|K$ is finite then so are $e_{B|K}$ and $f_{B|K}$, for any valuation ring B of L. The following statement holds even for an arbitrary field extension $L|K$.

(13.9) Let w be a Krull valuation and ζ a place of L corresponding to the same valuation ring B of L, and let Δ (resp. Γ) be the value group of w (resp. $w|K$) and $\mathcal{L}$ (resp. $\mathcal{K}$) the residue field of ζ (resp. $\zeta|K$). Let $x_1,\ldots,x_r \in L^*$ such that $wx_1 + \Gamma,\ldots,wx_r + \Gamma \in \Delta/\Gamma$ are distinct and let $y_1,\ldots,y_s \in B$ such that $\zeta y_1,\ldots,\zeta y_s \in \mathcal{L}$ are linearly independent over $\mathcal{K}$. Then the elements $x_i \cdot y_j$ $(i=1,\ldots,r;\ j=1,\ldots,s)$ are linearly independent over K.

Proof: Suppose there exist elements $a_{ij} \in K$ $(i=1,\ldots,r;j=1,\ldots,s)$, not all zero, such that $\sum_{i=1}^{r}\sum_{j=1}^{s} a_{ij}\cdot x_i \cdot y_j = 0$. We may assume that $w(a_{11}\cdot x_1) \leq w(a_{ij}\cdot x_i)$ for all $i=1,\ldots,r$ and $j=1,\ldots,s$. Then the strict inequality "<" holds for any $i \neq 1$, since otherwise $wx_1 + \Gamma = wx_i + \Gamma$, contradicting the choice of $x_1,\ldots,x_r$. Therefore $\sum_{j=1}^{s} \frac{a_{1j}\cdot x_1}{a_{11}\cdot x_1} \cdot y_j = - \sum_{i=2}^{r}\sum_{j=1}^{s} \frac{a_{ij}\cdot x_i}{a_{11}\cdot x_1} \cdot y_j \in \mathfrak{M}_B$, hence $\sum_{j=1}^{s} \zeta\left(\frac{a_{1j}}{a_{11}}\right)\cdot \zeta y_j = 0$, contradicting the choice of $y_1,\ldots,y_s$. □

We get as an easy consequence of (13.9):

(13.10) COROLLARY - Let $L|K$ be a finite extension. Then $e_{B|K}\cdot f_{B|K} \leq [L:K]$ for any valuation ring B of L.

In §17 we shall prove a "fundamental inequality" which at

the same time strengthens (13.7) and (13.10).

In the case of an infinite algebraic extension $L|K$, $e_{B|K}$ and $f_{B|K}$ may be finite or infinite. Nevertheless we have the following important result:

(13.11) COROLLARY - Let $L|K$ be an algebraic extension and let $w, \Delta, \Gamma, \zeta, \mathfrak{L}, \mathfrak{K}$ be as in (13.9). Then Δ/Γ is a torsion group and $\mathfrak{L}|\mathfrak{K}$ is an algebraic extension.

Proof: Let $\{L_\alpha \mid \alpha \in J\}$ be the set of all finite subextensions of $L|K$. We have $L = \bigcup_{\alpha\in J} L_\alpha$, $\Delta = \bigcup_{\alpha\in J} \Delta_\alpha$, and $\mathfrak{L} = \bigcup_{\alpha\in J} \mathfrak{L}_\alpha$, where Δ_α is the value group of $w|L_\alpha$ and $\mathfrak{L}_\alpha$ is the residue field of $\zeta|L_\alpha$. By (13.10), $(\Delta_\alpha:\Gamma)$ and $[\mathfrak{L}_\alpha:\mathfrak{K}]$ are finite for any $\alpha \in J$; therefore Δ/Γ is a torsion group and $\mathfrak{L}|\mathfrak{K}$ is algebraic. □

Let π be a place of K with residue field $\mathfrak{K}$ and let $\mathfrak{K}_c$ be an algebraic closure of $\mathfrak{K}$. For any algebraic field extension $L|K$ and any place ζ of L lying over π, the residue field of ζ can be imbedded in $\mathfrak{K}_c$, by (13.11); hence ζ is equivalent to a place ζ' of L which also lies over π and whose residue field is contained in $\mathfrak{K}_c$ (cf (8.3) and (8.5)). Therefore, when we consider extensions of π to algebraic field extensions of K, we may restrict ourselves to those places which have their residue field contained in $\mathfrak{K}_c$.

Similarly, let v be a Krull valuation of K with value group Γ and let Γ_c be the divisible closure of Γ, i.e., the quotient of $\Gamma \times (\mathbb{N}\setminus\{0\})$ by the equivalence relation $\sim$ defined by $[(\gamma,n) \sim (\delta,m) \Leftrightarrow n\delta = m\gamma]$, endowed with the following addition and total ordering:

$$\gamma/n + \delta/m = (m\gamma + n\delta)/m\cdot n \ ; \quad \gamma/n \leqq \delta/m \Leftrightarrow m\gamma \leqq n\delta$$

where γ/n denotes the equivalence class determined by (γ,n) and any $\gamma \in \Gamma$ is identified with $\gamma/1$. Obviously Γ_c is a totally ordered group containing Γ and such that Γ_c/Γ is a torsion group; on the other hand, for any totally ordered group Δ containing Γ and such that Δ/Γ is a torsion group there is an imbedding of Δ in Γ_c which induces ι_Γ. Therefore, when we consider extensions of v to algebraic field extensions of K, we may restrict ourselves to those Krull valuations which have their value group contained in Γ_c. Any such extension w of v is even uniquely determined by the corresponding valuation ring, as is shown in the following theorem. (Note that there is no analogous uniqueness statement for extensions of places; for a counter-example cf §14).

(13.12) THEOREM - Let v be a Krull valuation of K, with value group Γ, and $L|K$ an algebraic extension. For any valuation ring B of L lying over A_v there exists exactly one Krull valuation w of L which corresponds to B, lies over v, and whose value group is contained in Γ_c.

Proof: For any Krull valuation w' of L which corresponds to B and lies over v there is an isomorphism g from the value group of w' onto a subgroup Δ of Γ_c. Obviously $w = g \circ w'$ is a Krull valuation of L with the desired properties. Assume that w_1 is another Krull valuation of L with the desired properties. Then there is an isomorphism h of Δ onto the value group of w_1 such that $h \circ w = w_1$; in particular, $h\gamma = \gamma$ for all $\gamma \in \Gamma$. For any $\delta \in \Delta$ there is a non-zero $n \in N$ such that $n\delta \in \Gamma$, hence $n(h\delta) = h(n\delta) = n\delta$, hence $h\delta = \delta$. Therefore $w_1 = w$. □

From the fact that Δ/Γ is a torsion group it follows that Γ and Δ have the same rank; in fact:

(13.13) Let Γ and Δ be totally ordered (abelian) groups such that $\Gamma \subseteq \Delta$ and Δ/Γ is a torsion group. Let $\mathfrak{G}$ (resp. $\mathfrak{H}$)

be the set of all isolated subgroups of Γ (resp. Δ). Then $\Psi \mapsto \Psi \cap \Gamma$ ($\Psi \in \mathfrak{H}$) is an inclusion-preserving bijective mapping from $\mathfrak{H}$ onto $\mathfrak{G}$.

Proof: For any $\Phi \in \mathfrak{G}$ let $\sqrt{\Phi} = \{\delta \in \Delta \mid n\delta \in \Phi$ for some $n \geq 1\}$.
For any $\Psi \in \mathfrak{H}$ we have $\Psi \cap \Gamma \in \mathfrak{G}$ and $\Psi = \sqrt{\Psi \cap \Gamma}$, and for any $\Phi \in \mathfrak{G}$ we have $\sqrt{\Phi} \in \mathfrak{H}$ and $\Phi = \sqrt{\Phi} \cap \Gamma$, as is shown easily. □

From (13.13), (13.11), and (7.5) we conclude:

(13.14) THEOREM - Let $L|K$ be an algebraic extension and A a valuation ring of K. Then $\mathrm{rank}(B) = \mathrm{rank}(A)$ for any valuation ring B of L which lies over A.

In particular, $\mathrm{rank}(B) = 1$ if and only if $\mathrm{rank}(A) = 1$, and $B = L$ if and only if $A = K$. (Of course, this could be shown directly without using (13.14).) It follows that any Krull valuation w of L which lies over an exponential valuation v of K is equivalent to an exponential valuation of L (cf (7.6)). A similar statement for discrete valuations holds only in the case in which $L|K$ is a finite extension:

(13.15) THEOREM - Let $L|K$ be a finite extension and A a discrete valuation ring of K. Then any valuation ring B of L which lies over A is discrete.

Proof: Let v be the normalized exponential valuation of K corresponding to A (cf (4.2)) and let w be a Krull valuation of L, with value group Δ, which corresponds to B and lies over v. By (13.10), $\Delta/\mathbb{Z}$ is finite, hence there exists a non-zero $n \in \mathbb{N}$ such that the mapping $\delta \longmapsto n\delta$ ($\delta \in \Delta$) is an isomorphism from Δ onto a subgroup of $\mathbb{Z}$. Therefore w is discrete. □

We shall prove in §27 that algebraically closed fields

have no discrete valuation rings; therefore the statement of (13.15) is not generally true for infinite algebraic extensions. On the other hand, even for infinite algebraic extensions $L|K$ it is true that if the valuation ring B of L is discrete then so is $A = B \cap K$.

The following theorems will be used only in §18.

(13.16) THEOREM - Let $L|K$ be an algebraic extension, A a valuation ring of K, and $D = I_L(A)$. The following conditions are equivalent:

(i) D is a finite A-module.

(ii) $L|K$ is a finite extension and D is a free A-module.

In this case, the bases of the A-module D are its minimal systems of generators, and each of them consists of $[L:K]$ elements.

Proof: (i) $\Rightarrow$ (ii): Let $x_1,\dots,x_m$ be a minimal system of generators of the A-module D. We claim that these elements are linearly independent over A and therefore form a basis. In fact, suppose that there are $a_1,\dots,a_m \in A$, not all zero, such that $a_1 \cdot x_1 + \dots + a_m \cdot x_m = 0$. We may assume that $a_1 \cdot A \supseteq a_i \cdot A$ for all $i \in \{1,\dots,m\}$; then $-x_1 = a_1^{-1} \cdot a_2 \cdot x_2 + \dots + a_1^{-1} \cdot a_m \cdot x_m \in A \cdot x_2 + \dots + A \cdot x_m$, hence $x_2,\dots,x_m$ is a system of generators, too, contradicting the choice of $x_1,\dots,x_m$. For any $y \in L$ there is a non-zero $c \in A$ such that $c \cdot y \in D$, hence $L \subseteq K \cdot x_1 + \dots + K \cdot x_m$; therefore $[L:K] \leqq m$.
(ii) $\Rightarrow$ (i): Any basis of the A-module D is a minimal system of generators and has at most $[L:K]$ elements, since for any $y_1,\dots,y_k \in D$ with $k > [L:K]$ there exist $b_1,\dots,b_k \in A$, not all zero, and a non-zero $b_o \in A$ such that $b_o^{-1} \cdot b_1 \cdot y_1 + \dots + b_o^{-1} \cdot b_k \cdot y_k = 0$, hence $b_1 \cdot y_1 + \dots + b_k \cdot y_k = 0$. □

The conditions of theorem (13.16) are satisfied for any finite separable extension $L|K$ whenever A is a discrete valuation ring of K. In fact, any valuation ring of this type is an integrally closed noetherian subring of K , and we have the following theorem (cf Zariski & Samuel [36], Chap. V, §4):

(13.17) THEOREM - Let R be an integrally closed noetherian subring of its quotient field K . For any finite separable extension $L|K$, $I_L(R)$ is a finite R-module.

§14 The case of a normal field extension

Let $N|K$ be a normal field extension and G its Galois group, i.e., $G = \mathrm{Aut}(N|K)$, the group of all K-automorphisms of N. We denote by $\mathfrak{K}$ the set of all fields between K and N and by $\mathfrak{G}$ the set of all subgroups of G , both sets being ordered by inclusion. For any $L \in \mathfrak{K}$ we set $gL = \{\sigma \in G \mid \sigma x = x$ for all $x \in L\}$, and for any $H \in \mathfrak{G}$ we set $kH = \{x \in N \mid \sigma x = x$ for all $\sigma \in H\}$. Obviously $gL = \mathrm{Aut}(N|L) \in \mathfrak{G}$ and $kH \in \mathfrak{K}$; kH is called the fixed field of H . It is easily checked that the mappings $g: \mathfrak{K} \to \mathfrak{G}$ and $k: \mathfrak{G} \to \mathfrak{K}$ define a Galois connexion between $\mathfrak{K}$ and $\mathfrak{G}$, i.e., they are inclusion-inverting and satisfy $g(kH) \supseteq H$ for all $H \in \mathfrak{G}$ and $k(gL) \supseteq L$ for all $L \in \mathfrak{K}$. Their restrictions g' and k' to the image sets $\mathfrak{K}' = k\mathfrak{G}$ and $\mathfrak{G}' = g\mathfrak{K}$ define a Galois correspondence between these sets, i.e., $g': \mathfrak{K}' \to \mathfrak{G}'$ and $k': \mathfrak{G}' \to \mathfrak{K}'$ are bijective and inverse to each other [22].

By Infinite Galois Theory it is known that $\mathfrak{K}'$ consists of all subfields of N which contain $kG = K_{ins}$, the maximal purely inseparable subextension of $N|K$, and that $\mathfrak{G}'$ consists of all closed subgroups of G , where G is endowed with the "finite topo-

[22] The preceding statements hold even for arbitrary field extensions $N|K$.

logy" (i.e., considered as a subspace of the topological product $\bigtimes_{x \in N} N$ where N is discrete). With this topology, G is a compact and totally disconnected group, and a fundamental system of open neighborhoods of its unit element ι_N is given by $\{\mathrm{Aut}(N|N_\alpha) \mid \alpha \in I\}$, where $\{N_\alpha \mid \alpha \in I\}$ is the set of all finite normal subextensions of $N|K$. In particular, if $[N:K] < \infty$ then G is discrete and $\mathfrak{G}' = \mathfrak{G}$.

Moreover, let N_1 be any normal subextension of $N|K$ and $G_1 = \mathrm{Aut}(N_1|K)$. The mapping $\sigma \mapsto \sigma|N_1$ $(\sigma \in G)$ is a homomorphism from G onto G_1, called the <u>canonical homomorphism</u>. Its kernel equals $gN_1 = \mathrm{Aut}(N|N_1)$, it is continuous and induces a topological isomorphism from the topological quotient group $G/\mathrm{Aut}(N|N_1)$ onto G_1. In particular, $\mathrm{Aut}(K_{ins}|K)$ is the trivial group, and $N|K_{ins}$ is a Galois (i.e., separable and normal) extension with Galois group $\mathrm{Aut}(N|K_{ins}) = G$.

Note that $N = \bigcup_{\alpha \in I} N_\alpha$. Later on (in §19 and §20) we shall use the fact that G can be identified with the inverse limit of the Galois groups $G_\alpha = \mathrm{Aut}(N_\alpha|K)$.

Let A be a valuation ring of K and $\mathfrak{B}$ the set of all those valuation rings of N which lie over A. For any $B \in \mathfrak{B}$ we have obviously $\{\sigma B \mid \sigma \in G\} \subseteq \mathfrak{B}$ and therefore $I_N(A) \subseteq \bigcap_{\sigma \in G} \sigma B$, by (13.3 b). We claim that the equalities hold, and we prove them first in the case of a finite normal extension.

(14.1) <u>If the normal extension</u> $N|K$ <u>is finite then, for any</u> $B \in \mathfrak{B}$, <u>we have</u> $I_N(A) = \bigcap_{\sigma \in G} \sigma B$ <u>and</u> $\{\sigma B \mid \sigma \in G\} = \mathfrak{B}$.

<u>Proof</u>: a) Let $y \in \bigcap_{\sigma \in G} \sigma B$. Then $F = \prod_{\sigma \in G} (X - \sigma y) \in B[X] \cap K_{ins}[X]$, hence $F^{p^m} \in B[X] \cap K[X] = A[X]$ for some $m \in \mathbb{N}$, where p is the characteristic exponent of K; hence $y \in I_N(A)$.

b) Let $B_o \in \mathfrak{B}$. By a) and (13.3 b) we have $B_o \supseteq I_N(A) = \bigcap_{\sigma \in G} \sigma B$.

Since G is finite, it follows from (11.13 a) and (13.4 b) that $B_o = \sigma B$ for some $\sigma \in G$. □

The generalization of (14.1) to infinite normal extensions is a corollary of the following theorem. This theorem in its full strength will be needed in the proof of (15.4).

(14.2) THEOREM - <u>Let</u> $N|K$ <u>and</u> $N_o|K$ <u>be normal extensions with</u> $N_o \subseteq N$, <u>and let</u> B, B' <u>be valuation rings of</u> N <u>such that</u> $\sigma_o(B \cap N_o) = B' \cap N_o$ <u>for some</u> $\sigma_o \in \mathrm{Aut}(N_o|K)$. <u>Then there is a</u> $\sigma \in \mathrm{Aut}(N|K)$ <u>such that</u> $\sigma|N_o = \sigma_o$ <u>and</u> $\sigma B = B'$.

<u>Proof</u>: Let $\mathcal{S}$ be the set, ordered in a natural manner, of all pairs $(N_\gamma, \sigma_\gamma)$ consisting of a normal subextension N_γ of $N|K$ and an element $\sigma_\gamma \in \mathrm{Aut}(N_\gamma|K)$ such that $\sigma_\gamma|N_o = \sigma_o$ and $\sigma_\gamma(B \cap N_\gamma) = B' \cap N_\gamma$. Obviously $(N_o, \sigma_o) \in \mathcal{S}$. For any non-empty totally ordered subset $\mathcal{J} = \{(N_\beta;\sigma_\beta) \mid \beta \in J\}$ of $\mathcal{S}$ let $N_{\mathcal{J}} = \bigcup_{\beta \in J} N_\beta$ and $\sigma_{\mathcal{J}} \in \mathrm{Aut}(N_{\mathcal{J}}|K)$ be the common extension of all σ_β $(\beta \in J)$; obviously $(N_{\mathcal{J}}, \sigma_{\mathcal{J}}) \in \mathcal{S}$. Therefore $\mathcal{S}$ is inductively ordered and by Zorn's Lemma has a maximal element (M,ρ) (say). We claim that $M = N$. Otherwise let $x \in N \setminus M$ and $M_1 = M(\{\sigma x \mid \sigma \in G\})$. Since $M_1|K$ is normal, ρ extends to some $\rho_1 \in \mathrm{Aut}(M_1|K)$. Since $\rho_1(B \cap M_1)$ and $B' \cap M_1$ lie over $\rho(B \cap M) = B' \cap M$ and $M_1|M$ is finite, we conclude from (14.1) that $\tau(\rho_1(B \cap M_1)) = B' \cap M_1$ for some $\tau \in \mathrm{Aut}(M_1|M)$, hence $(M_1, \tau \circ \rho_1) \in \mathcal{S}$. Since $M_1 \supset M$ and $(\tau \circ \rho_1)|M = \rho$, this contradicts the maximality of (M,ρ). □

(14.3) COROLLARY - <u>The statement of</u> (14.1) <u>holds for arbitrary normal extensions</u> $N|K$.

<u>Proof</u>: We know already that a) $\{\sigma B \mid \sigma \in G\} \subseteq \mathcal{B}$ and b) $I_N(A) = \bigcap_{B' \in \mathcal{B}} B' \subseteq \bigcap_{\sigma \in G} \sigma B$. The equality in a) follows from (14.2), with $N_o = K$. The equality in b) follows from that in a). □

We conclude from (14.3) (or directly from the definition of $I_N(A)$):

(14.4) COROLLARY - <u>For any</u> $\sigma \in G$ <u>we have</u> $\sigma(I_N(A)) = I_N(A)$.

Let π be a place of K and ζ a place of N which lies over π . Let A (resp. B) be the valuation ring corresponding to π (resp. ζ), and let $\mathcal{K}$ (resp. $\mathfrak{h}$) be the residue field of π (resp. ζ) .

(14.5) THEOREM - $\mathfrak{h}|\mathcal{K}$ <u>is a normal extension and for any</u> $y \in I_N(A)$, <u>the roots of the polynomials</u> $\pi P_{y|K}$ <u>and</u> $P_{\zeta y|\mathcal{K}}$ <u>are in</u> $\{\zeta(\sigma y) \mid \sigma \in G\}$.

<u>Proof</u>: From $N = \bigcup_{\alpha\in I} N_\alpha$ we conclude $I_N(A) = \bigcup_{\alpha\in I} I_{N_\alpha}(A)$ and $\mathfrak{h} = \bigcup_{\alpha\in I} \mathfrak{h}_\alpha$, where $\mathfrak{h}_\alpha$ is the residue field of $\zeta|N_\alpha$. Moreover, for any $\alpha\in I$ and $\sigma_\alpha \in \mathrm{Aut}(N_\alpha|K)$, there is a $\sigma \in G$ such that $\sigma_\alpha = \sigma|N_\alpha$. Therefore it suffices to prove the theorem in the case of a finite normal extension, i.e., we may assume that $N|K$ is finite. Let $y \in I_N(A)$. By (14.4) we have $\sigma y \in I_N(A) \subseteq B$ for all $\sigma \in G$. Let $F = \prod_{\sigma\in G} (X-\sigma y)$; obviously $F \in K_{ins}[X] \cap I_N(A)[X]$, hence $F^{p^m} \in K[X] \cap I_N(A)[X] = A[X]$ for some $m \in \mathbb{N}$, where p is the characteristic exponent of K . By (10.6) and (10.10) we have $F^{p^m} = P_{y|K}\cdot Q$ for some $Q \in A[X]$, hence $\prod_{\sigma\in G} (X - \zeta(\sigma y))^{p^m} = \pi P_{y|K} \cdot \pi Q$; moreover, $\pi P_{y|K}$ is a multiple of $P_{\zeta y|\mathcal{K}}$. Therefore all roots of $\pi P_{y|K}$ and of $P_{\zeta y|\mathcal{K}}$ are in $\{\zeta(\sigma y) \mid \sigma \in G\} \subseteq \mathfrak{h}$. Since κ_B maps $I_N(A)$ onto $B/\mathfrak{M}_B$, by (13.4 c), and ζ induces an isomorphism from $B/\mathfrak{M}_B$ onto $\mathfrak{h}$, we have $\mathfrak{h} = \zeta(I_N(A))$; therefore, by the first part of this proof, $\mathfrak{h}|\mathcal{K}$ is a normal extension. □

Note that $\mathfrak{h}|\mathcal{K}$ may be inseparable even in the case in which $N|K$ is a Galois extension.

The following statement is obvious.

(14.6) Let w (resp. ζ) be a Krull valuation (resp. place) of N which corresponds to the valuation ring B of N. For any $\sigma \in G$, $w \circ \sigma^{-1}$ (resp. $\zeta \circ \sigma^{-1}$ [23]) is a Krull valuation (resp. place) of N which corresponds to the valuation ring σB and lies over $w|K$ (resp. $\zeta|K$). Moreover, the value groups of w and $w \circ \sigma^{-1}$ coincide, the residue fields of ζ and $\zeta \circ \sigma^{-1}$ coincide, hence $e_{\sigma B|K} = e_{B|K}$ and $f_{\sigma B|K} = f_{B|K}$.

From (13.12) and (14.6) we obtain the following statement of uniqueness:

(14.7) Let w, B be as in (14.6) and let $B' = \sigma B$ for some $\sigma \in G$. Then $w \circ \sigma^{-1}$ is the only Krull valuation of N which corresponds to B', lies over $w|K$ and has the same value group as w.

In particular, we have $w \circ \sigma = w$ for any $\sigma \in G$ such that $\sigma B = B$.

There is no similar statement for places. For example, let $N = \mathbb{Q}(X)$, $K = \mathbb{Q}(X^2)$, where X is an indeterminate over $\mathbb{Q}$, and σ the (only non-identical) K-automorphism of N, determined by $\sigma X = -X$. The place ζ of N, determined by $\zeta|\mathbb{Q} = \iota_{\mathbb{Q}}$ and $\zeta X = \sqrt{2}$, corresponds to the valuation ring $B = \mathbb{Q}[X]_{(X^2-2)}$; we have $\sigma B = B$ but $\zeta \circ \sigma \neq \zeta$.

In the following section we shall show, for an arbitrary normal extension $N|K$ and any valuation ring B of N, that the set of all $\sigma \in \mathrm{Aut}(N|K)$ such that $\sigma B = B$ (or, equivalently, $w \circ \sigma = w$) is a closed subgroup of G, called the decomposition group. In §19 we shall show that the set of all $\sigma \in \mathrm{Aut}(N|K)$ such

[23] For any $\sigma \in G$ we denote by $\zeta \circ \sigma$ the place $\zeta \circ \tilde{\sigma}: \tilde{N} \to \tilde{n}$, where $\tilde{\sigma}|N = \sigma$ and $\tilde{\sigma}\infty = \infty$.

that $\zeta \circ \sigma = \zeta$ is a closed normal subgroup of the decomposition group, called the inertia group.

§15 Decomposition group and decomposition field

Let $N|K$ be a normal field extension with Galois group G, B a valuation ring of N, $A = B \cap K$, and $\mathfrak{B}$ the set of all valuation rings of N which lie over A. By (14.3), a surjective mapping $G \to \mathfrak{B}$ is defined by $\sigma \mapsto \sigma B$ $(\sigma \in G)$. The set $G^Z(B|K) = \{\sigma \in G \mid \sigma B = B\}$ is obviously a subgroup of G and is called the <u>decomposition group</u> of B <u>over</u> K. If there is no ambiguity, we write G^Z instead of $G^Z(B|K)$.

Let $G/G^Z = \{\sigma \circ G^Z \mid \sigma \in G\}$ be the set of all left cosets of G^Z in G. We prove:

(15.1) a) <u>The mapping</u> $\sigma \mapsto \sigma B$ $(\sigma \in G)$ <u>induces a bijective mapping</u> $G/G^Z \to \mathfrak{B}$.

b) A <u>is indecomposed in</u> N <u>if and only if</u> $G^Z = G$.

<u>Proof</u>: a) follows from the fact that, for all $\sigma, \rho \in G$, we have $\rho B = \sigma B$ if and only if $\rho \circ G^Z = \sigma \circ G^Z$. b) is an immediate consequence of a). □

Conjugate valuation rings have conjugate decomposition groups. More precisely:

(15.2) $G^Z(\sigma B|K) = \sigma \circ G^Z \circ \sigma^{-1}$ <u>for any</u> $\sigma \in G$.

<u>Proof</u>: For any $\sigma \in G$ and $\rho \in G^Z$ we have $(\sigma \circ \rho \circ \sigma^{-1})(\sigma B) = \sigma B$, hence $\sigma \circ G^Z \circ \sigma^{-1} \subseteq G^Z(\sigma B|K)$. The inverse containment follows from $\sigma^{-1} \circ G^Z(\sigma B|K) \circ \sigma \subseteq G^Z(\sigma^{-1}(\sigma B)|K) = G^Z$. □

It is easy to determine the decomposition group of B over any intermediary field K' :

(15.3) $G^Z(B|K') = G^Z \cap \mathrm{Aut}(N|K')$ <u>for any field</u> K' <u>between</u> K <u>and</u> N .

Next we show that, for any normal subextension N' of $N|K$, the canonical homomorphism Θ from G onto the Galois group G' of $N'|K$ maps G^Z onto $G^Z((B \cap N')|K)$. More precisely:

(15.4) Θ <u>induces a surjective homomorphism</u> $\Theta^Z\colon G^Z \to G^Z((B \cap N')|K)$ <u>with kernel</u> $G^Z(B|N') = G^Z \cap \mathrm{Aut}(N|N')$.

<u>Proof</u>: For any $\sigma \in G^Z$ we have $\Theta\sigma \in G^Z((B \cap N')|K)$, since $(\Theta\sigma)(B \cap N') = \sigma B \cap \sigma N' = B \cap N'$. On the other hand, any $\sigma' \in G^Z((B \cap N')|K)$ extends to some $\sigma \in G^Z$, by (14.2); therefore the homomorphism $\Theta^Z\colon G^Z \to G^Z((B \cap N')|K)$ induced by Θ is surjective. Its kernel $G^Z \cap \mathrm{Aut}(N|N')$ equals $G^Z(B|N')$, by (15.3). □

The fixed field kG^Z of the decomposition group $G^Z = G^Z(B|K)$ is called the <u>decomposition field of</u> B <u>over</u> K and will be denoted by $K^Z(B|K)$ (or K^Z if there is no ambiguity). Obviously K^Z contains the maximal purely inseparable subextension K_{ins} of $N|K$. We prove:

(15.5) a) G^Z <u>is a closed subgroup of</u> G .

b) $N|K^Z$ <u>is a Galois extension with Galois group</u> $\mathrm{Aut}(N|K^Z) = G^Z$.

<u>Proof</u>: a) Let σ be an element of the topological closure of G^Z in G . For any finite normal subextension N_α of $N|K$ the set $(\sigma \circ \mathrm{Aut}(N|N_\alpha)) \cap G^Z$ is non-empty, and for any element ρ_α of this set we have $\sigma(B \cap N_\alpha) = \rho_\alpha(B \cap N_\alpha) = B \cap N_\alpha$. Since $B = \bigcup_{\alpha \in I} (B \cap N_\alpha)$, we conclude that $\sigma B = B$, hence $\sigma \in G^Z$.

b) Since $K_{ins} \subseteq K^Z \subseteq N$ and $N|K_{ins}$ is Galois, so is $N|K^Z$. Since

G^Z is closed in G we have $G^Z = g(kG^Z) = Aut(N|K^Z)$. □

The following statements are immediate consequences of (15.2), (15.3), and (15.4).

(15.6) a) $K^Z(\sigma B|K) = \sigma K^Z$ for any $\sigma \in G$.

b) $K^Z(B|K') = K^Z.K'$ for any field K' between K and N .

c) $K^Z((B \cap N')|K) = K^Z \cap N'$ for any normal subextension N' of $N|K$.

The decomposition field can be characterized by the following minimal property.

(15.7) THEOREM - For any field L between K_{ins} and N we have $K^Z \subseteq L$ if and only if $B \cap L$ is indecomposed in N .

Proof: If $K^Z \subseteq L$ then $G^Z(B|L) = G^Z \cap Aut(N|L) = Aut(N|L)$ by (15.3) and (15.5 b), hence $B \cap L$ is indecomposed in N , by (15.1 b). On the other hand, if $K_{ins} \subseteq L \subseteq N$ and L is indecomposed in N , then $gL = G^Z(B|L) \subseteq G^Z$, hence $L = k(gL) \supseteq kG^Z = K^Z$.

□

Let w be a Krull valuation and ζ a place of N, both corresponding to B . We denote by $\Delta, \Gamma, \Gamma^{\square}, \Gamma^Z$ the value groups of w , $w|K = v$, $w|K_{ins}$, $w|K^Z$ and by $\mathcal{K}, \mathcal{K}^{\square}, \mathcal{K}^Z$ the residue field of $\zeta|K = \pi$, $\zeta|K_{ins}$, $\zeta|K^Z$, respectively. It is easily seen that $\mathcal{K}^{\square}|\mathcal{K}$ is a purely inseparable extension and $\Gamma^{\square}|\Gamma$ is a p-group, where p is the characteristic exponent of K (cf Exercise III- 6). Of course $\mathcal{K}^{\square} = \mathcal{K}$ and $\Gamma^{\square} = \Gamma$ whenever $N|K$ is a Galois extension (for example, when K is perfect or $p = 1$).

The next theorem asserts that the ramification index and the residue degree of $(B \cap K^Z)|K_{ins}$ are equal to 1. We give a complete proof of this theorem only for the rank 1 case; the proof in the general case is based on Ribenboim's generalized independence theorem (mentioned at the end of §11) and Exercise III- 7 , which is also due to Ribenboim [30] .

(15.8) THEOREM - $\Gamma^Z = \Gamma^\square$ and $\mathcal{K}^Z = \mathcal{K}^\square$.

Proof: Since $K^Z(B|K_{ins}) = K^Z \cdot K_{ins} = K^Z$ by (15.6 b), we may assume that $K = K_{ins}$ or, in other words, that $N|K$ is a Galois extension; hence $\Gamma = \Gamma^\square$ and $\mathcal{K} = \mathcal{K}^\square$. For any finite Galois subextension N_α of $N|K$ we set $B_\alpha = B \cap N_\alpha$, $K_\alpha^Z = K^Z(B_\alpha|K)$ and denote by Γ_α^Z the value group of $w|K_\alpha^Z$ and by $\mathcal{K}_\alpha^Z$ the residue field of $\zeta|K_\alpha^Z$. Since $N = \bigcup_{\alpha\in I} N_\alpha$, we have $K^Z = \bigcup_{\alpha\in I} (K^Z \cap N_\alpha) = \bigcup_{\alpha\in I} K_\alpha^Z$, by (15.6 c), hence $\Gamma^Z = \bigcup_{\alpha\in I} \Gamma_\alpha^Z$ and $\mathcal{K}^Z = \bigcup_{\alpha\in I} \mathcal{K}_\alpha^Z$. It suffices to show that $\Gamma_\alpha^Z = \Gamma$ and $\mathcal{K}_\alpha^Z = \mathcal{K}$ for any $\alpha\in I$; therefore we may assume that $[N:K] < \infty$. Let $B_1 = B$, $B_2,\dots,B_r$ be the (finitely many) distinct valuation rings of N which lie over $A = B \cap K$ (cf (13.7)). We may assume $r > 1$, since otherwise A would be indecomposed in N and therefore $K^Z = K$, by (15.7), in which case the theorem is trivial. For any $j \in \{1,\dots,r\}$ let $A_j^Z = B_j \cap K^Z$; by (15.7) we have $A_1^Z \notin \{A_2^Z,\dots,A_r^Z\}$. We may assume that $A_2^Z,\dots,A_s^Z$ are distinct for some $s \in \{2,\dots,r\}$, whereas $A_{s+1}^Z,\dots, A_r^Z \in \{A_2^Z,\dots,A_s^Z\}$. By (13.2), (13.4 b), and (13.3 b), $A_1^Z,\dots,A_2^Z$ are all valuation rings of K^Z which lie over A , they are pairwise incomparable, and their intersection equals $I_{K^Z}(A)$.

a) Let $\alpha \in \mathcal{K}^Z$. By (11.14) there is an $x \in A_1^Z \cap\dots\cap A_s^Z$ such that $\zeta x = \alpha$, $x \in \mathfrak{M}_{A_2^Z}$,..., $x \in \mathfrak{M}_{A_s^Z}$; therefore $x \in I_N(A) \cap \mathfrak{M}_{B_2} \cap\dots\cap \mathfrak{M}_{B_r}$. Let $x, \sigma_1 x,\dots,\sigma_n x$ be the distinct conjugates of x over K , where $\sigma_1,\dots,\sigma_n \in G$. For any $i \in \{1,\dots,n\}$ we have $\sigma_i \notin \mathrm{Aut}(N|K^Z) = G^Z$, hence $\sigma_i^{-1} B \in \{B_2,\dots,B_r\}$, hence $\zeta(\sigma_i x) = 0$. We conclude that $\alpha = \zeta x = \zeta(x + \sigma_1 x +\dots+ \sigma_r x) \in \mathcal{K}$, since $x + \sigma_1 x +\dots+ \sigma_r x$ is the trace of x with respect to the extension $K(x)|K$. Therefore $\mathcal{K}^Z = \mathcal{K}$.

b) For any $j \in \{1,\dots,r\}$ let w_j be the unique Krull valuation of N, with value group Δ, which corresponds to B_j and lies over v (cf (14.7)), and let $v_j^Z = w_j|K^Z$. We claim that, for any $\gamma \in \Gamma^Z$, there is an $x \in K^Z$ such that $v_1^Z x = \gamma$, $v_2^Z x > \gamma$, $\dots$, $v_s^Z x > \gamma$. If $A_1^Z,\dots,A_s^Z$ are pairwise independent (which occurs, in particular, whenever $\operatorname{rank}(A) = 1$), this claim follows from (11.17). In the general case, it is an immediate consequence of Ribenboim's general independence theorem and Exercise III- 7 . Let $x, \sigma_1 x,\dots, \sigma_n x$ be the distinct conjugates of x over K, where $\sigma_1,\dots,\sigma_n \in G$. For any $i \in \{1,\dots,n\}$ we have $\sigma_i \notin \operatorname{Aut}(N|K^Z) = G^Z$, hence $\sigma_i^{-1} B \in \{B_2,\dots,B_r\}$, hence $w_1 \circ \sigma_1 \in \{w_2,\dots,w_r\}$. Therefore $(w_1 \circ \sigma_i)|K^Z$ is equivalent to, and by (13.12) is even equal to, one of the Krull valuations $v_2^Z,\dots,v_s^Z$; hence $(w_1 \circ \sigma_1)x > \gamma = v_1^Z x = w_1 x$. We conclude that $\gamma = \min\{w_1 x, w_1(\sigma_1 x),\dots,w_1(\sigma_n x)\} = w_1(x + \sigma_1 x + \dots + \sigma_n x) \in \Gamma$, since $x + \sigma_1 x + \dots + \sigma_n x$ is the trace of x with respect to the extension $K(x)|K$. Therefore $\Gamma^Z = \Gamma$. □

§16 Henselian valuation rings

We first consider an arbitrary normal field extension $N|K$ and characterize those valuation rings A of K which are indecomposed in N. We show that it suffices to consider only the finite Galois subextensions of $N|K$; in fact:

(16.1) A _is indecomposed in_ N _if and only if_ A _is indecomposed in any finite Galois subextension of_ $N|K$.

Proof: If A is indecomposed in N then also in any subextension of $N|K$. On the other hand, if A is not indecomposed in N then, by (13.6 b), A is not indecomposed in some finite subextension L

of $N|K$. There is a finite Galois subextension N' of $N|K$ which contains the maximal separable subextension L_{sep} of $L|K$ and, by (13.8) and (13.6 a), A is not indecomposed in N' . □

For any $y \in N$ we denote by $c_i(y)$ the i-th coefficient of the minimal polynomial $P_{y|K}$ of y over K . More precisely, we set

$$P_{y|K} = X^{n(y)} + c_1(y)\cdot X^{n(y)-1} +\dots+ c_{n(y)-1}(y)\cdot X + c_{n(y)}(y)$$

and write sometimes $c(y)$ instead of $c_{n(y)}(y)$. We recall that $(-1)^{n(y)}\cdot c(y)$ (resp. $-c_1(y)$) is the norm (resp. trace) of y with respect to the extension $K(y)|K$.

In case A is indecomposed in N , the unique extension w of v to N is explicitely given in the following theorem.

(16.2) THEOREM - Let A be indecomposed in N and let w be the unique Krull valuation of N lying over v and with value group $\Delta \subseteq \Gamma_c$. Then for any $y \in N$ and $j \in \{1,\dots,n(y)\}$ we have $wy = v(c(y))/n(y) \leqq v(c_j(y))/j$.

Proof: There exist $\sigma_1,\dots,\sigma_{n(y)} \in G = \mathrm{Aut}(N|K)$ such that $P_{y|K} = (X - \sigma_1 y)\cdot\ldots\cdot(X - \sigma_{n(y)}y)$. For any $j \in \{1,\dots,n(y)\}$ we have $c_j(y) = (-1)^j \cdot \sum_{(i_1,\dots,i_j)\in I_j} \sigma_{i_1}y \cdot \ldots \cdot \sigma_{i_j}y$, where $I_j = \{(i_1,\dots,i_n) \mid 1 \leqq i_1 < \dots < i_j \leqq n\}$; therefore $v(c_j(y)) \geqq \min\{w(\sigma_{i_1}y) +\dots+ w(\sigma_{i_j}y) \mid (i_1,\dots,i_j) \in I_j\}$ and in particular $v(c(y)) = w(\sigma_1 y) +\dots+ w(\sigma_n y)$. Since $w\circ\sigma = w$ for all $\sigma \in G$ by (14.7), we conclude $v(c_j(y)) \geqq j \cdot wy$ for $j = 1,\dots,n(y)$ and $v(c(y)) = n(y)\cdot wy$. □

We conclude from (16.2) that if A is indecomposed in N then $y^{n(y)} \cdot c(y)^{-1}$ is a unit in $I_N(A)$ for all non-zero $y \in N$. That this condition (and even a weaker one) is also sufficient for

A to be indecomposed in N is one of the statements of the following theorem.

(16.3) THEOREM - Let A be a valuation ring of K, π a place of K corresponding to A, and $\mathcal{K}$ its residue field. The following conditions are equivalent:

(i) A is indecomposed in N.

(ii) For any $y \in I_N(A)$, $\pi P_{y|K}$ is a power of some irreducible polynomial in $\mathcal{K}[X]$.

(iii) $y^{n(y)} \cdot c(y)^{-1} \in I_N(A)$ for any non-zero $y \in N$.

(iv) $c_j(y)^{n(y)} \in c(y)^j \cdot A$ for any $y \in N$ and $j \in \{1,\ldots,n(y)\}$.

(v) $c(y) \in \mathfrak{M}_A$ implies $\{c_1(y),\ldots,c_{n(y)}(y)\} \subseteq \mathfrak{M}_A$ for any $y \in I_N(A)$.

(vi) $c(y) \in A$ implies $\{c_1(y),\ldots,c_{n(y)}(y)\} \subseteq A$ for any $y \in N$.

(vii) $\{c_2(y),\ldots,c_{n(y)}(y)\} \subseteq \mathfrak{M}_A$ implies $c_1(y) \in \mathfrak{M}_A$ for any $y \in I_N(A) \setminus K$.

(viii) $\{c_2(y),\ldots,c_{n(y)}(y)\} \subseteq A$ implies $c_1(y) \in A$ for any $y \in N \setminus K$.

Proof: (i) ⇒ (iii) and (i) ⇒ (iv) are immediate consequences of (16.2). (v) ⇒ (vii) and (vi) ⇒ (viii) are trivial (note that $y \in N \setminus K$ implies $c(y) \in \{c_2(y),\ldots,c_{n(y)}(y)\} \neq \emptyset$) and so is (iv) ⇒ (vi).

(iii) ⇒ (vi): Let $y \in N$ be such that $c(y) \in A$; then $y^{n(y)} \in I_N(A)$, hence $y \in I_N(A)$, hence $\{c_1(y),\ldots,c_{n(y)}(y)\} \subseteq A$ by (10.6) and (10.10 b).

(i) ⇒ (ii): Let ζ be a place of N which lies over π; obviously ζ corresponds to the valuation ring $I_N(A)$. From (14.4) and (14.6) it follows that $\zeta \circ \sigma$ is equivalent to ζ for any $\sigma \in G = \mathrm{Aut}(N|K)$, hence $\zeta \circ \sigma = \bar{\sigma} \circ \zeta$ for some trivial place $\bar{\sigma}$ of the residue field $\mathfrak{n}$ of ζ; obviously $\bar{\sigma}|\mathfrak{n} \in \bar{G} = \mathrm{Aut}(\mathfrak{n}|\mathcal{K})$. Let $y \in I_N(A)$. By (14.5) all roots of $\pi P_{y|K}$ are in $\{\zeta(\sigma y) \mid \sigma \in G\} \subseteq$

$\subseteq \{\bar{\sigma}(\zeta y) \mid \bar{\sigma} \in \bar{G}\}$. We conclude that $P_{\zeta y|\mathcal{K}}$ is the only irreducible and monic polynomial in $\mathcal{K}[X]$ which divides $\pi\, P_{y|K}$, hence $\pi\, P_{y|K}$ is a power of $P_{\zeta y|\mathcal{K}}$.

(ii) ⇒ (v): Suppose there is a $y \in I_N(A)$ such that $c_k(y) \notin \mathfrak{M}_A$ and $\{c_{k+1}(y),\ldots,c_{n(y)}(y)\} \subseteq \mathfrak{M}_A$ for some $k \in \{1,\ldots,n(y)-1\}$. Then $\pi\, P_{y|K} = X^{n(y)-k} \cdot (X^k + \pi(c_1(y))\cdot X^{k-1} +\ldots+ \pi(c_k(y)))$ is not a power of an irreducible polynomial in $\mathcal{K}[X]$.

(vii) ⇒ (viii): Suppose there exists a $y \in N \setminus K$ such that $c_1(y) \notin A$ and $\{c_2(y),\ldots,c_{n(y)}(y)\} \subseteq A$. Let $z = c_1(y)^{-1} \cdot y$; then $n(z) = n(y)$ and $c_j(z) = c_j(y) \cdot c_1(y)^{-j}$ for all $j \in \{1,\ldots,n(y)\}$, hence $c_1(z) = 1 \notin \mathfrak{M}_A$, $\{c_2(z),\ldots,c_{n(z)}(z)\} \subseteq \mathfrak{M}_A$, and $z \in I_N(A) \setminus K$.

(viii) ⇒ (i): By (16.1) it suffices to prove that A is indecomposed in any finite Galois subextension L of $N|K$; therefore we may assume that $N|K$ is finite Galois. Suppose A is not indecomposed in N and let $B_1,\ldots,B_r$ be the distinct valuation rings of N which lie over A . Moreover let $K^Z = K^Z(B_1|K)$ and $B_i' = B_i \cap K^Z$ $(i = 1,\ldots,r)$. By (15.7) we have $B_1' \notin \{B_2',\ldots,B_r'\}$, by (13.4 b) any two distinct valuation rings in $\{B_1',\ldots,B_r'\}$ are incomparable, and by (11.14) there is a $y \in B_1' \cap\ldots\cap B_r'$ such that $y - 1 \in \mathfrak{M}_{B_1'}$ and $y \in \mathfrak{M}_{B_2'} \cap\ldots\cap \mathfrak{M}_{B_r'}$. Let w be a Krull valuation of N corresponding to B_1 and let $G = \mathrm{Aut}(N|K)$, $G^Z = G^Z(B_1|K)$. Since, for any $\sigma \in G \setminus G^Z$, $w \circ \sigma$ corresponds to some valuation ring in $\{B_2,\ldots,B_r\}$, we have $\delta = \min\{w(\sigma y) \mid \sigma \in G \setminus G^Z\} > 0$, and by (13.11) there exist $d \in K$ and $k \in \mathbb{N} \setminus \{0\}$ such that $vd = k \cdot \delta$, where $v = w|K$. Let $z = y^{2k} \cdot d^{-1}$; then $wz = 2k \cdot wy - vd = -k\cdot\delta$ and $w(\sigma z) = 2k\cdot w(\sigma y) - vd \geq k\cdot\delta$ for all $\sigma \in G \setminus G^Z$; hence $z \notin K$. Let $\sigma_2,\ldots,\sigma_n \in G$ such that $z, \sigma_2 z,\ldots,\sigma_n z$ are the distinct conjugates of z over K , and let $\sigma_1 = \iota_N$; obviously $n = n(z) \geqq 2$ and $\sigma_2,\ldots,\sigma_n \notin G^Z$. Therefore $v(c_1(z)) = w(z + \sigma_2 z +\ldots+ \sigma_n z) =$

$= -k\cdot\delta < 0$ and $v(c_j(z)) \geqq \min\{w(\sigma_{i_1}z) + \ldots + w(\sigma_{i_j}z) \mid (i_1,\ldots,i_j) \in I_j\}$ $\geqq 0$ for all $j \in \{2,\ldots,n\}$ (where I_j is as in (16.2)), contradicting (viii). □

We are particularly interested in the case in which $N = ac(K)$ or $N = sc(K)$ where $ac(K)$ (resp. $sc(K)$) is an algebraic (resp. separable) closure of the field K . A valuation ring A of K is called henselian if it satisfies one of the following equivalent conditions:

(16.4) The following conditions are equivalent:

(i) A is indecomposed in $ac(K)$.

(ii) A is indecomposed in $sc(K)$.

(iii) A is indecomposed in any finite Galois subextension of $ac(K)|K$.

In this case, A is indecomposed in any algebraic extension L of K .

Proof: (i) ⇔ (iii) and (ii) ⇔ (iii) follow from (16.1). The last assertion follows from (13.6 a) and the fact that any algebraic extension L of K is K-isomorphic to some subfield of $ac(K)$. □

Note that any valuation ring of any separably closed field is henselian. From (15.7) we conclude that $B \cap K^Z(B|K)$ is a henselian valuation ring of $K^Z(B|K)$, where B is any valuation ring of $ac(K)$ or $sc(K)$.

In the case $N = ac(K)$, theorem (16.3) may be reformulated in terms of irreducible polynomials. In fact, let $Irr(A)$ be the set of all monic and irreducible polynomials in $A[X]$ and $Irr^*(A) = \{P \in Irr(A) \mid \deg P > 1\}$; then $Irr(K) = \{P_{y|K} \mid y \in ac(K)\}$, $Irr^*(K) = \{P_{y|K} \mid y \in ac(K) \setminus K\}$, $Irr(A) = Irr(K) \cap A[X] =$

$= \{P_{y|K} | y \in I_N(A)\}$, and $Irr^*(A) = \{P_{y|K} | y \in I_N(A) \setminus K\}$. Setting

$$F = X^{n(F)} + c_1(F) \cdot X^{n(F)-1} + \ldots + c_{n(F)-1}(F) \cdot X + c_{n(F)}(F)$$

and $c(F) = c_{n(F)}(F)$ for any monic $F \in K[X]$, we get as an immediate consequence of (16.3):

(16.5) COROLLARY - In the conditions (iv) through (viii) of (16.3) replace y, N, $I_N(A)$, $N \setminus K$, $I_N(A) \setminus K$ by P, $Irr(K)$, $Irr(A)$, $Irr^*(K)$, $Irr^*(A)$, respectively. Any of these modified conditions is necessary and sufficient for A to be henselian.

Condition (iii) of the following corollary is sometimes referred to as "Hensel's condition".

(16.6) COROLLARY - Let A, π, and $\mathcal{K}$ be as in (16.3). The following conditions are equivalent:

(i) A is henselian.

(ii) For any $P \in Irr(A)$, πP is a power of some irreducible polynomial in $\mathcal{K}[X]$.

(iii) For any monic polynomials $F \in A[X]$ and Φ_1, $\Phi_2 \in \mathcal{K}[X]$ such that Φ_1, Φ_2 are relatively prime and $\pi F = \Phi_1 \cdot \Phi_2$, there exist monic polynomials F_1, $F_2 \in A[X]$ such that $\pi F_1 = \Phi_1$, $\pi F_2 = \Phi_2$, and $F = F_1 \cdot F_2$.

(iv) For any monic polynomial $F \in A[X]$ and any $\beta \in \mathcal{K}$ such that β is a simple root of πF, there is a $b \in A$ such that $\pi b = \beta$ and $F(b) = 0$.

(v) For any monic polynomial $F \in A[X]$ such that $c_1(F) \notin \mathfrak{M}_A$ and $\{c_2(F), \ldots, c_{n(F)}(F)\} \subseteq \mathfrak{M}_A$, there is a $b \in A$ such that $\pi b = -\pi(c_1(F))$ and $F(b) = 0$.

Proof: (i) ⇒ (ii) follows from (16.3).

(ii) ⇒ (iii): Let $F = P_1 \cdot \ldots \cdot P_k$ be the factorization of

F in (not necessarily distinct) polynomials $P_1,\ldots,P_k \in \operatorname{Irr}(K)$. By (10.6) and (10.10 a) we have $P_1,\ldots,P_k \in \operatorname{Irr}(A)$. By (ii) we have $\pi P_i = \Psi_i^{n_i}$ for some $\Psi_i \in \operatorname{Irr}(\mathcal{K})$ $(i = 1,\ldots k)$, hence $\pi F = \Psi_1^{n_1} \cdot \ldots \cdot \Psi_k^{n_k}$. If $\Phi_1 , \Phi_2 \in \mathcal{K}[X]$ are monic and relatively prime and $\pi F = \Phi_1 \cdot \Phi_2$, then $\Phi_1 = \prod_{i\in I} \Psi_i^{n_i}$ and $\Phi_2 = \prod_{j\in J\setminus I} \Psi_j^{n_j}$ for some subset I of $J = \{1,\ldots,k\}$; hence $F_1 = \prod_{i\in I} P_i$ and $F_2 = \prod_{j\in J\setminus I} P_j$ have the desired properties.

(iii) ⇒ (iv): If $\beta \in \mathcal{K}$ is a simple root of πF , then $\pi F = (X-\beta) \cdot \Phi$ for some monic polynomial $\Phi \in \mathcal{K}[X]$, and Φ , $X-\beta$ are relatively prime. By (iii) there exist $F_1 \in A[X]$ and $b \in A$ such that $\pi F_1 = \Phi$, $\pi(X-b) = X-\beta$, and $F = F_1 \cdot (X-b)$; therefore $\pi b = \beta$ and $F(b) = 0$.

(iv) ⇒ (v): Assume that F satisfies the hypotheses of (v) and let $\beta = -\pi(c_1(F))$. Then $\pi F = X^n - \beta \cdot X^{n-1} = X^{n-1} \cdot (X-\beta)$, hence β is a simple root of πF . Therefore (v) follows from (iv).

(v) ⇒ (i): Suppose that A is not henselian. By (16.5) there is a $P \in \operatorname{Irr}^*(A)$ such that $\{c_2(P),\ldots,c_{n(P)}(P)\} \subseteq \mathfrak{M}_A$ and $c_1(P) \notin \mathfrak{M}_A$. By (v) there exist $b \in A$ and $F_1 \in K[X]$ such that $P = F_1 \cdot (X-b)$, contradicting the irreducibility of P . □

There exist various modifications of Hensel's condition. For example, non-monic polynomials may be admitted, and the condition "relatively prime" may be weakened.

We recall that, for any exponential valuation v of K such that (K,v) is complete, the corresponding valuation ring A_v is henselian (cf (2.8) and §13). This fact was originally proved by means of Hensel's Lemma, which states that A_v satisfies condition (iii) of (16.6) (and which gave rise to the denotations "henselian" and "Hensel's condition"). We prove here a version of Hensel's lemma in which non-monic polynomials are admitted; condition (iii) of (16.6) can be obtained by setting $g = 1$ and assuming F to be monic.

(16.7) HENSEL'S LEMMA - Let (K,v) be a complete valued field, where v is an exponential valuation, let A be its valuation ring and π a corresponding place with residue field $\mathcal{K}$. Then for any $F \in A[X]$, $g \in A$, and $\Phi, \Psi \in \mathcal{K}[X]$ such that Φ, Ψ are relatively prime, with $\pi F = \Phi \cdot \Psi$ and πg as the leading coefficient of Φ, there exist polynomials $G, H \in A[X]$ such that $\pi G = \Phi$, $\pi H = \Psi$, $F = G \cdot H$, $\deg G = \deg \Phi$, and g is the leading coefficient of G.

Proof: Let $s = \deg F$ and $r = \deg \Phi$; then $\deg \Psi = \deg(\pi F) - \deg \Phi \leqq s - r$. Let $G_o, H_o \in A[X]$ such that $\pi G_o = \Phi$, $\pi H_o = \Psi$, $\deg G_o = r$, $\deg H_o \leqq s - r$, and $lc(G_o) = g$ (where $lc(\ )$ denotes the leading coefficient of a polynomial). Let $C, D \in A[X]$ such that $\pi C \cdot \Phi + \pi D \cdot \Psi = 1$ and let $\gamma = \min\{v(F - G_o \cdot H_o), v(C \cdot G_o + D \cdot H_o - 1\}$[24]; then obviously $\gamma > 0$. We choose $z \in A$ such that $0 < vz \leqq \gamma$ and set $W_o = z^{-1} \cdot (F - G_o \cdot H_o)$. We prove by induction that for any $i \in \mathbb{N}$ there exist $G_i, H_i, W_i \in A[X]$ with the following properties:

---

24 We set $v(\sum_{i=0}^{n} a_i \cdot X^i) = \min\{v\,a_o, \ldots, v\,a_n\}$ for any $a_o, \ldots, a_n \in K$, $n \in \mathbb{N}$.

1) $\pi G_i = \Phi$ and $\pi H_i = \Psi$.

2) $\deg G_i = r$, $\deg H_i \leqq s-r$, and $lc(G_i) = g$.

3) $G_i - G_{i-1} \in z^i \cdot A[X]$ and $H_i - H_{i-1} \in z^i \cdot A[X]$.

4) $F - G_i \cdot H_i = z^{i+1} \cdot W_i$.

This is obviously true for $i = 0$ (condition 3 being void). We assume that it is true for $i = 0,1,\ldots,n-1$ and construct polynomials G_n , H_n , $W_n \in A[X]$ which satisfy these conditions for $i = n$. Since $lc(G_o) \in U_A$, there exist Q_n , $U_n \in A[X]$ such that $W_{n-1} \cdot D = Q_n \cdot G_o + U_n$ and $\deg U_n < \deg G_o = r$. Let $V_n \in A[X]$ be a polynomial of least degree such that $W_{n-1} \cdot C + Q_n \cdot H_o - V_n \in z \cdot A[X]$; then $V_n \cdot G_o + U_n \cdot H_o - W_{n-1} = (C \cdot G_o + D \cdot H_o - 1) \cdot W_{n-1} - (W_{n-1} \cdot C + Q_n \cdot H_o - V_n) \cdot G_o \in z \cdot A[X]$. We claim that $\deg V_n \leqq s-r$. In fact, otherwise $\deg(U_n \cdot H_o - W_{n-1}) \leqq \max \{\deg(U_n \cdot H_o), \deg W_{n-1}\} \leqq s < \deg(V_n \cdot G_o)$, hence $lc(V_n) \cdot lc(G_o) = lc(V_n \cdot G_o) \in z \cdot A$, hence $lc(V_n) \in z \cdot A$, contradicting the choice of V_n . The polynomials $G_n = G_{n-1} + z^n \cdot U_n$ and $H_n = H_{n-1} + z^n \cdot V_n$ obviously satisfy 1), 2), and 3) for $i = n$. Since $G_{n-1} - G_o$, $H_{n-1} - H_o$, and $V_n \cdot G_o + U_n \cdot H_o - W_{n-1}$ are in $z \cdot A[X]$, so is $V_n \cdot G_{n-1} + U_n \cdot H_{n-1} - W_{n-1}$; hence $F - G_n \cdot H_n = -z^{2n} \cdot U_n \cdot V_n - z^n \cdot (V_n \cdot G_{n-1} + U_n \cdot H_{n-1} - W_{n-1}) = z^{n+1} \cdot W_n$ for some $W_n \in A[X]$. Therefore also 4) is satisfied for $i = n$.

Let $G_i = \sum_{j=0}^{r} g_{ij} \cdot X^j$ and $H_i = \sum_{j=0}^{s-r} h_{ij} \cdot X^j$. From 3) follows that the sequences $(g_{i0})_{i \in \mathbb{N}}, \ldots, (g_{ir})_{i \in \mathbb{N}}$, $(h_{i0})_{i \in \mathbb{N}}, \ldots, (h_{i\ s-r})_{i \in \mathbb{N}}$ are v-Cauchy, hence v-convergent. (It is only here that we use the completeness of (K,v).) Let $g_0,\ldots,g_r$, $h_0,\ldots,h_{s-r}$ be the respective v-limits and $G = \sum_{j=0}^{r} g_j \cdot X^j$, $H = \sum_{j=0}^{s-r} h_j \cdot X^j$. Since $G - G_n$, $H - H_n$, and $F - G_n \cdot H_n$ are in $z^n \cdot A[X]$ for any $n \in \mathbb{N}$, we have $F - G \cdot H \in \bigcap_{n \in \mathbb{N}} z^n \cdot A[X]$, hence $F = G \cdot H$. Obviously G and H have all the desired properties. □

Note that Hensel's Lemma is not true for complete valued fields (K,v) , where v is a Krull valuation of rank > 1; in fact, the corresponding valuation ring is generally not henselian (see Bourbaki [5], Chap. 6, §8). However, there exist different notions of "completeness" for Krull valuations, which imply Hensel's condition (see Krull [20] or Ribenboim [30]).

The following statement, known as Krasner's Lemma, asserts that $K(x) = K(y)$ for any two elements $x, y \in sc(K)$ which are sufficiently close to each other in the topology defined by the unique extension of a henselian Krull valuation of K. For any $x \in ac(K)$ we denote by C_x the (finite) set of all K-conjugates of x, i.e. $C_x = \{\sigma x \mid \sigma \in G\}$, where $G = Aut(ac(K)|K)$.

(16.8) KRASNER'S LEMMA - Let A be a henselian valuation ring of K and w a Krull valuation of $ac(K)$ such that $w|K$ corresponds to A. Moreover, let $x, y \in ac(K)$ such that $w(y-x) > \max\{w(x'-x) \mid x' \in C_x \setminus \{x\}\}$. Then $K(x,y)$ is purely inseparable over $K(y)$, and if $x \in sc(K)$ then $K(x) \subseteq K(y)$.

Proof: We recall that the separability degree $[K(x,y):K(y)]_{sep}$ is equal to the number of $K(y)$-monomorphisms $\rho: K(x,y) \to ac(K)$ and that any such ρ is the restriction of some $K(y)$-automorphism of $ac(K)$. Therefore it suffices to prove that any $\sigma \in Aut(ac(K)|K(y))$ induces the identity $\iota_{K(x,y)}$ of $K(x,y)$. In fact, we have $w \circ \sigma = w$ by (14.7), hence $w(\sigma x - x) = w((y-x) - (\sigma y - \sigma x)) \geqq w(y-x) > w(x'-x)$ for all $x' \in C_x \setminus \{x\}$, hence $\sigma x = x$; therefore $\sigma|K(x,y) = \iota_{K(x,y)}$ for any $\sigma \in Aut(ac(K)|K(y))$. □

REMARK - The statement of Krasner's lemma holds also under the following hypotheses: Let φ be a valuation of K (in the sense of §1) which admits only one extension ψ to $ac(K)$, and let $x, y \in ac(K)$ such that $\psi(y-x) < \frac{1}{2} \cdot \min\{\psi(x'-x) \mid x' \in C_x \setminus \{x\}\}$. In fact, in the preceding proof we need only replace "w" by "ψ" and

"$\geq w(y-x) >$" by "$\leq 2\cdot\psi(y-x) <$" .

§17 Extensions of valuation rings and henselization

We recall that, for any finite separable field extension $L|K$, the valuations of L which extend a given valuation φ of K can be obtained by means of the completion $(\hat{K},\hat{\varphi})$ of (K,φ) (cf (2.12)). Similarly, for a given valuation ring A of K , we shall use a "henselization" of (K,A) to determine the valuation rings of L which lie over A ; here $L|K$ may be even an arbitrary algebraic field extension.

Let Ω be an algebraically closed field. For any subfield K of Ω we denote by $ac(K)$ (resp. $sc(K)$) the algebraic (resp. separable) closure of K contained in Ω . For any algebraic field extension $L|K$ (which need not be a subextension of $\Omega|K$) we denote by $Mon(L|K)$ the set of all K-monomorphisms from L into Ω . Obviously $\lambda L \subseteq ac(K)$ for all $\lambda \in Mon(L|K)$, and if $L|K$ is finite then $\# Mon(L|K) = [L:K]_{sep}$. If L is a subextension of $\Omega|K$, the identical imbedding $\iota = \iota_{L,\Omega}: L \to \Omega$ is in $Mon(L|K)$. If N is a normal subextension of $\Omega|K$ then $Mon(N|K)$ can be identified with the Galois group $Aut(N|K)$.

Let $\bar{K}$ be a subextension of $\Omega|K$. We define in $Mon(L|K)$ an equivalence relation $\sim_{(\bar{K};L)}$ by

$$\lambda' \sim_{(\bar{K};L)} \lambda \Leftrightarrow [\lambda' = \sigma \circ \lambda \text{ for some } \sigma \in Aut(\Omega|\bar{K})] ,$$

for all $\lambda, \lambda' \in Mon(L|K)$. If no ambiguity is possible we write $\sim_{\bar{K}}$ instead of $\sim_{(\bar{K};L)}$. If $\lambda' \sim_{\bar{K}} \lambda$ we say that λ , λ' are $\bar{K}$-conjugate to each other. The following statement is obvious.

(17.1) <u>Let</u> $K \subseteq \bar{K}_1 \subseteq \bar{K}_2 \subseteq \Omega$. <u>For any algebraic extension</u> $L|K$ <u>and any</u> $\lambda, \lambda' \in \mathrm{Mon}(L|K)$ <u>we have</u>:

a) $\lambda \sim_K \lambda'$.

b) $\lambda \sim_\Omega \lambda'$ <u>if and only if</u> $\lambda = \lambda'$.

c) $\lambda \sim_{\bar{K}_2} \lambda'$ <u>implies</u> $\lambda \sim_{\bar{K}_1} \lambda'$.

The relations $\sim_{(\bar{K}_1;L)}$, $\sim_{(\bar{K}_2;L)}$ may coincide for different fields $\bar{K}_1$, $\bar{K}_2$; in fact:

(17.2) <u>Let</u> $K \subseteq \bar{K}_i \subseteq \Omega$ $(i=1,2)$. <u>The following conditions are equivalent</u>:

(i) $\sim_{(\bar{K}_1;L)}$ <u>and</u> $\sim_{(\bar{K}_2;L)}$ <u>coincide for any algebraic extension</u> $L|K$.

(ii) $\sim_{(\bar{K}_1;L)}$ <u>and</u> $\sim_{(\bar{K}_2;L)}$ <u>coincide for any finite separable subextension</u> L <u>of</u> $\Omega|K$.

(iii) $\bar{K}_1 \cap sc(K) = \bar{K}_2 \cap sc(K)$.

<u>Proof</u>: (i) $\Rightarrow$ (ii) is trivial.

(ii) $\Rightarrow$ (iii): Suppose $\bar{K}_1 \cap sc(K) \not\subseteq \bar{K}_2 \cap sc(K)$ and let $y \in (\bar{K}_1 \cap sc(K)) \setminus \bar{K}_2$, $L = K(y)$; then L is a finite separable subextension of $\Omega|K$. Since y is separable over $\bar{K}_2$ and $y \notin \bar{K}_2$, we have $\sigma y \neq y$ for some $\sigma \in \mathrm{Aut}(\Omega|\bar{K}_2)$. Let $\lambda = \sigma|L$; obviously $\lambda \in \mathrm{Mon}(L|K)$ and $\lambda \sim_{\bar{K}_2} \iota$. On the other hand, we have $\lambda \not\sim_{\bar{K}_1} \iota$ since $\lambda y \neq y = (\rho \circ \iota)y$ for any $\rho \in \mathrm{Aut}(\Omega|\bar{K}_1)$.

(iii) $\Rightarrow$ (i): The canonical homomorphism $\mathrm{Aut}(\Omega|\bar{K}_i) \to \mathrm{Aut}(sc(K)|(\bar{K}_i \cap sc(K)))$, defined by $\sigma \longmapsto \sigma|sc(K)$, is surjective; therefore for any $\lambda, \lambda' \in \mathrm{Mon}(L|K)$ we have $\lambda' \sim_{\bar{K}_i} \lambda$ if and only if $\lambda' = \rho \circ \lambda$ for some $\rho \in \mathrm{Aut}(sc(K)|(\bar{K}_i \cap sc(K)))$. Hence (iii) implies (i). □

We denote by $\mathrm{Mon}_{\bar{K}}(L|K)$ the set of all $\sim_{(\bar{K};L)}$-classes of $\mathrm{Mon}(L|K)$. Obviously $\mathrm{Mon}_\Omega(L|K) = \mathrm{Mon}(L|K)$, $\# \mathrm{Mon}_K(L|K) = 1$, and for $K \subseteq \bar{K}_1 \subseteq \bar{K}_2 \subseteq \Omega$ we have a canonical surjective mapping ϑ_L : $\mathrm{Mon}_{\bar{K}_2}(L|K) \to \mathrm{Mon}_{\bar{K}_1}(L|K)$. In particular, ϑ_L is bijective for any algebraic extension $L|K$ if and only if $\bar{K}_1$, $\bar{K}_2$ satisfy the conditions of (17.2); in fact, in this case ϑ_L is the identity of $\mathrm{Mon}_{\bar{K}_2}(L|K) = \mathrm{Mon}_{\bar{K}_1}(L|K)$.

For any <u>finite</u> extension $L|K$ and any $\lambda \in \mathrm{Mon}(L|K)$ we denote by g_λ the quotient $[L:K]_{insep} \cdot ([\lambda L\cdot\bar{K} : \bar{K}]_{insep})^{-1}$ of inseparability degrees [25]. Note that g_λ depends not only on λ but also on the choice of K and $\bar{K}$, and that all $g_\lambda = 1$ whenever $L|K$ is separable.

(17.3) <u>Let</u> $L|K$ <u>be a finite extension. Then</u>

a) <u>For any</u> $\lambda \in \mathrm{Mon}(L|K)$, g_λ <u>is a power of the characteristic exponent of</u> K .

b) <u>For any</u> $\lambda_1 , \lambda_2 \in \mathrm{Mon}(L|K)$, $\lambda_1 \sim_{\bar{K}} \lambda_2$ <u>implies</u> $g_{\lambda_1} = g_{\lambda_2}$.

c) <u>Let</u> $\lambda_1,\dots,\lambda_r \in \mathrm{Mon}(L|K)$ <u>be a set of representatives of</u> $\mathrm{Mon}_{\bar{K}}(L|K)$; <u>then</u> $[L:K] = \sum_{i=1}^{r} [\lambda_i L\cdot\bar{K} : \bar{K}]\cdot g_{\lambda_i}$.

<u>Proof</u>: a) We have $[L:K]_{insep} = [L : L_{sep}] \geq [\lambda L\cdot\bar{K} : \lambda L_{sep}\cdot\bar{K}] = [\lambda L\cdot\bar{K} : \bar{K}]_{insep}$, since $\lambda L_{sep}\cdot\bar{K}$ is the maximal separable subextension of $\lambda L\cdot\bar{K} \mid \bar{K}$. Since both inseparability degrees are powers of the characteristic exponent of K , so is g_λ .

b) If $\lambda_1 \sim_{\bar{K}} \lambda_2$ then $\lambda_1 L\cdot\bar{K}$, $\lambda_2 L\cdot\bar{K}$ are $\bar{K}$-isomorphic, hence $g_{\lambda_1} = g_{\lambda_2}$.

c) Since $L_{sep}|K$ is finite and separable, we have $L_{sep} = K(y)$ for some $y \in L_{sep}$. Obviously $\lambda_i L_{sep}\cdot\bar{K} = \bar{K}(\lambda_i y)$ $(i=1,\dots,r)$ and $P_{y|K} = \prod_{i=1}^{r} P_{\lambda_i y|\bar{K}}$; therefore $[L:K]_{sep} = \deg P_{y|K} = \sum_{i=1}^{r} \deg P_{\lambda_i y|\bar{K}} = \sum_{i=1}^{r} [\lambda_i L\cdot\bar{K} : \bar{K}]_{sep}$, hence $[L:K] = \sum_{i=1}^{r} [\lambda_i L\cdot\bar{K}:\bar{K}]\cdot g_{\lambda_i}$. □

[25] The inseparability degree of $L|K$ is defined as $[L:K]\cdot[L:K]_{sep}^{-1}$.

If $L|K$ is finite and normal, then all fields λL coincide ($\lambda \in \mathrm{Mon}(L|K)$) ; therefore $[\lambda L\cdot\bar{K} : \bar{K}]$ and g_λ do not depend on λ , and the equality in (17.3 c) can be written as $[L:K] =$ $= [\lambda L\cdot\bar{K} \cdot \bar{K}]\cdot g_\lambda \cdot \# \mathrm{Mon}_{\bar{K}}(L|K)$.

Note that the preceding considerations are essentially independent of the choice of the algebraically closed field Ω . In fact, let φ be a K-monomorphism from Ω into another algebraically closed field Ω' . It is easy to check that $\lambda \mapsto \varphi\circ\lambda$ is a bijection from $\mathrm{Mon}(L|K)$ onto $\mathrm{Mon}'(L|K)$, and $\lambda_1 \sim_{\bar{K}} \lambda_2$ if and only if $\varphi\circ\lambda_1 \sim'_{\varphi\bar{K}} \varphi\circ\lambda_2$, where $\mathrm{Mon}'(\ )$ and $\sim'$ are defined with respect to Ω'.

As in §2, it is practical to define a <u>valued field</u> as a pair (K,A) , consisting of a field K and a valuation ring A of K . If (K,A) and (L,B) are valued fields such that $K \subseteq L$ and $A = B \cap L$, we write $(K,A) \subseteq (L,B)$ and call (L,B) an <u>extension of</u> (K,A) . Such an extension is called <u>henselian</u> if B is a henselian valuation ring. It is called <u>immediate</u> if $e_{B|K} = f_{B|K} = 1$. Note that, for any valuation ring C of $sc(K)$, $(sc(K),C)$ is a henselian extension of (K,A), where $A = C \cap K$, and by (15.7) and (15.8), $(K^Z(C|K),\ B \cap K^Z(C|K))$ is even an immediate henselian extension of (K,A). There are also immediate henselian extensions $(\bar{K},\bar{A})$ such that $\bar{K}|K$ is transcendental; in fact, if $\mathrm{rank}(A) = 1$ then the completion of (K,A) [26] is of this type, and in the general case, any maximal immediate extension of (K,A) has this property (cf Ribenboim [30], D,F).

Let $L|K$ and $L'|K$ be field extensions. By a K-<u>imbedding</u> $\lambda: (L,B) \to (L',B')$ we mean a K-monomorphism $\lambda: L \to L'$ such that

[26] We call $(\hat{K},\hat{A})$ the completion of (K,A) if $(\hat{K},\hat{v})$ is the completion of (K,v) , where $\hat{v}$ (resp. v) is a Krull valuation corresponding to $\hat{A}$ (resp. A).

$(\lambda L, \lambda B) \subseteq (L', B')$; in this case $B \cap K = B' \cap K$. A bijective K-imbedding will be called a K-<u>isomorphism</u>; its inverse is a K-isomorphism too.

For the rest of this section, we assume that $(\bar{K},\bar{A})$ is a henselian extension of (K,A) such that $\bar{K} \subseteq \Omega$, and we denote by $\bar{D}$ the unique valuation ring of $ac(\bar{K})$ which lies over $\bar{A}$. For any algebraic field extension $L|K$ we denote by $\mathcal{B}(L;A)$ the set of all valuation rings of L which lie over A .

(17.4) THEOREM - <u>Let</u> $L|K$ <u>be an algebraic extension and, for any</u> $\lambda \in Mon(L|K)$, <u>let</u> $B_\lambda = \lambda^{-1}\bar{D}$. <u>Then any</u> $\lambda \in Mon(L|K)$ <u>is a K-imbedding</u> $\lambda: (L,B_\lambda) \to (ac(\bar{K}),\bar{D})$, <u>we have</u> $\mathcal{B}(L;A) = \{B_\lambda \mid \lambda \in Mon(L|K)\}$, <u>and</u> $\lambda \longmapsto B_\lambda$ <u>induces a mapping from</u> $Mon_{\bar{K}}(L|K)$ <u>onto</u> $\mathcal{B}(L;K)$.

<u>Proof</u>: For any $\lambda \in Mon(L|K)$, $B_\lambda = \{x \in L \mid \lambda x \in \bar{D}\}$ is a valuation ring of L such that $\lambda B_\lambda = \bar{D} \cap \lambda L$; therefore $\lambda: (L,B_\lambda) \to (ac(\bar{K}),\bar{D})$ is a K-imbedding. It follows that $B_\lambda \cap K = D \cap K = \bar{A} \cap K = A$, hence $B_\lambda \in \mathcal{B}(L;A)$. On the other hand, we claim that any $B \in \mathcal{B}(L;A)$ equals B_λ for some $\lambda \in Mon(L|K)$. In fact, choose $\lambda_o \in Mon(L|K)$ and $C \in \mathcal{B}(ac(K); \lambda_o B)$. Since C and $\bar{D} \cap ac(K)$ lie over A , we have $\rho C = \bar{D} \cap ac(K)$ for some $\rho \in Aut(ac(K)|K)$, by (14.3) . Let $\lambda = \rho \circ \lambda_o$; then $\lambda \in Mon(L|K)$ and $\lambda B = \rho(\lambda_o B) = \rho(C \cap \lambda_o L) = (\bar{D} \cap ac(K)) \cap \lambda L = \bar{D} \cap \lambda L = \lambda B_\lambda$, hence $B = B_\lambda$. Finally, we have $\sigma\bar{D} = \bar{D}$ for any $\sigma \in Aut(\Omega|\bar{K})$, since $\bar{A}$ is henselian. Therefore $\lambda \sim_{\bar{K}} \lambda'$ implies $B_\lambda = B_{\lambda'}$, i.e., the last statement of the theorem holds. □

Assuming furthermore that the henselian extension $(\bar{K},\bar{A})$ of (K,A) is immediate, we get easily the following "fundamental inequality":

(17.5) COROLLARY - For any finite field extension $L|K$ we have

$$[L:K] \geq \sum_{i=1}^{r} e_i \cdot f_i \cdot g_{\lambda_i} \geq \sum_{i=1}^{r} e_i \cdot f_i \geq \sum_{B \in \mathcal{B}(L;A)} e_{B|K} \cdot f_{B|K} ,$$

where $(\bar{K},\bar{A})$ is an immediate henselian extension of (K,A) ; $\lambda_1,\ldots,\lambda_r \in \mathrm{Mon}(L|K)$ is any set of representatives of $\mathrm{Mon}_{\bar{K}}(L|K)$; and $e_i = e_{B_{\lambda_i}|K}$, $f_i = f_{B_{\lambda_i}|K}$ for $i = 1,\ldots,r$.

Proof: Let $\bar{w}$ (resp. $\bar{\zeta}$) be a Krull valuation (resp. place) of $ac(\bar{K})$ corresponding to $\bar{D}$, and let $\bar{L}_i = \lambda_i L\cdot\bar{K}$, $\bar{B}_i = \bar{D} \cap \bar{L}_i$ $(i = 1,\ldots,r)$. Let Γ , $\bar{\Gamma}$, $\bar{\Delta}_i$, Δ_i be the value groups of the Krull valuations $v = \bar{w}|K$, $\bar{v} = \bar{w}|\bar{K}$, $\bar{w}_i = \bar{w}|\bar{L}_i$, $w_i = \bar{w}\circ\lambda_i$, and let $\mathcal{K}$, $\bar{\mathcal{K}}$, $\bar{\mathcal{L}}_i$, $\mathcal{L}_i$ be the residue fields of the places $\pi = \bar{\zeta}|K$, $\bar{\pi} = \bar{\zeta}|\bar{K}$, $\bar{\zeta}_i = \bar{\zeta}|\bar{L}_i$, $\zeta_i = \bar{\zeta}\circ\lambda_i$; these Krull valuations and places correspond to the valuation rings A , $\bar{A}$, $\bar{B}_i$, B_{λ_i} of K , $\bar{K}$, $\bar{L}_i$, L , respectively $(i = 1,\ldots,r)$. By hypothesis we have $\Gamma = \bar{\Gamma}$ and $\mathcal{K} = \bar{\mathcal{K}}$. For $i = 1,\ldots,r$ we have $\Delta_i \subseteq \bar{\Delta}_i$ and $\mathcal{L}_i \subseteq \bar{\mathcal{L}}_i$, hence $e_i \leq e_{\bar{B}_i|\bar{K}}$, $f_i \leq f_{\bar{B}_i|\bar{K}}$, hence $e_i\cdot f_i \leq e_{\bar{B}_i|\bar{K}} \cdot f_{\bar{B}_i|\bar{K}} \leq [\bar{L}_i:\bar{K}]$, by (13.10); therefore the first two inequalities follow from (17.3). Since $\mathcal{B}(L;A) = \{B_{\lambda_1},\ldots,B_{\lambda_r}\}$ by (17.4), the last inequality holds. □

If $L|K$ is a normal extension, then the inequalities of (17.5) may be written as

$$[L:K] \geq e\cdot f\cdot g\cdot \# \mathrm{Mon}_{\bar{K}}(L|K) \geq e\cdot f\cdot \# \mathrm{Mon}_{\bar{K}}(L|K) \geq e\cdot f\cdot \# \mathcal{B}(L;A) ,$$

where e (resp. f) is the common ramification index (resp. residue degree) of all $B \in \mathcal{B}(L;A)$ over K (see §14) and $g = g_\lambda$ for all $\lambda \in \mathrm{Mon}(L|K)$.

Let $L|K$ be an algebraic extension. The henselian extension $(\bar{K},\bar{A})$ of (K,A) is called L-allowable if the surjective mapping $\mathrm{Mon}_{\bar{K}}(L|K) \to \mathcal{B}(L;A)$ in (17.4) is bijective, i.e., if $B_\lambda = B_{\lambda'}$ implies

$\lambda \sim_{\bar{K}} \lambda'$ for any $\lambda, \lambda' \in \mathrm{Mon}(L|K)$. It is easy to give examples of valued fields (K,A) and henselian extensions $(\bar{K},\bar{A})$ of (K,A) which are not L-allowable for some algebraic extension $L|K$. This is possible even under the additional hypothesis that $(\bar{K},\bar{A})$ be an immediate extension of (K,A) (see Ribenboim [30], G, Example 2).

(17.6) Let $(\bar{K},\bar{A})$ be an L-allowable extension of (K,A) , and let

a) L' be a subextension of $L|K$ or

b) $L'|L$ be a purely inseparable extension.

Then $(\bar{K},\bar{A})$ is an L'-allowable extension of (K,A) .

Proof: Let $\lambda_1', \lambda_2' \in \mathrm{Mon}(L'|K)$ such that $(\lambda_1')^{-1}\bar{D} = (\lambda_2')^{-1}\bar{D} = B'$ (say). We have to show that $\lambda_1' \sim_{\bar{K}} \lambda_2'$.

a) Suppose $K \subseteq L' \subseteq L$. Let B be a valuation ring of L which lies over B' , and for $i = 1,2$ let $\lambda_i \in \mathrm{Mon}(L|K)$ be an extension of λ_i' . Since $(\lambda_i'L'\cdot\bar{K}, \bar{D} \cap \lambda_i'L'\cdot\bar{K})$ is a henselian extension of $(\lambda_i'L', \lambda_i'B')$ such that $ac(\lambda_i'L'\cdot\bar{K}) = ac(\bar{K})$, we conclude from (17.4) (applied on λ_iL , $\lambda_i'L'$, $\lambda_i'B$ instead of L, K , A) that $\mu_i(\lambda_iB) = \bar{D} \cap \mu_i(\lambda_iL)$ for some $\mu_i \in \mathrm{Mon}(\lambda_iL|\lambda_i'L')$. Let $\nu_i = \mu_i \circ \lambda_i$; obviously $\nu_i \in \mathrm{Mon}(L|K)$ and $B = \nu_i^{-1}\bar{D}$ $(i=1,2)$. By hypothesis we have $\nu_1 = \sigma \circ \nu_2$ for some $\sigma \in \mathrm{Aut}(\Omega|\bar{K})$, hence $\lambda_1' = \nu_1|L' = \sigma \circ \nu_2|L' = \lambda_2'$, hence $\lambda_1' \sim_{\bar{K}} \lambda_2'$.

b) Suppose that $L'|L$ is purely inseparable. Let $B = B' \cap L$ and $\lambda_i = \lambda_i'|L$. Obviously $B = \lambda_i^{-1}\bar{D}$ for $i = 1,2$, hence $\lambda_1 = \sigma \circ \lambda_2$ for some $\sigma \in \mathrm{Aut}(\Omega|\bar{K})$, by hypothesis. Since λ_1' is the only extension of λ_1 , we have $\lambda_1' = \sigma \circ \lambda_2'$, hence $\lambda_1' \sim_{\bar{K}} \lambda_2'$. □

We get as an immediate consequence of (17.6):

(17.7) The following conditions are equivalent:

(i) $(\bar{K},\bar{A})$ is an $sc(K)$-allowable extension of (K,A) .

(ii) $(\bar{K},\bar{A})$ <u>is an</u> ac(K)-<u>allowable extension of</u> (K,A) .

(iii) $(\bar{K},\bar{A})$ <u>is an</u> L-<u>allowable extension of</u> (K,A) <u>for any algebraic extension</u> $L|K$.

We call $(\bar{K},\bar{A})$ an <u>allowable extension</u> of (K,A) if it satisfies one of the equivalent conditions of (17.7).

(17.8) <u>Let</u> $(\bar{K}_1,\bar{A}_1)$ <u>and</u> $(\bar{K}_2,\bar{A}_2)$ <u>be henselian extensions of</u> (K,A) <u>such that</u> $(\bar{K}_1,\bar{A}_1) \subseteq (\bar{K}_2,\bar{A}_2)$. <u>The following conditions are equivalent</u>:

(i) $(\bar{K}_2,\bar{A}_2)$ <u>is an allowable extension of</u> (K,A) .

(ii) $(\bar{K}_1,\bar{A}_1)$ <u>is an allowable extension of</u> (K,A) <u>and</u> $\bar{K}_2 \cap sc(K) \subseteq \bar{K}_1$.

<u>Proof</u>: For $i = 1,2$ and any algebraic extension $L|K$ let $\eta_{i,L}$ be the surjective mapping $Mon_{\bar{K}_i}(L|K) \to \mathcal{B}(L;A)$ defined in (17.4). Obviously $\eta_{2,L} = \eta_{1,L} \circ \vartheta_L$, and $\eta_{2,L}$ is bijective if and only if $\eta_{1,L}$ and ϑ_L are bijective. Moreover, by (17.2) ϑ_L is bijective for all algebraic extension $L|K$ if and only if $\bar{K}_1 \cap sc(K) = \bar{K}_2 \cap sc(K)$, if and only if $\bar{K}_2 \cap sc(K) \subseteq \bar{K}_1$. □

In order to decide whether $(\bar{K},\bar{A})$ is allowable or not, it suffices to consider the relative separable closure of K in $\bar{K}$. In fact:

(17.9) THEOREM - <u>Let</u> $(K',A') = (\bar{K} \cap sc(K), \bar{A} \cap sc(K))$. <u>Then</u>:

a) A' <u>is a henselian valuation ring of</u> K' .

b) $(\bar{K},\bar{A})$ <u>is an allowable extension of</u> (K',A') .

c) $(\bar{K},\bar{A})$ <u>is an allowable extension of</u> (K,A) <u>if and only if</u> (K',A') <u>is an allowable extension of</u> (K,A) .

<u>Proof</u>: Obviously $sc(K') = sc(K)$. Any $\lambda \in Mon(sc(K)|K') = Aut(sc(K)|K')$ is the restriction of some $\sigma \in Aut(\Omega|\bar{K})$;

therefore $\# \operatorname{Mon}_{\bar{K}}(sc(K)|K') = 1$. From (17.4) we conclude that $\# \mathfrak{B}(sc(K);A') = 1$ and the mapping $\operatorname{Mon}_{\bar{K}}(sc(K)|K') \to \mathfrak{B}(sc(K);A')$ is bijective; therefore a) and b) are true. Statement c) is a consequence of (17.8). □

Next we show that any valued field (K,A) has a "smallest" henselian extension, the so-called henselization. It will turn out to be an allowable immediate extension of (K,A) . We define the henselization by a universal property: A henselian extension $(\tilde{K},\tilde{A})$ of (K,A) is called a <u>henselization</u> of (K,A) if, for any valued field $(\bar{K}_o,\bar{A}_o)$ such that $\bar{A}_o$ is henselian and any imbedding $\mu:(K,A) \to (\bar{K}_o,\bar{A}_o)$, there exists exactly one imbedding $\tilde{\mu}: (\tilde{K},\tilde{A}) \to (\bar{K}_o,\bar{A}_o)$ which extends μ . The following statements are obvious.

(17.10) a) <u>If</u> $(\tilde{K}_1,\tilde{A}_1)$ <u>and</u> $(\tilde{K}_2,\tilde{A}_2)$ <u>are henselizations of</u> (K,A), <u>then there is exactly one</u> K-<u>isomorphism</u> $(\tilde{K}_1,\tilde{A}_1) \to (\tilde{K}_2,\tilde{A}_2)$.

b) <u>Any henselian extension of</u> (K,A) <u>contains at most one henselization of</u> (K,A) .

We hasten to show that "at most" can be replaced by "exactly". Recalling the notation preceding Theorem (17.4), we have:

(17.11) THEOREM - <u>Let</u> $C = \bar{D} \cap sc(K)$ <u>and</u> $K^Z = K^Z(C|K)$. <u>Then</u> $(K^Z, C \cap K^Z)$ <u>is the unique henselization of</u> (K,A) <u>contained in the henselian extension</u> $(\bar{K},\bar{A})$ <u>of</u> (K,A) .

<u>Proof</u>: By (15.7), $(K^Z, C \cap K^Z)$ is a henselian extension of (K,A) .

Let $(K',A') = (\bar{K} \cap sc(K), \bar{A} \cap sc(K))$. Since $(K',A') \subseteq (sc(K),C)$ and A' is henselian by (17.9), we have $(K^Z,C \cap K^Z) \subseteq (K',A') \subseteq (\bar{K},\bar{A})$, by (15.7). To prove that $(K^Z,C \cap K^Z)$ is a henselization of (K,A) , it suffices to show that for any henselian extension $(\bar{K}_o,\bar{A}_o)$ of (K,A) there is exactly one K-imbedding $(K^Z,C \cap K^Z) \to (\bar{K}_o,\bar{A}_o)$. Let Ω_o be an algebraic closure of $\bar{K}_o$,

$D_o = I_{\Omega_o}(\bar{A}_o)$, $sc_o(K)$ the separable closure of K contained in Ω_o, and $(K_o', A_o') = (\bar{K}_o \cap sc_o(K), \bar{A}_o \cap sc_o(K))$. By (17.4) there exists a K-imbedding $\lambda: (sc(K),C) \to (\Omega_o, D_o)$. Since $(K_o', A_o') \subseteq (sc_o(K), \lambda C)$ and A_o' is henselian by (17.9), we have $(K^Z(\lambda C|K), \lambda C \cap K^Z(\lambda C|K)) \subseteq$ $\subseteq (K_o', A_o') \subseteq (\bar{K}_o, \bar{A}_o)$ by (15.7). Obviously $\lambda K^Z = K^Z(\lambda C|K)$, hence λ induces a K-imbedding $(K^Z, C \cap K^Z) \to (\bar{K}_o, \bar{A}_o)$. On the other hand, for $i = 1,2$ let $\mu_i: (K^Z, C \cap K^Z) \to (\bar{K}_o, \bar{A}_o)$ be a K-imbedding and $\lambda_i: sc(K) \to \Omega_o$ a K-monomorphism which extends $\mu_i: K^Z \to \bar{K}_o$; obviously $\lambda_i\ sc(K) = sc_o(K)$. As $C \cap K^Z$ is henselian, so is $\mu_i(C \cap K^Z)$, and therefore $\lambda_i C = D_o \cap sc_o(K)$ $(i=1,2)$. It follows that $\lambda_1 =$ $= \lambda_2 \circ \sigma$ for some $\sigma \in G^Z(C|K) = \mathrm{Aut}(sc(K)|K^Z)$, hence $\mu_1 = \lambda_1|K^Z =$ $= \lambda_2|K^Z = \mu_2$. □

We conclude from (17.11) that, for any henselization $(\tilde{K}, \tilde{A})$ of (K,A) , the field $\tilde{K}$ is separable over K . It is easy to determine all henselizations $(\tilde{K}, \tilde{A})$ of (K,A) such that $\tilde{K} \subseteq \Omega$ or, equivalently, $\tilde{K} \subseteq sc(K)$; in fact:

(17.12) COROLLARY - <u>A bijective mapping from</u> $\mathfrak{B}(sc(K);A)$ <u>onto the set of all henselizations</u> $(\tilde{K}, \tilde{A})$ <u>of</u> (K,A) <u>such that</u> $\tilde{K} \subseteq \Omega$ <u>is given by</u>

$$C \mapsto (K^Z(C|K), C \cap K^Z(C|K)) \text{ , } \underline{\text{for all}}\ C \in \mathfrak{B}(sc(K);A) .$$

<u>Proof</u>: From (17.11) we conclude that $(K^Z(C|K), C \cap K^Z(C|K))$ is a henselization of (K,A) , for all $C \in \mathfrak{B}(sc(K);A)$, and that any henselization of (K,A) is of this form. If $C_1, C_2 \in \mathfrak{B}(sc(K);A)$ such that $(K^Z(C_i|K), C_i \cap K^Z(C_i|K))$ coincide $(i = 1,2)$, then $C_1 =$ $= C_2$ by (15.7). □

Note that the valuation rings in $\mathfrak{B}(sc(K);A)$ are K-conjugate, by (14.3), and so are the corresponding henselizations, by (15.6 a).

(17.13) COROLLARY - Any henselization of (K,A) is an allowable extension of (K,A) .

Proof: Because of (17.10), it suffices to consider the henselization $(K^Z, C \cap K^Z)$ indicated in (17.11) . Let $\lambda_1, \lambda_2 \in \mathrm{Mon}(sc(K)|K) = \mathrm{Aut}(sc(K)|K)$ such that $\lambda_1^{-1}\bar{D} = \lambda_2^{-1}\bar{D}$ and set $C = \bar{D} \cap sc(K)$. Since $\lambda^{-1}\bar{D} = \lambda^{-1}C$ for any $\lambda \in \mathrm{Aut}(sc(K)|K)$, we have $\lambda_2 \circ \lambda_1^{-1} \in G^Z(C|K) = \mathrm{Aut}(sc(K)|K^Z)$ (cf (15.5 b)), hence $\lambda_2 = \sigma \circ \lambda_1$ for some $\sigma \in \mathrm{Aut}(\Omega|K^Z)$. Therefore $\lambda_1 \sim_{K^Z} \lambda_2$. □

Using again the notations of (17.11), it is obvious that $K^Z \subseteq \bar{K} \cap sc(K) \subseteq sc(K)$. As to the first containment, we show:

(17.14) COROLLARY - $(\bar{K},\bar{A})$ is an allowable extension of (K,A) if and only if $\bar{K} \cap sc(K) = K^Z$.

Proof: If $\bar{K} \cap sc(K) = K^Z$ then $(\bar{K},\bar{A})$ is an allowable extension of (K,A) , by (17.9 c) and (17.13). Conversely, if $(\bar{K},\bar{A})$ is an allowable extension of (K,A) then $\bar{K} \cap sc(K) \subseteq K^Z$, by (17.8). □

The following corollary states several conditions, each of which is equivalent to the equality $K^Z = sc(K)$ (i.e., K^Z separably closed). We need the following definitions, which will be studied in detail in the next section: Let $L|K$ be a finite extension and A a valuation ring of K ; A is called defectless in L if $[L:K] = \sum_{B \in \mathcal{B}(L;A)} e_{B|L} \cdot f_{B|L}$, i.e., if the equality sign holds in the fundamental inequality.

(17.15) COROLLARY - Let $C \in \mathcal{B}(sc(K);A)$ and $K^Z = K^Z(C|K)$. The following conditions are equivalent:

(i) $K^Z = sc(K)$.

(ii) $(sc(K),C)$ is a henselization of (K,A) .

(iii) $(sc(K),C)$ is an allowable extension of (K,A) .

(iv) $\# \mathcal{B}(L;A) = [L:K]$ <u>for any finite separable extension</u> $L|K$.

(v) <u>For any finite separable extension</u> $L|K$, A <u>is defectless in</u> L <u>and</u> $e_{B|K} = f_{B|K} = 1$ <u>for all</u> $B \in \mathcal{B}(L;A)$.

<u>Proof</u>: (i) ⇒ (ii) follows from (17.11). (ii) ⇒ (iii) follows from (17.13). (iii) ⇒ (iv): For any finite separable extension $L|K$ we have $\mathrm{Mon}_{sc(K)}(L|K) = \mathrm{Mon}_{\Omega}(L|K) = \mathrm{Mon}(L|K)$, by (17.2), hence $\# \mathcal{B}(L;A) = \# \mathrm{Mon}(L|K) = [L:K]_{sep} = [L:K]$. (iv) ⇒ (i): Suppose $sc(K) \not\subseteq K^Z$ and let $y \in sc(K) \smallsetminus K^Z$, $L = K(y)$; then $\sigma y \neq y$ for some $\sigma \in \mathrm{Aut}(\Omega|K^Z)$, hence $\sigma|L \neq \iota$, but $\sigma|L \sim_{K^Z} \iota$. Therefore $\# \mathcal{B}(L;A) \leq \# \mathrm{Mon}_{K^Z}(L|K) < \# \mathrm{Mon}(L|K) = [L:K]$, in contradiction to (iv). (iv) ⇔ (v) follows from the inequalities $[L:K] \geq \sum_{B \in \mathcal{B}(L;A)} e_{B|K} \cdot f_{B|K} \geq \# \mathcal{B}(L;A)$ (cf (17.5)) and $e_{B|K} \geq 1$, $f_{B|K} \geq 1$ for all $B \in \mathcal{B}(L;A)$. □

It is obvious that the equalities $e_{B|K} = f_{B|K} = 1$ hold for any finite separable extension $L|K$ and any $B \in \mathcal{B}(L;A)$ whenever the value group Γ of A is a divisible group (i.e., $\Gamma = \Gamma_c$) and its residue field $\mathcal{K}$ is algebraically closed. Actually these properties of Γ and $\mathcal{K}$ are even equivalent to these equalities, as will follow from theorem (27.1) (cf Exercise IV-11).

It is natural and often convenient to choose for $(\bar{K},\bar{A})$ a henselization of (K,A) . In this case we have $(K^Z, C \cap K^Z) = (\bar{K},\bar{A})$, by (17.11), and the mapping $\mathrm{Mon}_{\bar{K}}(L|K) \to \mathcal{B}(L;A)$ in (17.4) is bijective, as follows from (17.13). Moreover, using the notations of (17.4), we prove:

(17.16) <u>Let</u> $(\bar{K},\bar{A})$ <u>be a henselization of</u> (K,A). <u>For any separable extension</u> $L|K$ <u>and any</u> $\lambda \in \mathrm{Mon}(L|K)$, $(\lambda L\cdot\bar{K}, \bar{D} \cap \lambda L\cdot\bar{K})$ <u>is a henselization of</u> $(\lambda L, \lambda B_\lambda)$.

<u>Proof</u>: Since $K \subseteq \lambda L \subseteq sc(K) = sc(\lambda L)$ and $(\lambda L, \lambda B_\lambda) \subseteq (ac(\bar{K}),\bar{D})$,

$(K^Z(C|\lambda L),\ C \cap K^Z(C|\lambda L))$ is a henselization of $(\lambda L, \lambda B_\lambda)$, by (17.11), where $C = \bar{D} \cap sc(K)$. By (15.6 b) we have $K^Z(C|\lambda L) = \lambda L \cdot K^Z = \lambda L \cdot \bar{K}$. □

The following theorem, which is an analogue to theorem (2.12), shows that, for any finite separable extension $L|K$ and any primitive element y of $L|K$, the valuation rings of L which lie over A are in 1-1 correspondence with the irreducible factors in $\bar{K}[X]$ of the minimal polynomial $P_{y|K}$ of y over K . More precisely:

(17.17) THEOREM - Let $(\bar{K},\bar{A})$ be a henselization of (K,A) , $L = K(y)$ a finite separable extension of K , of degree n, and let $P_{y|K} = \bar{P}_1 \cdot \ldots \cdot \bar{P}_r$ be the factorization of $P_{y|K}$ in (necessarily distinct) irreducible monic polynomials in $\bar{K}[X]$. For any $i \in \{1,\ldots,r\}$ choose $\lambda_i \in \mathrm{Mon}(L|K)$ such that $\bar{P}_i(\lambda_i y) = 0$. Then $\mathcal{B}(L;A)$ consists of exactly r valuation rings, namely $B_i = \lambda_i^{-1}\bar{D}$ $(i=1,\ldots,r)$ and, for any $i = 1,\ldots,r$, $(\bar{K}(\lambda_i y), \bar{D} \cap \bar{K}(\lambda_i y))$ is a henselization of $(\lambda_i L, \lambda_i B_i)$. Moreover, $\sum_{i=1}^{r} [\bar{K}(\lambda_i y):\bar{K}] = n$.

Proof: Obviously $\lambda_1,\ldots,\lambda_r$ is a set of representatives of $\mathrm{Mon}_{\bar{K}}(L|K)$, and $\lambda_i L \cdot \bar{K} = \bar{K}(\lambda_i y)$ for $i = 1,\ldots,r$. Therefore the theorem follows from (17.4), (17.13), and (17.16). □

In the case of a rank 1 valuation ring A of K , one can conclude from (2.12) that the completion $(\hat{K},\hat{A})$ of (K,A) is L-allowable for any finite separable extension $L|K$. More generally, we prove by means of Krasner's lemma:

(17.18) THEOREM - Let A be a valuation ring of K of rank 1. Then the completion $(\hat{K},\hat{A})$ of (K,A) is an allowable henselian extension of (K,A) , and $(\hat{K} \cap sc(K),\ \hat{A} \cap sc(K))$ is the unique henselization of (K,A) contained in $(\hat{K},\hat{A})$.

Proof: By (16.7) and (16.6), $(\hat{K},\hat{A})$ is henselian; therefore we may assume $(\bar{K},\bar{A}) = (\hat{K},\hat{A})$. Because of (17.14) and (17.11), it suffices to show that $\bar{K} \cap sc(K) \subseteq K^Z$. Let w be a Krull valuation of $ac(K)$ corresponding to $\bar{D} \cap ac(K)$; then $w|K^Z$ corresponds to the valuation ring $\bar{D} \cap K^Z = C \cap K^Z$, which is henselian by (15.7). Since K is dense in $\bar{K}$, for any $x \in \bar{K} \cap sc(K)$ there is an $a \in K$ such that $w(a-x) > \max\{w(x'-x) \mid x' \in C_x \setminus \{x\}\}$, hence $x \in K^Z(x) \subseteq \subseteq K^Z(a) = K^Z$; by (16.8). □

We showed in (3.12) that, for any valuation ring A of rank 1, the completion of (K,A) is an immediate extension of (K,A) [27]. The same is true for any henselization of any valued field (K,A). In fact, we conclude from (17.11) and (15.8):

(17.19) THEOREM - Any henselization of (K,A) is an immediate extension of (K,A).

§18 The equality $\Sigma\, e_i \cdot f_i = n$

Given a finite extension $L|K$, we have called a valuation ring A of K defectless in L if $\Sigma\, e_{B|K} \cdot f_{B|K} = [L:K]$, where B ranges over the set $\mathcal{B}(L;A)$ of all valuation rings of L which lie over A. Defectlessness is transitive in the following sense:

(18.1) Let L' be a subextension of the finite extension $L|K$. The following conditions are equivalent:

---

[27] This is even true for the completion of any valued field (K,A), where A may have any rank (cf Bourbaki [ ], Chap. 6, §5, Proposition 5).

(i) A <u>is defectless in</u> L .

(ii) A <u>is defectless in</u> L' <u>and any</u> $B' \in \mathcal{B}(L';A)$ <u>is defectless in</u> L .

<u>Proof</u>: We set $\mathcal{B}(L';A) = \{B_1',\ldots,B_r'\}$ and $\mathcal{B}(L;B_i') = \{B_{i,1},\ldots,B_{i,s_i}\}$ for $i = 1,\ldots,r$; then $\mathcal{B}(L;A) = \{B_{i,j} \mid j=1,\ldots,s_i; i=1,\ldots,r\}$. Let $e_i' = e_{B_i'|K}$, $e_{i,j} = e_{B_{i,j}|K}$, $e''_{i,j} = e_{B_{i,j}|L'}$, and let f_i' , $f_{i,j}$, $f''_{i,j}$ have a similar meaning. Obviously $\sum_{j=1}^{s_i} e_{i,j} \cdot f_{i,j} = e_i' \cdot f_i' \cdot \sum_{j=1}^{s_i} e''_{i,j} \cdot f''_{i,j}$ for any $i \in \{1,\ldots,r\}$. By (17.5) we have $[L:K] \geq \sum_{i=1}^{r}\sum_{j=1}^{s_i} e_{i,j} \cdot f_{i,j} = \sum_{i=1}^{r} e_i' \cdot f_i' \cdot \sum_{j=1}^{s_i} e''_{i,j} \cdot f''_{i,j}$, $[L':K] \geq \sum_{i=1}^{r} e_i' \cdot f_i'$, and $[L:L'] \geq \sum_{j=1}^{s_i} e''_{i,j} \cdot f''_{i,j}$ for any $i \in \{1,\ldots,r\}$. Since $[L:K] = [L:L']\cdot[L':K]$, the equality sign holds in the first inequality if and only if it holds in the other inequalities. Therefore (i) $\Leftrightarrow$ (ii). □

The question whether A is defectless in L can be reduced to the question whether, for all $\lambda \in \mathrm{Mon}_{\bar K}(L|K)$, the valuation ring $\bar A$ of a henselization $(\bar K,\bar A)$ of (K,A) is defectless in $\lambda L\cdot\bar K$, i.e., whether $e_{\bar B_\lambda|\bar K} \cdot f_{\bar B_\lambda|\bar K} = [\lambda L\cdot\bar K:\bar K]$, where $\bar B$ is the unique valuation ring of $\lambda L\cdot\bar K$ which lies over $\bar A$. Actually, it suffices to consider only a set of representatives of $\mathrm{Mon}_{\bar K}(L|K)$.

(18.2) THEOREM - <u>Let</u> $(\bar K,\bar A)$ <u>be a henselization of</u> (K,A) , $L|K$ <u>a finite separable extension, and</u> $\lambda_1,\ldots,\lambda_r$ <u>a set of representatives of</u> $\mathrm{Mon}_{\bar K}(L|K)$. <u>The following conditions are equivalent</u>:

(i) A <u>is defectless in</u> L .

(ii) $\bar A$ <u>is defectless in</u> $\lambda_i L\cdot\bar K$ <u>for all</u> $i \in \{1,\ldots,r\}$.

<u>Proof</u>: We use the notations of the proof of (17.5). By (17.19) and (17.17), $(\bar L_i,\bar B_i)$ and $(\bar K,\bar A)$ are immediate extensions of $(\lambda_i L, \lambda_i B_{\lambda_i})$ and (K,A) , respectively; therefore $e_{B_{\lambda_i}|K} = e_{\bar B_i|\bar K}$

and $f_{B_{\lambda_i}|K} = f_{\bar{B}_i|\bar{K}}$ $(i = 1,\ldots,r)$. Since $(\bar{K},\bar{A})$ is an allowable extension of (K,A) by (17.13), we have $\sum_{B\in\mathfrak{B}(L;A)} e_{B|K} \cdot f_{B|K} = \sum_{i=1}^{r} e_{B_{\lambda_i}|K} \cdot f_{B_{\lambda_i}|K}$. Since $L|K$ is separable, we have $[L:K] = \sum_{i=1}^{r} [\bar{L}_i:\bar{K}]$ by (17.3); moreover, $[\bar{L}_i:\bar{K}] \geq e_{\bar{B}_i|K} \cdot f_{\bar{B}_i|\bar{K}}$ $(i=1,\ldots,r)$, by (13.10). Therefore A is defectless in L if and only if $\sum_{i=1}^{r}[\bar{L}_i:\bar{K}] = \sum_{i=1}^{r} e_{\bar{B}_i|\bar{K}} \cdot f_{\bar{B}_i|\bar{K}}$, if and only if, for all $i\in\{1,\ldots,r\}$, $[\bar{L}_i:\bar{K}] = e_{\bar{B}_i|\bar{K}} \cdot f_{\bar{B}_i|\bar{K}}$, i.e., $\bar{A}$ is defectless in $\lambda_i L\cdot\bar{K}$. □

It is easy to generalize this theorem using, instead of a henselization of (K,A), any immediate L-allowable henselian extension $(\bar{K},\bar{A})$ of (K,A) such that $(\lambda L\cdot\bar{K}\,,\ \bar{D}\cap\lambda L\cdot\bar{K})$ is an immediate extension of $(\lambda L\,,\ \lambda B_\lambda)$ for any $\lambda \in \mathrm{Mon}(L|K)$ (cf Exercises III-14 and III-15). For example, the completion of any valued field (K,A) such that $\mathrm{rank}(A) = 1$ has this property, as follows from (2.12) and (3.12). We do not know whether all immediate henselian extensions of any valued field (K,A) have this property with respect to every finite separable extension $L|K$.

We are going to show that the integral closure $D = I_L(A)$ of A in L can be used for characterizing those valuation rings A which are defectless in L. We set $\mathfrak{B} = \mathfrak{B}(L;A)$ and, for any $B \in \mathfrak{B}$, we define the <u>initial index</u> $\epsilon_{B|K}$ as the number of elements of the set $E = \{\delta \in \Delta \mid 0 \leq \delta < \gamma$ for all positive $\gamma \in \Gamma\}$, where Δ is the value group of a Krull valuation w corresponding to B and Γ the value group of $w|K$; obviously $\epsilon_{B|K}$ does not depend on the choice of w.

(18.3) a) <u>For any</u> $B \in \mathfrak{B}$ <u>we have</u> $1 \leq \epsilon_{B|K} \leq e_{B|K}$.

b) <u>If</u> B <u>is discrete then</u> $\epsilon_{B|K} = e_{B|K}$.

c) <u>If</u> $\mathfrak{M}_B$ <u>is not a principal ideal of</u> B <u>then</u> $\epsilon_{B|K} = 1$.

Proof: a) We have $\epsilon_{B|K} \geqq 1$ since $0 \in E$. Let $\delta_1, \delta_2 \in E$, $\delta_1 < \delta_2$. For all positive $\gamma \in \Gamma$ we have $0 < \delta_2 - \delta_1 \leqq \delta_2 < \gamma$, hence $\delta_2 - \delta_1 \notin \Gamma$, $\delta_2 + \Gamma \neq \delta_1 + \Gamma$. Therefore $\epsilon_{B|K} \leqq \# \Delta/\Gamma = e_{B|K}$.

b) If B is discrete, then Δ has a least positive element δ_o, and we have $\Delta = \mathbb{Z}\cdot\delta_o$, $\Gamma = e_{B|K}\cdot\Delta$. Therefore $E = \{0, \delta_o, \ldots, (e_{B|K}-1)\cdot\delta_o\}$, hence $\epsilon_{B|K} = e_{B|K}$.

c) If $\epsilon_{B|K} > 1$ then E has a least positive element δ_o, and δ_o is the least positive element of Δ. Let $t \in B$ such that $wt = \delta_o$; then $\mathfrak{M}_B = t\cdot B$. □

(18.4) a) <u>If</u> A <u>is discrete then</u> $\epsilon_{B|K} = e_{B|K}$ <u>for all</u> $B \in \mathfrak{B}$.

b) <u>If</u> A <u>is of rank</u> 1 <u>and non-discrete then</u> $\epsilon_{B|K} = 1$ <u>for all</u> $B \in \mathfrak{B}$.

Proof: If $\text{rank}(A) = 1$ then $\text{rank}(B) = 1$ for all $B \in \mathfrak{B}$, by (13.14). If A is discrete then so is any $B \in \mathfrak{B}$, by (13.15). If A is of rank 1 and non-discrete then so is any $B \in \mathfrak{B}$, hence $\mathfrak{M}_B$ is not a principal ideal, as follows easily from (4.1). Therefore (18.4) follows from (18.3). □

Given an A-module M and an A-submodule M' of M such that $\mathfrak{M}_A\cdot M \subseteq M'$, the quotient M/M' can be considered, in a natural way, as a vector space over the residue field $\mathcal{K} = A/\mathfrak{M}_A$; its dimension will be denoted by $[M/M':\mathcal{K}]$. In particular, for any $B \in \mathfrak{B}$, $B/\mathfrak{M}_B$ and $B/\mathfrak{M}_A\cdot B$ are vector spaces over $\mathcal{K}$ and so is $D/\mathfrak{M}_A\cdot D$. Obviously $[B/\mathfrak{M}_B : \mathcal{K}] = f_{B|K}$. As to the dimension of the other vector spaces we prove:

(18.5) a) $[B/\mathfrak{M}_A\cdot B : \mathcal{K}] = \epsilon_{B|K}\cdot f_{B|K}$ <u>for any</u> $B \in \mathfrak{B}$.

b) $[D/\mathfrak{M}_A\cdot D : \mathcal{K}] \leqq \sum_{B\in\mathfrak{B}} e_{B|K}\cdot f_{B|K}$.

Proof: a) We set $f = f_{B|K}$ and $e = e_{B|K}$. Let w be a Krull valuation corresponding to B, Δ (resp. Γ) the value group of w (resp. $w|K$), and $E = \{\delta \in \Delta \mid 0 \leqq \delta < \gamma$ for all positive $\gamma \in \Gamma\}$, say $E = \{\delta_1, \ldots, \delta_e\}$, where $0 = \delta_1 < \ldots < \delta_e$. We consider the A-modules $M_i = \{x \in B \mid wx \geqq \delta_i\}$ for $i = 1, \ldots, e$, and $M_{e+1} = \{x \in B \mid wx \geqq \gamma$ for some positive $\gamma \in \Gamma\}$; obviously $B = M_1 \supset M_2 \supset \ldots \supset M_e \supset M_{e+1} = \mathfrak{M}_A \cdot B$. Since $\mathfrak{M}_A \cdot M_i \subseteq \mathfrak{M}_A \cdot B \subseteq M_{i+1}$, M_i/M_{i+1} is a vector space over $\mathcal{K}$. We claim that $[M_i/M_{i+1} : \mathcal{K}] = f$. In fact, let $y \in B$ such that $wy = \delta_i$ and let $z_1, \ldots, z_f$ be a set of representatives of some basis of $B/\mathfrak{M}_B$, considered as a vector space over $\mathcal{K}$. Then $z_1 \cdot y, \ldots, z_f \cdot y \in M_i$, and for any elements $a_1, \ldots, a_f \in A$, not all in $\mathfrak{M}_A$, we have $w(a_1 \cdot z_1 + \ldots + a_f \cdot z_f) = 0$, hence $a_1 \cdot z_1 \cdot y + \ldots + a_f \cdot z_f \cdot y \notin M_{i+1}$. Therefore $[M_i/M_{i+1} : \mathcal{K}] \geqq f$. On the other hand, for any $x_1, \ldots, x_{f+1} \in M_i$ we have $x_1 \cdot y^{-1}, \ldots, x_{f+1} \cdot y^{-1} \in B$ and there exist elements $a_1, \ldots, a_{f+1} \in A$, not all in $\mathfrak{M}_A$, such that $w(a_1 \cdot x_1 \cdot y^{-1} + \ldots + a_{f+1} \cdot x_{f+1} \cdot y^{-1}) > 0$; hence $w(a_1 \cdot x_1 + \ldots + a_{f+1} \cdot x_{f+1}) > wy = \delta_i$, hence $a_1 \cdot x_1 + \ldots + a_{f+1} \cdot x_{f+1} \in M_{i+1}$. Therefore $[M_i/M_{i+1} : \mathcal{K}] \leqq f$. We conclude that
$[B/\mathfrak{M}_A \cdot B : \mathcal{K}] = [M_1/M_{e+1} : \mathcal{K}] = \sum_{i=1}^{e} [M_i/M_{i+1} : \mathcal{K}] = e \cdot f$.

b) The kernel of the ring homomorphism $D \to \bigtimes_{B \in \mathcal{B}} B/\mathfrak{M}_A \cdot B$, defined by $x \mapsto (x + \mathfrak{M}_A \cdot B)_{B \in \mathcal{B}}$, obviously contains $\mathfrak{M}_A \cdot D$. It is even equal to $\mathfrak{M}_A \cdot D$; in fact, since $\mathcal{B}$ is finite, for any $x \in \bigcap_{B \in \mathcal{B}} \mathfrak{M}_A \cdot B$ there exists an element $c \in \mathfrak{M}_A$ such that $\frac{x}{c} \in \bigcap_{B \in \mathcal{B}} B = D$, hence $x \in \mathfrak{M}_A \cdot D$. Therefore the above-mentioned homomorphism induces an injective homomorphism from $D/\mathfrak{M}_A \cdot D$ into $\bigtimes_{B \in \mathcal{B}} B/\mathfrak{M}_A \cdot B$, which is $\mathcal{K}$-linear. So $D/\mathfrak{M}_A \cdot D$ is isomorphic to some subspace of the product space $\bigtimes_{B \in \mathcal{B}} B/\mathfrak{M}_A \cdot B$, which has dimension $\sum_{B \in \mathcal{B}} e_{B|K} \cdot f_{B|K}$, by a). □

Actually, the homomorphism $D/\mathfrak{M}_A \cdot D \to \bigtimes_{B \in \mathcal{B}} B/\mathfrak{M}_A \cdot B$ is even an isomorphism and therefore in (18.5 b) the equality holds (see Bourbaki [5], Chap.6, §8). But this equality will not be used in the proof of the following theorem, except for the implication (iv) ⇒ (iii).

(18.6) THEOREM - The following conditions are equivalent:

(i) D is a finite A-module.

(ii) D is a free A-module.

(iii) $[D/\mathfrak{M}_A \cdot D : \mathcal{K}] = [L:K]$.

(iv) A is defectless in L , and $e_{B|K} = e_{B|K}$ for all $B \in \mathfrak{B}$.

Proof: (i) ⇔ (ii) follows from (13.16).

(ii) ⇒ (iii): By (13.16) there exists a basis $x_1,\dots,x_n$ of the A-module D such that $n = [L:K]$. Let $\bar{x}_i = x_i + \mathfrak{M}_A \cdot D$ $(i=1,\dots,n)$; obviously $\bar{x}_1,\dots,\bar{x}_n$ is a system of generators of $D/\mathfrak{M}_A \cdot D$ considered as a vector space over $\mathcal{K}$. We claim that $\bar{x}_1,\dots,\bar{x}_n$ are linearly independent over $\mathcal{K}$. In fact, for any $a_1,\dots,a_n \in A$ such that $a_1 \cdot x_1 + \dots + a_n \cdot x_n \in \mathfrak{M}_A \cdot D$ we have $c \in \mathfrak{M}_A$ and $b_1,\dots,b_n \in A$ such that $a_1 \cdot x_1 + \dots + a_n \cdot x_n = c \cdot (b_1 \cdot x_1 + \dots + b_n \cdot x_n)$, hence $a_i = c \cdot b_i \in \mathfrak{M}_A$ for $i = 1,\dots,n$. Therefore $[D/\mathfrak{M}_A \cdot D : \mathcal{K}] = n = [L:K]$.

(iii) ⇒ (i): Let $x_1,\dots,x_r \in D$ such that $\bar{x}_1,\dots,\bar{x}_r$ is a basis of the vector space $D/\mathfrak{M}_A \cdot D$. We claim that $x_1,\dots,x_r$ are linearly independent over K . In fact, otherwise there exist $c_1,\dots,c_r \in K$, not all zero, such that $c_1 \cdot x_1 + \dots + c_r \cdot x_r = 0$, and we may assume $c_1 \neq 0$ and $c_i \cdot c_1^{-1} \in A$ for $i = 1,\dots,r$; then $x_1 \in A \cdot x_2 + \dots + A \cdot x_r$, hence $\bar{x}_1 \in \mathcal{K} \cdot \bar{x}_2 + \dots + \mathcal{K} \cdot \bar{x}_r$, contradicting the linear independence of $\bar{x}_1,\dots,\bar{x}_r$. From (iii) follows that $x_1,\dots,x_r$ is even a basis of $L|K$, hence $D \subseteq L = K \cdot x_1 + \dots + K \cdot x_r$. Let $y = a_1 \cdot x_1 + \dots + a_r \cdot x_r \in D$, $y \neq 0$, where $a_1,\dots,a_r \in K$. We may assume that $a_1 \neq 0$ and $a_i \cdot a_1^{-1} \in A$ for $i = 1,\dots,r$. We claim that $a_1,\dots,a_r \in A$. In fact, otherwise we would have $a_1 \notin A$, hence $a_1^{-1} \in \mathfrak{M}_A$, $x_1 + a_2 \cdot a_1^{-1} \cdot x_2 + \dots + a_r \cdot a_1^{-1} \cdot x_r = a_1^{-1} \cdot y \in \mathfrak{M}_A \cdot D$, contradicting the linear independence of $\bar{x}_1,\dots,\bar{x}_r$. Therefore $D =$

$= A \cdot x_1 + \ldots + A \cdot x_r$.

(iii) ⇔ (iv): By (18.3 a), (18.5 b), and (17.5), we have $[D/\mathfrak{M}_A \cdot D : \mathcal{K}] \leq \sum_{B \in \mathfrak{B}} e_{B|K} \cdot f_{B|K} \leq \sum_{B \in \mathfrak{B}} e_{B|K} \cdot f_{B|K} \leq [L:K]$. Obviously, (iii) implies that, in all these inequalities, the equality sign holds; therefore (iv) is true. On the other hand, if (iv) holds, then the equality sign holds in the second and third inequality. Using a remark preceding this theorem, we conclude that (iii) holds. □

For discrete valuation rings, the second statement of condition (iv) can be dropped, because of (18.4 a). Moreover, we conclude from (13.17):

(18.7) COROLLARY - Let A be a discrete valuation ring of K and $L|K$ a finite separable extension. Then the equivalent conditions of (18.6) hold.

This corollary can also be obtained from (18.2), (18.4 a), and the following theorem.

Note that, in corollary (18.7), the assumption that $L|K$ be separable cannot be dropped, in general. In fact, there is an example, due to F.K. Schmidt, of a discrete valuation ring A of a field K of characteristic $p \neq 0$ and a purely inseparable extension $L|K$ of degree p such that $e_{B|K} = f_{B|K} = 1$ for the unique valuation ring B of L which lies over A (cf Ribenboim [30], G, Example 1). However, in the case in which (K,A) is complete, we have the following theorem.

(18.8) THEOREM - Let (K,A) be complete, where A is a discrete valuation ring of K . Then, for any finite extension $L|K$, $D = I_L(A)$ is a discrete valuation ring of L , (L,D) is complete, and the equivalent conditions of (18.6) hold. In particular, we have $[D/\mathfrak{M}_A \cdot D : \mathcal{K}] = [L:K] = e_{D|K} \cdot f_{D|K}$.

Proof: By (2.7) and (13.5), D is the unique valuation ring of L which lies over A and (L,D) is complete. Moreover, D is discrete, by (13.15). Let $e = e_{D|K}$, $f = f_{D|K}$, and let w (resp. v) be the normalized exponential valuation corresponding to D (resp. A); then $w|K = e\cdot v$. Let $z \in D$ such that $wz = 1$ and let $y_1,\dots,y_f \in D$ be a set of representatives of a basis of $D/\mathfrak{M}_D$, considered as a vector space over $\mathcal{K}$. Obviously, it suffices to prove the containment $D \subseteq \sum_{j=1}^{f}\sum_{i=o}^{e-1} A\cdot z^i\cdot y_j$. Let S be a set of representatives for (K,v) (in the sense of §5); then $U = S\cdot y_1 +\dots+ S\cdot y_f$ is a set of representatives for (L,w). Let $c \in A$ such that $vc = 1$ and set $t_{i+e\cdot h} = z^i\cdot c^h$ $(i=0,\dots,e-1;\ h \in \mathbb{N})$; then $wt_k = k$ for any $k \in \mathbb{N}$. By (5.2), for any $x \in D$ there is a sequence $(u_k)_{k\in\mathbb{N}}$ of elements $u_k \in U$ such that $x = \sum_{k=0}^{\infty} u_k\cdot t_k$. Since $u_k = s_{k,1}\cdot y_1 +\dots+ s_{k,f}\cdot y_f$ for appropriate elements $s_{k,j}\in S$, we get $x = \sum_{h=0}^{\infty}\sum_{i=0}^{e-1} s_{i+e\cdot h,j}\cdot y_j\cdot z^i\cdot c^h = \sum_{i=0}^{e-1}\sum_{j=1}^{f} a_{i,j}\cdot y_j\cdot z^i$, where $a_{i,j} = \sum_{h=0}^{\infty} s_{i+e\cdot h,j}\cdot c^h \in A$ $(i=0,\dots,e-1;j=1,\dots,f)$, by (5.2). □

Note that, in theorem (18.8), the assumption that A be a discrete valuation ring cannot be dropped, in general. In fact, there is an example, due to Ostrowski, of a complete valued field (K,A), where A is a non-discrete valuation ring of rank 1, and a finite separable extension $L|K$ such that A is not defectless in L (cf Ribenboim [30], G, Example 2). Moreover, for any non-discrete valuation ring A of rank 1, $I_L(A)$ is an infinite A-module unless $e_{B|K} = 1$ for any $B \in \mathcal{B}(L;A)$. In fact, we get as an immediate consequence of (18.6) and (18.4 b):

(18.9) <u>For any non-discrete valuation ring</u> A <u>of rank</u> 1, <u>the equivalent conditions of</u> (18.6) <u>hold if and only if</u> A <u>is defectless in</u> L <u>and</u> $e_{B|K} = 1$ <u>for all</u> $B \in \mathcal{B}$.

It is easy to indicate examples of non-discrete valuation rings A of rank 1 such that A is defectless in L, but $I_L(A)$ is not a finite A-module. An interesting example of a valuation ring A of rank 2 with this property is given by Bourbaki [5], Chap. 6, §8, Exercise 4).

In §20, we shall discuss the question under which conditions an arbitrary valuation ring A of K is defectless in a given finite normal extension. In particular, we shall show that any valuation ring whose residue field has characteristic zero is defectless in any finite extension. In §22 we shall generalize defectlessness to infinite algebraic extensions.

§19 Inertia group and inertia field

Let $N|K$ be a normal extension with Galois group G and B a valuation ring of N. In §15 we defined the decomposition field K^Z of B over K as the fixed field of the decomposition group $G^Z = \{\sigma \in G \mid \sigma B = B\}$ and proved that $(K^Z, B \cap K^Z)$ is an immediate extension of $(K_{ins}, B \cap K_{ins})$.

Similarly, in this section we define the inertia field K^T of B over K as the fixed field of the inertia group G^T which, in turn, is defined as the kernel of a natural homomorphism $\Phi: G^Z \to \mathrm{Aut}(\mathfrak{K}|\kappa)$, where $\mathfrak{K}$ is the residue field of a place ζ corresponding to B and κ that of its restriction to K. We shall prove that Φ is surjective and therefore induces an isomorphism $\mathrm{Aut}(K^T|K^Z) \cong \mathrm{Aut}(\mathfrak{K}|\kappa)$. Another main result of this section will be the fact that, passing from K^Z to K^T, the value group remains unaltered whereas the residue field suffers the largest possible separable extension within $\mathfrak{h}$.

Moreover, we shall characterize K^T by a minimal property and indicate a 1-1 correspondence between the subextensions of $K^T|K^Z$ and those of the corresponding residue field extensions.

(19.1) THEOREM - a) <u>For any</u> $\sigma \in G^Z(B|K)$ <u>there is one and only one</u> $\bar{\sigma} \in \mathrm{Aut}(\mathfrak{h}|\mathfrak{K})$ <u>such that</u> $\bar{\sigma}\circ\zeta = \zeta\circ\sigma$ [28].

b) <u>The mapping</u> $\sigma \mapsto \bar{\sigma}$ $(\sigma \in G^Z(B|K))$ <u>is a continuous homomorphism</u> $\Phi_{\zeta;K}: G^Z(B|K) \to \mathrm{Aut}(\mathfrak{h}|\mathfrak{K})$.

c) <u>The kernel of</u> $\Phi_{\zeta;K}$ <u>is the set of all</u> $\sigma \in G$ <u>such that</u> $\sigma x - x \in \mathfrak{M}_B$ <u>for any</u> $x \in B$.

<u>Proof</u>: a) Let $G^Z = G^Z(B|K)$. For any $\sigma \in G^Z$, $\zeta\circ\sigma$ is a place of N, with residue field $\mathfrak{h}$, which corresponds to $\sigma^{-1}B = B$ and therefore is equivalent to ζ. By (8.5), we have $\zeta\circ\sigma = \bar{\sigma}\circ\zeta$ for a unique automorphism $\bar{\sigma}$ of $\mathfrak{h}$. Since $\bar{\sigma}(\zeta a) = \zeta(\sigma a) = \zeta a$ for all $a \in K$, we have $\bar{\sigma} \in \mathrm{Aut}(\mathfrak{h}|\mathfrak{K})$.

b) The mapping $\Phi: G^Z \to \mathrm{Aut}(\mathfrak{h}|\mathfrak{K})$ defined by $\sigma \mapsto \bar{\sigma}$ is a homomorphism since $\Phi(\sigma\circ\rho)\circ\zeta = \zeta\circ(\sigma\circ\rho) = (\Phi\sigma\circ\zeta)\circ\rho = (\Phi\sigma\circ\Phi\rho)\circ\zeta$ for all $\sigma, \rho \in G^Z$. As to the continuity of Φ, it suffices to show that, for any finite normal subextension $\mathfrak{h}'$ of $\mathfrak{h}|\mathfrak{K}$, there is a finite normal subextension N' of $N|K$ such that $\Phi(G^Z \cap \mathrm{Aut}(N|N')) \subseteq \mathrm{Aut}(\mathfrak{h}|\mathfrak{h}')$. Let $y_1,\dots,y_r \in B$ such that $\mathfrak{h}' = \mathfrak{K}(\zeta y_1,\dots,\zeta y_r)$ and let N' be the smallest normal subextension of $N|K$ which contains $K(y_1,\dots,y_r)$. Then $N'|K$ is finite, $\mathfrak{h}'$ is contained in the residue field $\mathfrak{h}''$ of $\zeta|N'$, and for any $\sigma \in G^Z \cap \mathrm{Aut}(N|N')$ we have $\Phi\sigma \in$

---

[28] For simplification, we identify σ (resp. $\bar{\sigma}$) with the place of N (resp. $\mathfrak{h}$) defined by it.

$\mathrm{Aut}(\mathfrak{h}|\mathfrak{h}'') \subseteq \mathrm{Aut}(\mathfrak{h}|\mathfrak{h}')$, since $(\Phi\sigma)(\zeta x) = \zeta(\sigma x) = \zeta x$ for any $x \in$ $\in B \cap N'$.

c) Let $\sigma \in G$ such that $\sigma x - x \in \mathfrak{M}_B$ for all $x \in B$. Then $\sigma B \subseteq$ $\subseteq B$ and even $\sigma B = B$, by (13.4 b), hence $\sigma \in G^Z$. Moreover, for any $x \in B$ we have $\zeta(\sigma x - x) = 0$, hence $(\Phi\sigma)(\zeta x) = \zeta x$; therefore $\Phi\sigma = \iota_{\mathfrak{h}}$. On the other hand, if $\sigma \in G^Z$ such that $\Phi\sigma =$ $= \iota_{\mathfrak{h}}$ then, for any $x \in B$, we have $\zeta(\sigma x) = \zeta x$, hence $\zeta(\sigma x - x) =$ $= 0$, $\sigma x - x \in \mathfrak{M}_B$. □

The kernel of $\Phi_{\zeta;K}$ is called the <u>inertia group</u> of B <u>over</u> K and will be denoted by $G^T(B|K)$; it is obviously independent of the choice of ζ . We shall write G^Z , G^T , Φ instead of $G^Z(B|K)$, $G^T(B|K)$, $\Phi_{\zeta;K}$, if there is no ambiguity. Before proving that Φ is surjective, we state some elementary facts about the inertia group. Since $\mathrm{Aut}(\mathfrak{h}|\mathfrak{K})$ is Hausdorff, we conclude from (19.1) and (15.5 a):

(19.2) COROLLARY - G^T <u>is a closed normal subgoup of</u> G^Z <u>and a closed subgroup of</u> G.

Let K' be a field between K and N . From (19.1) and (15.3) we conclude:

(19.3) $\Phi_{\zeta;K'}$ <u>coincides with the restriction of</u> Φ <u>to</u> $G^Z(B|K')$, <u>and</u> $G^T(B|K') = G^T \cap \mathrm{Aut}(N|K')$.

Conjugate valuation rings have conjugate inertia groups. More precisely:

(19.4) $G^T(\sigma B|K) = \sigma \circ G^T \circ \sigma^{-1}$ <u>for any</u> $\sigma \in G$.

<u>Proof</u>: For any $\sigma \in G$, $\rho \in G^T$, $x \in B$ we have $(\sigma\circ\rho\circ\sigma^{-1})(\sigma x) - \sigma x =$

$\sigma(\rho x - x) \in \sigma\mathfrak{M}_B = \mathfrak{M}_{\sigma B}$, hence $\sigma \circ G^T \circ \sigma^{-1} \subseteq G^T(\sigma B|K)$. On the other hand, $\sigma^{-1} \circ G^T(\sigma B|K) \circ \sigma \subseteq G^T(\sigma^{-1}(\sigma B)|K) = G^T$. □

Let N' be a normal subextension of $N|K$ with Galois group G' and let $\mathfrak{h}'$ be the residue field of $\zeta' = \zeta|N'$. We recall that the canonical homomorphism from G onto G' induces a surjective homomorphism $\Theta^Z: G^Z \to G^Z((B \cap N')|K)$ and prove:

(19.5) a) $\Phi_{\zeta';K} \circ \Theta^Z = \bar{\Theta} \circ \Phi$, where $\bar{\Theta}$ is the canonical homomorphism from $\mathrm{Aut}(\mathfrak{h}|\mathcal{K})$ onto $\mathrm{Aut}(\mathfrak{h}'|\mathcal{K})$.

b) Θ^Z <u>induces a homomorphism</u> $\Theta^T: G^T \to G^T((B \cap N')|K)$ <u>with kernel</u> $G^T(B|N') = G^T \cap \mathrm{Aut}(N|N')$.

<u>Proof</u>: a) Let $\Phi' = \Phi_{\zeta';K}$. For any $\sigma \in G^Z$ and any $x \in B \cap N'$ we have $(\Phi'(\Theta^Z \sigma))(\zeta' x) = (\zeta'(\Theta^Z\sigma))x = \zeta(\sigma x) = (\Phi\sigma)(\zeta x) =$ $= (\bar{\Theta} \circ \Phi'\sigma)(\zeta' x)$, hence $\Phi' \circ \Theta^Z = \bar{\Theta} \circ \Phi$.

b) For any $\sigma \in G^T$ we have $\Phi'(\Theta^Z \sigma) = \bar{\Theta}(\Phi\sigma) = \iota_{\mathfrak{h}'}$, hence $\Theta^Z\sigma \in$ $\in G^T((B \cap N')|K)$. Therefore Θ^Z induces a homomorphism $\Theta^T: G^T \to G^T((B \cap N')|K)$. Since Θ^Z has kernel $G^Z \cap \mathrm{Aut}(N|N')$, by (15.4), Θ^T has kernel $G^T \cap (G^Z \cap \mathrm{Aut}(N|N'))$, which is equal to $G^T(B|N')$, by (19.3). □

We prove now the surjectivity of Φ and will obtain, as a consequence, that of Θ^T .

(19.6) THEOREM - <u>The homomorphism</u> $\Phi: G^Z \to \mathrm{Aut}(\mathfrak{h}|\mathcal{K})$ <u>is surjective and induces a topological isomorphism from</u> G^Z/G^T <u>onto</u> $\mathrm{Aut}(\mathfrak{h}|\mathcal{K})$.

<u>Proof</u>: a) We first assume that $N|K$ is finite. By (15.3) and (15.5b) we have $G^Z(B|K^Z) = G^Z = \mathrm{Aut}(N|K^Z)$, and by (19.3) the homomorphisms Φ and $\Phi_{\zeta;K^Z}$ coincide. Moreover, since the residue

field $\mathcal{K}^Z$ of $\zeta|K^Z$ is purely inseparable over $\mathcal{K}$, by (15.8) and Exercise III-6, we have $\mathrm{Aut}(\mathfrak{h}|\mathcal{K}^Z) = \mathrm{Aut}(\mathfrak{h}|\mathcal{K})$. Therefore we may assume $K^Z = K$, hence $G^Z = G$, and $B = I_N(B \cap K)$ by (15.7) and (13.5). Since $\mathfrak{h}|\mathcal{K}$ is finite by (13.10), there exists a $y \in B$ such that $\mathcal{K}(\zeta y)$ is the separable closure of $\mathcal{K}$ in $\mathfrak{h}$. For any $\bar{\sigma} \in \mathrm{Aut}(\mathfrak{h}|\mathcal{K})$, $\bar{\sigma}(\zeta y)$ is a root of $P_{\zeta y|\mathcal{K}}$ and therefore is equal to $\zeta(\sigma y) = (\Phi\sigma)(\zeta y)$ for some $\sigma \in G$, by (14.5). We conclude that $\bar{\sigma}$ and $\Phi\sigma$ coincide on $\mathcal{K}(\zeta y)$. Since $\mathfrak{h}|\mathcal{K}(\zeta y)$ is purely inseparable, we have $\bar{\sigma} = \Phi\sigma$. Therefore Φ is surjective.

b) For an arbitrary normal extension $N|K$, let $\{N_\alpha \mid \alpha \in I\}$ be the set of all finite normal subextensions of $N|K$. We may assume that I is ordered such that $\alpha \leqq \beta$ if and only if $N_\alpha \subseteq N_\beta$. Let $G_\alpha = \mathrm{Aut}(N_\alpha|K)$ and $\Theta_{\alpha\beta}$ be the canonical homomorphism from G_β onto G_α, where $\alpha, \beta \in I$, $\alpha \leqq \beta$. Then $(G_\alpha, \Theta_{\alpha\beta})_I$ is an inverse system of finite groups, its inverse limit $\varprojlim (G_\alpha, \Theta_{\alpha\beta})_I$ is the subgroup $\{(\sigma_\alpha)_{\alpha\in I} \mid \Theta_{\alpha\beta}\,\sigma_\beta = \sigma_\alpha$ for all $\alpha, \beta \in I, \alpha \leqq \beta\}$ of the product $\times_{\alpha\in I} G_\alpha$ of discrete topological groups G_α, and the mapping $\sigma \mapsto (\sigma|N_\alpha)_{\alpha\in I}$ $(\sigma \in G)$ is a topological isomorphism Θ from G (endowed with the finite topology) onto $\varprojlim (G_\alpha, \Theta_{\alpha\beta})_I$ [29]. Let $G_\alpha^Z = G^Z((B \cap N_\alpha)|K)$. For any $\alpha, \beta \in I$, $\alpha \leqq \beta$, $\Theta_{\alpha\beta}$ induces a homomorphism $\Theta_{\alpha\beta}^Z$ from G_β^Z onto G_α^Z, by (15.4), and it is easily seen that Θ maps G^Z onto the subgroup $\varprojlim (G_\alpha^Z, \Theta_{\alpha\beta}^Z)_I$ of $\varprojlim (G_\alpha, \Theta_{\alpha\beta})_I$.

Let $\{\mathfrak{h}_\gamma \mid \gamma \in J\}$ be the set of all finite normal subextensions of $\mathfrak{h}|\mathcal{K}$, where J is ordered such that $\gamma \leqq \delta$ if and only if $\mathfrak{h}_\gamma \subseteq \mathfrak{h}_\delta$. Let $\bar{G}_\gamma = \mathrm{Aut}(\mathfrak{h}_\gamma|\mathcal{K})$ and $\bar{\Theta}_{\gamma\delta}$ be the canonical homo-

---

[29] For inverse systems and inverse limits, in particular those of Galois groups, see for example Goldhaber and Ehrlich [11], Ch.5, Sec. 6, Bourbaki [4], Chap.3, §7, or Serre [34], §1. A proof of (19.6) without inverse limits can be found in Endler [6].

morphism from $\bar{G}_\delta$ onto $\bar{G}_\gamma$, where $\gamma, \delta \in J$, $\gamma \leqq \delta$. For any $\alpha \in I$, the residue field of $\zeta_\alpha = \zeta | N_\alpha$ is equal to $\mathfrak{h}_{\bar{\alpha}}$ for a uniquely determined $\bar{\alpha} \in J$, by (13.10) and (14.5). Since $\alpha \mapsto \bar{\alpha}$ is an order-preserving mapping from I onto a cofinal subset of J (i.e. for any $\gamma \in J$ there is an $\alpha \in I$ such that $\gamma \leqq \bar{\alpha}$), the canonical mapping $\bigtimes_{\gamma \in J} \bar{G}_\gamma \to \bigtimes_{\alpha \in I} \bar{G}_{\bar{\alpha}}$ induces a topological isomorphism from $\varprojlim (\bar{G}_\gamma, \bar{\Theta}_{\gamma\delta})_J$ onto $\varprojlim (\bar{G}_{\bar{\alpha}}, \bar{\Theta}_{\bar{\alpha}\bar{\beta}})_I$. We conclude that the mapping $\bar{\sigma} \mapsto (\bar{\sigma}|\mathfrak{h}_{\bar{\alpha}})_{\alpha \in I}$ is a topological isomorphism $\bar{\Theta}$ from $\mathrm{Aut}(\mathfrak{h}|\mathfrak{K})$ onto $\varprojlim (\bar{G}_{\bar{\alpha}}, \bar{\Theta}_{\bar{\alpha}\bar{\beta}})_I$.

By (19.5), the family $(\Phi_\alpha)_{\alpha \in I}$ of homomorphisms $\Phi_\alpha = \Phi_{\zeta_\alpha;K}: G^Z_\alpha \to \bar{G}_{\bar{\alpha}}$ satisfies $\Phi_\alpha \circ \Theta^Z_{\alpha\beta} = \bar{\Theta}_{\bar{\alpha}\bar{\beta}} \circ \Phi_\beta$ for all $\alpha, \beta \in I$, $\alpha \leqq \beta$, and therefore defines a homomorphism $\varprojlim (\Phi_\alpha)_I : \varprojlim (G^Z_\alpha, \Theta^Z_{\alpha\beta})_I \to \varprojlim (\bar{G}_{\bar{\alpha}}, \bar{\Theta}_{\bar{\alpha}\bar{\beta}})_I$ such that $\bar{\Theta} \circ \Phi = \varprojlim (\Phi_\alpha)_I \circ \Theta | G^Z$. From the fact that, for all $\alpha \in I$, G^Z_α and $\bar{G}_{\bar{\alpha}}$ are finite and Φ_α is surjective (as was shown in part (a) of this proof), it follows that $\varprojlim (\Phi_\alpha)_I$ is surjective (see Bourbaki [4], Chap. 3, §7, Prop.5, Cor. 1), hence so is $\Phi = \bar{\Theta}^{-1} \circ \varprojlim (\Phi_\alpha)_I \circ \Theta | G^Z$. The last statement of the theorem follows from the fact that G^Z is compact and $\mathrm{Aut}(\mathfrak{h}|\mathfrak{K})$ is Hausdorff. □

(19.7) COROLLARY - <u>For any normal subextension</u> N' <u>of</u> $N|K$, <u>the homomorphism</u> $\Theta^T: G^T \to G^T((B \cap N')|K)$ <u>is surjective</u>.

<u>Proof</u>: From (19.3), (19.5), (19.6), and (15.4) it follows that the diagram

$$\begin{array}{ccccccc}
 & & & & E & & \\
 & & & & \downarrow & & \\
G^T(B|N') & \xrightarrow{\iota} & G^T & \xrightarrow{\Theta^T} & G^T((B \cap N')|K) & & \\
\downarrow{\scriptstyle \iota} & & \downarrow{\scriptstyle \iota} & & \downarrow{\scriptstyle \iota} & & \\
G^Z(B|N') & \xrightarrow{\iota} & G^Z & \xrightarrow{\Theta^Z} & G^Z((B \cap N')|K) & \longrightarrow & E \\
\downarrow{\scriptstyle \Phi_{\zeta;N'}} & & \downarrow{\scriptstyle \Phi} & & \downarrow{\scriptstyle \Phi_{\zeta';K}} & & \\
\mathrm{Aut}(\mathfrak{n}|\mathfrak{n}') & \xrightarrow{\iota} & \mathrm{Aut}(\mathfrak{n}|\mathfrak{K}) & \xrightarrow{\bar{\Theta}} & \mathrm{Aut}(\mathfrak{n}'|\mathfrak{K}) & & \\
\downarrow & & & & & & \\
E & & & & & &
\end{array}$$

is commutative and has exact rows and columns (where E is the trivial group, $\bar{\Theta}$ as in (19.5), and the arrows $\xrightarrow{\iota}$ denote identical injections). By diagram chasing one proves that Θ^T is surjective (see Exercise III- 17). □

The fixed field $k\, G^T$ of the inertia group $G^T = G^T(B|K)$ is called the <u>inertia field</u> of B <u>over</u> K and will be denoted by $K^T(B|K)$. We shall write K^Z, K^T instead of $K^Z(B|K)$, $K^T(B|K)$ if there is no ambiguity.

(19.8) a) $N|K^T$ <u>is a Galois extension with Galois group</u> $\mathrm{Aut}(N|K^T) = G^T$.

b) $K^T|K^Z$ <u>is a Galois extension, and the canonical homomorphism from</u> $G^Z = \mathrm{Aut}(N|K^Z)$ <u>onto</u> $\mathrm{Aut}(K^T|K^Z)$ <u>induces a topological isomorphism from</u> G^Z/G^T <u>onto</u> $\mathrm{Aut}(K^T|K^Z)$.

<u>Proof</u>: a) Obviously $G^T \subseteq g(k\, G^T) = \mathrm{Aut}(N|K^T)$. Since G^T is closed in G^Z , by (19.2), we have $G^T = g(k\, G^T)$.

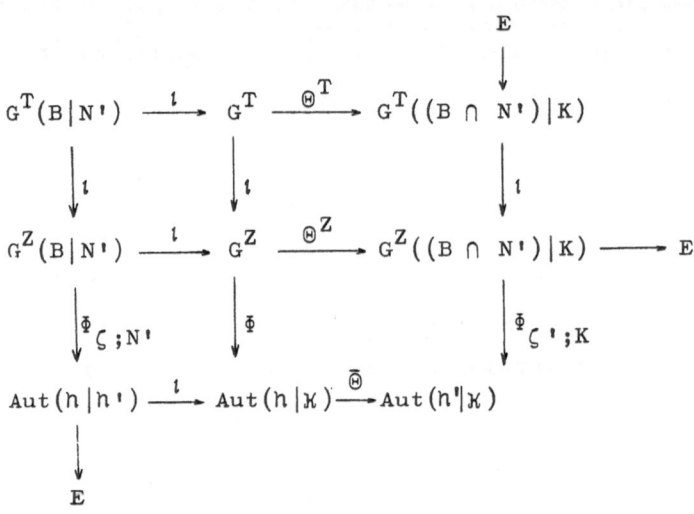

b) By (15.5 b), $N|K^Z$ is a Galois extension with Galois group G^Z and, by (19.2), G^T is a normal subgroup of G^Z ; therefore $K^T|K^Z$ is a Galois extension. The last statement of b) holds by Infinite Galois Theory. □

From (19.6) and (19.8 b) we conclude:

(19.9) COROLLARY - *The topological groups* $\mathrm{Aut}(K^T|K^Z)$ *and* $\mathrm{Aut}(\mathfrak{h}|\mathfrak{K})$ *canonically isomorphic*.

The following statements are immediate consequences of (19.3), (19.4), and (19.7).

(19.10) a) $K^T(\sigma B|K) = \sigma K^T$ *for any* $\sigma \in G$.

b) $K^T(B|K') = K^T \cdot K'$ *for any field* K' *between* K *and* N.

c) $K^T((B \cap N')|K) = K^T \cap N'$ *for any normal subextension* N' *of* $N|K$.

For any subfield L of N, let $\mathfrak{K}_L$ be the residue field of $\zeta|L$ and Γ_L the value group of $w|L$, where w is a Krull valuation of N corresponding to B . In particular, $\mathfrak{K}_N = \mathfrak{h}$ and $\Gamma_N = \Delta$. Instead of $\mathfrak{K}_{K^Z}$, $\mathfrak{K}_{K^T}$, Γ_{K^Z} , Γ_{K^T} we write $\mathfrak{K}^Z$, $\mathfrak{K}^T$, Γ^Z , Γ^T .

The inertia field can be characterized by the following minimal property:

(19.11) THEOREM - *For any field* L *between* K^Z *and* N *we have* $K^T \subseteq L$ *if and only if* $\mathfrak{h}|\mathfrak{K}_L$ *is purely inseparable*.

Proof: If $K^T \subseteq L$ then $G^T(B|L) = G^T \cap \mathrm{Aut}(N|L) = \mathrm{Aut}(N|L) \supseteq G^Z(B|L)$ $\supseteq G^T(B|L)$, by (19.3) and (19.8 a). By (19.6), $\mathrm{Aut}(\mathfrak{h}|\mathfrak{K}_L)$ is the trivial group, hence $\mathfrak{h}|\mathfrak{K}_L$ is purely inseparable. On the other hand, let L be a field between K^Z and N such that $\mathfrak{h}|\mathfrak{K}_L$ is purely inseparable. Then $K^Z(B|L) = K^Z \cdot L = L$ by (15.6 b), hence $\mathrm{Aut}(N|L) = G^Z(B|L)$ by (15.5 b). Since $\mathrm{Aut}(\mathfrak{h}|\mathfrak{K}_L)$ is the trivial group, we have $G^T(B|L) = G^Z(B|L)$ by (19.6). We conclude that $L =$

$= k(\mathrm{Aut}(N|L)) = k(G^T(B|L)) = K^T(B|L) = K^T \cdot L \supseteq K^T$, by (19.10 b). □

Moreover, we prove:

(19.12) THEOREM - $\Gamma^T = \Gamma^Z$, $\mathcal{K}^T$ *is the separable closure of* $\mathcal{K}^Z$ *in* $\mathfrak{h}$, *and* $\mathcal{K}^T|\mathcal{K}^Z$ *is a Galois extension.*

Proof: By (19.8 b), $K^T|K^Z$ is a Galois extension. By (15.6) and (19.10), we have $K^T((B \cap K^T)|K^Z) = K^T(B|K^Z) \cap K^T = K^T \cdot K^Z \cap K^T = K^T$ and $K^Z((B \cap K^T)|K^Z) = K^Z(B|K^Z) \cap K^T = K^Z \cdot K^Z \cap K^T = K^Z$; therefore we may assume $N = K^T$ and $K = K^Z$, hence $\Delta = \Gamma^T$, $\Gamma = \Gamma^Z$, $\mathfrak{h} = \mathcal{K}^T$, $\mathcal{K} = \mathcal{K}^Z$, and $G = \mathrm{Aut}(K^T|K^Z)$ is isomorphic to $\mathrm{Aut}(\mathfrak{h}|\mathcal{K})$, by (19.9). If $[N:K] = n < \infty$ then $n = \# G = \# \mathrm{Aut}(\mathfrak{h}|\mathcal{K}) \leq [\mathfrak{h}:\mathcal{K}] = f_{B|K} \leq e_{B|K} \cdot f_{B|K} \leq n$, by (13.10), hence $f_{B|K} = n$, $e_{B|K} = 1$, and $\mathfrak{h}|\mathcal{K}$ is a Galois extension. If $N|K$ is infinite then $N = \bigcup_{\alpha \in I} N_\alpha$, where $\{N_\alpha \mid \alpha \in I\}$ is the set of all finite normal subextensions of $N|K$, hence $\Delta = \bigcup_{\alpha \in I} \Gamma_{N_\alpha}$ and $\mathfrak{h} = \bigcup_{\alpha \in I} \mathcal{K}_{N_\alpha}$. Since $\Gamma_{N_\alpha} = \Gamma$ for all $\alpha \in I$, we have $\Delta = \Gamma$. Since $\mathcal{K}_{N_\alpha}|\mathcal{K}$ is a Galois extension for all $\alpha \in I$, so is $\mathfrak{h}|\mathcal{K}$. Therefore $\Gamma^T = \Gamma^Z$ and $\mathcal{K}^T|\mathcal{K}^Z$ is a Galois extension. From (19.11) it follows that $\mathcal{K}^T$ is the separable closure of $\mathcal{K}^Z$ in $\mathfrak{h}$. □

Let $\mathfrak{R}^{ZT}$ (resp. $\mathfrak{D}^{ZT}$) be the set of all subextensions of $K^T|K^Z$ (resp. $\mathcal{K}^T|\mathcal{K}^Z$). We prove that there is an inclusion-preserving 1-1 correspondence between $\mathfrak{R}^{ZT}$ and $\mathfrak{D}^{ZT}$ and that the Galois groups of corresponding fields are canonically isomorphic. For any $L \in \mathfrak{R}^{ZT}$ we set $\Phi_L = \Phi_{\zeta|K^T;L}$.

(19.13) THEOREM - a) *For any* $L \in \mathfrak{R}^{ZT}$, Φ_L *is an isomorphism from* $\mathrm{Aut}(K^T|L)$ *onto* $\mathrm{Aut}(\mathcal{K}^T|\mathcal{K}_L)$ *and equals the restriction of* Φ_{K^Z} *to* $\mathrm{Aut}(K^T|L)$.

b) *The mapping* $\mathfrak{R}^{ZT} \to \mathfrak{D}^{ZT}$ *defined by* $L \mapsto \mathcal{K}_L$ *is bijective and*

<u>inclusion-preserving and so is its inverse.</u>

<u>Proof</u>: a) For any $L \in \mathfrak{K}^{ZT}$, $K^T|L$ is a Galois extension and, by (15.4) and (19.7), the canonical homomorphism from $\mathrm{Aut}(N|L)$ onto $\mathrm{Aut}(K^T|L)$ maps $G^Z(B|L)$ onto $G^Z(B^T|L)$ and $G^T(B|L)$ onto $G^T(B^T|L)$, where $B^T = B \cap K^T$. Since $G^Z(B|L) = \mathrm{Aut}(N|L)$, by (15.3) and (15.5 b), and $G^T(B|L) = \mathrm{Aut}(N|K^T)$, by (19.3) and (19.8 a), we conclude that $G^Z(B^T|L) = \mathrm{Aut}(K^T|L)$ and $G^T(B^T|L) = \{\iota_{K^T}\}$; therefore Φ_L is a topological isomorphism from $\mathrm{Aut}(K^T|L)$ onto $\mathrm{Aut}(\mathcal{K}^T|\mathcal{K}_L)$, by (19.6). By (19.3), Φ_L coincides with the restriction of Φ_{K^Z} to $\mathrm{Aut}(K^T|L)$.

b) Let $\mathfrak{G}^{ZT}$ (resp. $\mathfrak{H}^{ZT}$) be the set of all closed subgroups of $\mathrm{Aut}(K^T|K^Z)$ (resp. $\mathrm{Aut}(\mathcal{K}^T|\mathcal{K}^Z)$). By Infinite Galois Theory, the mapping $L \mapsto \mathrm{Aut}(K^T|L)$ (resp. $\mathfrak{L} \mapsto \mathrm{Aut}(\mathcal{K}^T|\mathfrak{L})$) is a bijection from $\mathfrak{K}^{ZT}$ onto $\mathfrak{G}^{ZT}$ (resp. from $\mathfrak{D}^{ZT}$ onto $\mathfrak{H}^{ZT}$), which is inclusion-inverting, as is its inverse. By a), Φ_{K^Z} defines a bijection $\tilde{\Phi}$ from $\mathfrak{G}^{ZT}$ on $\mathfrak{H}^{ZT}$, which is inclusion-preserving, as is its inverse, and for any $L \in \mathfrak{K}^{ZT}$ we have $\tilde{\Phi}\,\mathrm{Aut}(K^T|L) = \mathrm{Aut}(\mathcal{K}^T|\mathcal{K}_L)$. We conclude that $L \mapsto \mathcal{K}_L$ $(L \in \mathfrak{K}^{ZT})$ is a bijective mapping from $\mathfrak{K}^{ZT}$ onto $\mathfrak{D}^{ZT}$, which is inclusion-preserving, as is its inverse. □

It is obvious that $\mathfrak{K}^{ZT}$ and $\mathfrak{D}^{ZT}$ are lattices, the join (resp. meet) of any two elements being their composition (resp. intersection). It follows that the mapping $L \mapsto \mathcal{K}_L$ in (19.13 b) is a lattice isomorphism from $\mathfrak{K}^{ZT}$ onto $\mathfrak{D}^{ZT}$, i.e., for all $L, L' \in \mathfrak{K}^{ZT}$ we have $\mathcal{K}_{L \cap L'} = \mathcal{K}_L \cap \mathcal{K}_{L'}$ and $\mathcal{K}_{L \cdot L'} = \mathcal{K}_L \cdot \mathcal{K}_{L'}$.

(19.14) COROLLARY - <u>Let</u> $L, M \in \mathfrak{K}^{ZT}$ <u>such that</u> $L \subseteq M$. <u>Then</u> $\mathcal{K}_M|\mathcal{K}_L$ <u>is separable</u>, $f_{(B \cap M)|L} = [M:L]$, <u>and</u> $e_{(B \cap M)|L} = 1$.

<u>Proof</u>: By (19.12) we have $\Gamma_M = \Gamma_L$, hence $e_{(B \cap M)|L} = 1$, and $\mathcal{K}_M|\mathcal{K}_L$ is separable. To prove $f_{(B \cap M)|L} = [M:L]$, we first

assume that $M|L$ is finite. Let N' be a finite Galois subextension of $K^T|L$ which contains M. Since $K^T((B \cap N')|L) = K^T(B|L) \cap N' = K^T \cdot L \cap N' = N'$ and $K^Z((B \cap N')|L) = K^Z(B|L) \cap N' = K^Z \cdot L \cap N' = L$, by (19.10) and (15.6), we may assume $N = N' = K^T(B|L)$ and $L = K^Z(B|L)$. From (19.13 a) we conclude that $[N:L] = [\mathfrak{h}:\mathcal{K}_L]$ and $[N:M] = [\mathfrak{h}:\mathcal{K}_M]$, hence $[M:L] = [\mathcal{K}_M:\mathcal{K}_L] = f_{(B \cap M)|L}$. If $M|L$ is infinite then, for any $n \in \mathbb{N}$, there is a finite subextension M' of $M|L$ such that $[M':L] > n$. Since $[\mathcal{K}_{M'}:\mathcal{K}_L] = [M':L]$ and $\mathcal{K}_{M'} \subseteq \mathcal{K}_M$, the extension $\mathcal{K}_M|\mathcal{K}_L$ is also infinite. □

§20 Ramification group and ramification field

In §19 we defined the inertia group G^T as the kernel of the homomorphism $\Phi: G^Z \to \mathrm{Aut}(\mathfrak{h}|\mathcal{K})$ and the inertia field K^T as the fixed field of G^T. Similarly we shall define the ramification group G^V as the kernel of a natural homomorphism Ψ from G^T into a "$\bar{p}$-character group" of the torsion group Δ/Γ, where $\bar{p}$ is the characteristic exponent of the residue fields, and the ramification field K^V will be defined as the fixed field of G^V.

The homomorphism Ψ has properties similar to those of Φ. In fact, we shall prove that Ψ is surjective and that, passing from K^T to K^V, the residue field remains unaltered, whereas the value group suffers the largest possible $\bar{p}$-free extension within Δ. We shall also characterize K^T by a minimal property and indicate a 1-1 correspondence between the subextensions of $K^V|K^T$ and those of the corresponding value group extensions. Moreover, we shall show that

G^V is the only $\bar{p}$-Sylow group of G^T (in particular, $G^V = \{\iota_N\}$ if $\bar{p} = 1$) and give several sufficient conditions for $A = B \cap K$ to be defectless in N.

We need some preliminaries on character groups. Let X be an additively written abelian torsion group, Ω an algebraically closed field, and p its characteristic exponent ($p = 1$ or a prime number). The set of all homomorphisms χ from X into the multiplicative group Ω^* of Ω, endowed with the multiplication defined by

$$(\chi \cdot \chi')x = \chi x \cdot \chi' x \quad \text{for all} \quad x \in X ,$$

is an abelian group $H_\Omega(X) = \mathrm{Hom}(X,\Omega^*)$ and will be called a <u>p-character group</u> of X; its unit element is the trivial homomorphism 1_X (defined by $1_X x = 1$ for all $x \in X$). Since any $\chi \in H_\Omega(X)$ maps X into the group of all roots of unity of Ω, which depends only on the characteristic exponent p of Ω, any two p-character groups of X are isomorphic.

For any homomorphism $\lambda: X \to X'$ between abelian torsion groups X, X', the mapping $\chi \mapsto \chi \circ \lambda$ $(\chi \in H_\Omega(X'))$ is a homomorphism $H_\Omega(\lambda): H_\Omega(X') \to H_\Omega(X)$. Moreover, it is well known that H_Ω is a left-exact additive contravariant functor; since Ω^* is injective, the functor H_Ω is even exact. In particular, for any subgroup X' of X, the canonical exact sequence $0 \to X' \overset{\iota}{\to} X \overset{\pi}{\to} X/X' \to 0$ gives rise to a surjective homomorphism $H_\Omega(\iota): H_\Omega(X) \to H_\Omega(X')$ and an injective homomorphism $H_\Omega(\pi): H_\Omega(X/X') \to H_\Omega(X)$, and the kernel of $H_\Omega(\iota)$ is equal to the image $H_\Omega(X|X') = \{\chi \in H_\Omega(X) \mid \chi X' = \{1\}\}$ of $H_\Omega(\pi)$. The homomorphism $H_\Omega(\iota)$ will be called the <u>canonical homomorphism</u> from $H_\Omega(X)$ onto $H_\Omega(X')$ and be denoted by $\Xi_{X',X}$. It assigns to each $\chi \in H_\Omega(X)$ the restriction $\chi|X'$ of χ to X'.

Let X be an abelian torsion group and, for any $x \in X$, let n_x be the order of x. If n_x is a p-power for any $x \in X$

then X is called a p-<u>group</u>. If p, n_x are relatively prime for any $x \in X$ then X is called p-<u>free</u>. For an arbitrary abelian torsion group X we have $X = X_p \oplus X^{(p)}$ where $X_p = \{x \in X \mid n_x \text{ a p-power}\}$ (resp. $X^{(p)} = \{x \in X \mid p, n_x \text{ relatively prime}\}$) is the largest subgroup of X which is a p-group (resp. p-free). In particular, $X_p = 0$ and $X^{(p)} = X$ for $p = 1$.

We shall use the following statements:

(20.1) a) $H_\Omega(X) = \{1_X\}$ <u>if an only if</u> X <u>is a</u> p-<u>group</u>.

b) $\Xi_{X^{(p)},X}: H_\Omega(X) \to H_\Omega(X^{(p)})$ <u>is an isomorphism</u>.

c) <u>If</u> X_1, X_2 <u>are subgroups of</u> X <u>such that</u> $H_\Omega(X|X_2) \subseteq$ $\subseteq H_\Omega(X|X_1)$ <u>then</u> $(X_1 \cdot X_2)/X_2$ <u>is a</u> p-<u>group</u>.

<u>Proof</u>: a) Let X be a p-group. For any $x \in X$, 1 is the only n_x-th root of unity in Ω, hence $\chi x = 1$ for all $\chi \in H_\Omega(X)$; therefore $H_\Omega(X) = \{1_X\}$. Conversely, if X is not a p-group then there is a non-zero $x \in X$ such that p, n_x are relatively prime. Let $\eta \in \Omega$ be a primitive n_x-th root of unity in Ω. Since $\Xi_{\mathbb{Z}x,X}$ is surjective, the homomorphism $\chi' \in H_\Omega(\mathbb{Z}x)$, defined by $\chi'(mx) = \eta^m$ for all $m \in \mathbb{N}$, extends to a homomorphism $\chi \in H_\Omega(X)$, hence $H_\Omega(X) \neq \{1_X\}$.

b) The kernel of $\Xi_{X^{(p)},X}$ is isomorphic to $H_\Omega(X/X^{(p)}) \cong$ $\cong H_\Omega(X_p)$, and $H_\Omega(X_p) = \{1_{X_p}\}$, by a).

c) Assuming that $(X_1 \cdot X_2)/X_2$ is not a p-group, we can choose $x \in X_1$ such that the order of $x \cdot X_2$ is divisible by some prime number $q \neq p$. Let X_3 be the subgroup of X generated by $\{x\} \cup X_2$. Since X_3/X_2 is not a p-group, there is a $\chi \in H_\Omega(X_3/X_2)$ such that $\chi(x \cdot X_2) \neq 1$, by a). Since $\Xi_{X_3/X_2,X/X_2}$ is surjective, χ is the restriction of some $\chi' \in H_\Omega(X/X_2)$. Let $\chi'' = \chi' \circ \pi$, where π is the canonical homorphism $X \to X/X_2$. Then $\chi'' \in H_\Omega(X|X_2)$, but $\chi'' \notin H_\Omega(X|X_1)$ since $\chi'' x = \chi(x \cdot X_2) \neq 1$. Therefore $H_\Omega(X|X_2) \not\subseteq H_\Omega(X|X_1)$. □

(20.2) <u>If</u> X <u>is finite then</u> $H_\Omega(X)$ <u>is isomorphic to</u> $X^{(p)}$.

<u>Proof</u>: Since $H_\Omega(X)$ is isomorphic to $H_\Omega(X^{(p)})$, by (20.1 b), we may assume that X is p-free. Since X is the direct product of p-free cyclic subgroups $Z_1,\dots,Z_r$, the p-character group $H_\Omega(X)$ is the direct product of subgroups $H_1,\dots,H_r$ which are isomorphic to $H_\Omega(Z_1),\dots,H_\Omega(Z_r)$, respectively. Therefore it suffices to prove that $H_\Omega(Z) \cong Z$ for any cyclic group Z or order n such that p , n are relatively prime. Let z be a generator of Z and $\eta \in \Omega$ a primitive n-th root of unity. Then the mapping $mz \mapsto \eta^m$ $(m \in \mathbb{N})$ is an element $\chi \in H_\Omega(Z)$ or order n , and $H_\Omega(Z) = \{\chi^k \mid k = 0,\dots,n-1\}$. Therefore $H_\Omega(Z)$ is isomorphic to Z . □

For an arbitrary abelian torsion group X we endow $H_\Omega(X)$ with the "finite topology", i.e. consider it as a subspace of the product $(\Omega^*)^X = \bigtimes_{x\in X} \Omega^*$, where Ω^* is discrete. With this topology, $H_\Omega(X)$ is a compact, totally disconnected topological group, and a fundamental system of neighborhoods of its unit element 1_X is given by $\{H_\Omega(X|X_\gamma) \mid \gamma \in J\}$, where $\{X_\gamma \mid \gamma \in J\}$ is the set of all finite subgroups of X . For any subgroup X' of X , $H_\Omega(X|X')$ is a closed subgroup of $H_\Omega(X)$. If X is finite then the finite topology of $H_\Omega(X)$ is discrete. (These facts are proven similarly as the analogous statements in Infinite Galois Theory; cf Endler [6], vol. I, §14.)

After these preliminary remarks, we consider again a normal extension $N|K$, with Galois group G . For any $\sigma \in G$, the mapping $x \mapsto \frac{\sigma x}{x}$ $(x \in N^*)$ is an endomorphism of N^* and its kernel contains K^* ; therefore it induces a homomorphism from N^*/K^* into N^* which will be denoted by $\partial\sigma$. It is easy to check that

(20.3) $$\partial(\rho \circ \sigma) = \partial\rho \cdot (\rho \circ \partial\sigma) \quad \underline{\text{for any}} \quad \rho, \sigma \in G .$$

Moreover, let B be a valuation ring of N ; then:

(20.4) a) $(\partial\rho)(N^*/K^*) \subseteq U_B$ <u>for any</u> $\rho \in G^Z(B|K)$.

b) $(\partial\rho)((U_B \cdot K^*)/K^*) \subseteq 1 + \mathfrak{M}_B$ <u>for any</u> $\rho \in G^T(B|K)$.

<u>Proof</u>: a) For any $\rho \in G^Z$ and $x \in N^*$ we have $(\partial\rho)(x\cdot K^*) = \frac{\rho x}{x} \in U_B$.

b) For any $\rho \in G^T$ and any $u \in U_B$ we have $(\partial\rho)(u\cdot K^*) = \frac{\rho u}{u} = 1 + u^{-1}(\rho u - u) \in 1 + \mathfrak{M}_B$. □

Let w be a Krull valuation of N corresponding to B and let Δ (resp. Γ) be the value group of w (resp. $w|K$) . Obviously w induces a surjective homomorphism $\overset{*}{w}: N^*/K^* \to \Delta/\Gamma$. Let ζ be a place of N corresponding to B , $\mathfrak{h}$ its residue field, and Ω an algebraic closure of $\mathfrak{h}$. By (8.1 f), ζ induces a homomorphism $\overset{*}{\zeta}: U_B \to \Omega^*$ with image $\mathfrak{h}^*$ and kernel $1 + \mathfrak{M}_B$. Using these homomorphisms, we prove the existence of a natural homomorphism from G^T into the $\bar{p}$-character group $H_\Omega(\Delta/\Gamma)$, where $\bar{p}$ denotes the characteristic exponent of Ω (and of $\mathfrak{h}$) . Note that the characteristic exponent p of N equals $\bar{p}$ or 1 (cf §8).

(20.5) THEOREM - a) <u>For any</u> $\sigma \in G^T(B|K)$ <u>there is one and only one</u> $\tilde{\sigma} \in H_\Omega(\Delta/\Gamma)$ <u>such that</u> $\tilde{\sigma} \circ \overset{*}{w} = \overset{*}{\zeta} \circ \partial\sigma$.

b) <u>The mapping</u> $\sigma \mapsto \tilde{\sigma}$ $(\sigma \in G^T(B|K))$ <u>is a continuous homomorphism</u> $\Psi_{w,\zeta;K}: G^T(B|K) \to H_\Omega(\Delta/\Gamma)$.

c) <u>The kernel of</u> $\Psi_{w,\zeta;K}$ <u>is the set of all</u> $\sigma \in G$ <u>such that</u> $\frac{\sigma x}{x} - 1 \in \mathfrak{M}_B$ <u>for any</u> $x \in N^*$.

<u>Proof</u>: a) For any $\sigma \in G^T = G^T(B|K)$, the kernel of $\overset{*}{\zeta} \circ \partial\sigma$ contains the kernel $\{u\cdot K^* \mid u \in U_B\}$ of $\overset{*}{w}$, by (20.4 b), hence

there is a $\tilde{\sigma} \in H_\Omega(\Delta/\Gamma)$ such that $\tilde{\sigma} \circ \overset{*}{w} = \overset{*}{\zeta} \circ \partial\sigma$. Since $\overset{*}{w}: N^*/K^* \to \Delta/\Gamma$ is surjective, $\tilde{\sigma}$ is uniquely determined.

b) The mapping $\Psi: G^T \to H_\Omega(\Delta/\Gamma)$ defined by $\sigma \mapsto \tilde{\sigma}$ is a homomorphism, since $\Psi(\rho \circ \sigma) \circ \overset{*}{w} = \overset{*}{\zeta} \circ \partial(\rho \circ \sigma) = (\overset{*}{\zeta} \circ \partial\rho) \cdot (\overset{*}{\zeta} \circ \rho \circ \partial\sigma) = (\overset{*}{\zeta} \circ \partial\rho) \cdot (\overset{*}{\zeta} \circ \partial\sigma) = (\Psi\rho \circ \overset{*}{w}) \cdot (\Psi\sigma \circ \overset{*}{w})$ for all $\rho, \sigma \in G^T$, by (20.3) and (19.1). As to the continuity of Ψ, it suffices to show that, for any finite subgroup Δ'/Γ of Δ/Γ, there is a finite normal subextension N' of $N|K$ such that $(\Psi\sigma)(\Delta'/\Gamma) = \{1\}$ for all $\sigma \in G^T \cap \mathrm{Aut}(N|N')$. Let $z_1, \dots, z_s \in N^*$ such that $\Delta'/\Gamma = \{wz_1 + \Gamma, \dots, wz_s + \Gamma\}$ and N' be the smallest normal subextension of $N|K$ which contains $K(z_1, \dots, z_s)$. Then $N'|K$ is finite, Δ' is contained in the value group Δ'' of $w|N'$, and for any $\sigma \in G^T \cap \mathrm{Aut}(N|N')$ we have $(\Psi\sigma)(\Delta'/\Gamma) \subseteq (\Psi\sigma)(\Delta''/\Gamma) = \{1\}$, since $(\Psi\sigma)(wx + \Gamma) = \zeta(\frac{\sigma x}{x}) = \zeta 1 = 1$ for all non-zero $x \in N'$.

c) Let $\sigma \in G$ such that $\frac{\sigma x}{x} - 1 \in \mathfrak{M}_B$ for all $x \in N^*$. Then obviously $\sigma \in G^T$ and $(\Psi\sigma)(wx + \Gamma) = \zeta(\frac{\sigma x}{x}) = \zeta 1 = 1$ for all $x \in N^*$, hence $\Psi\sigma = 1_{\Delta/\Gamma}$. On the other hand, if $\sigma \in G^T$ such that $\Psi\sigma = 1_{\Delta/\Gamma}$ then, for any $x \in N^*$, $\zeta(\frac{\sigma x}{x}) = (\overset{*}{\zeta} \circ \partial\sigma)(x \cdot K^*) = 1_{\Delta/\Gamma}(wx + \Gamma) = 1$, hence $\frac{\sigma x}{x} - 1 \in \mathfrak{M}_B$. □

The kernel of $\Psi_{w,\zeta;K}$ is called the <u>ramification group of</u> B <u>over</u> K and will be denoted by $G^V(B|K)$; it is obviously independent of the choice of w and ζ. We shall write G^V and Ψ instead of $G^V(B|K)$ and $\Psi_{w,\zeta;K}$ is there is no ambiguity. From (20.5), (19.2), and the fact that $H_\Omega(\Delta/\Gamma)$ is Hausdorff, we conclude:

(20.6) COROLLARY - G^V <u>is a closed normal subgroup of</u> G^T <u>and a closed subgroup of</u> G^Z <u>and</u> G.

We shall see in §21 that G^V is also a normal subgroup of G^Z.

Let K' be any field between K and N and let Γ' be the value group of $w|K'$. The canonical surjective homomorphism $\pi: \Delta/\Gamma \to \Delta/\Gamma'$ gives rise to an injective homomorphism $H_\Omega(\pi): H_\Omega(\Delta/\Gamma') \to H_\Omega(\Delta/\Gamma)$, which will be denoted by $I_{K'}$; its image is equal to the kernel $H_\Omega((\Delta/\Gamma)|(\Gamma'/\Gamma))$ of $\Xi_{\Gamma'/\Gamma;\Delta/\Gamma}$. The following statement is analogous to (19.3).

(20.7) <u>The restriction of</u> Ψ <u>to</u> $G^T(B|K')$ <u>is equal to</u> $I_{K'} \circ \Psi_{w,\zeta;K'}$, <u>and</u> $G^V(B|K') = G^V \cap \operatorname{Aut}(N|K')$.

<u>Proof</u>: Let $\Psi' = \Psi_{w,\zeta;K'}$ and $G'^T = G^T(B|K')$. For any $\sigma \in G'^T$ and $x \in N^*$ we have $(\Psi\sigma)(wx + \Gamma) = \zeta(\frac{\sigma x}{x}) = (\Psi'\sigma)(wx + \Gamma') = (\Psi'\sigma \circ \pi)(wx + \Gamma) = (I_{K'}(\Psi'\sigma))(wx + \Gamma)$, hence $\Psi|G'^T = I_{K'} \circ \Psi'$. The kernel $G^V(B|K')$ of Ψ' coincides with the kernel of $I_{K'} \circ \Psi' = \Psi|G'^T$, which is equal to $G^V \cap G'^T = G^V \cap (G^T \cap \operatorname{Aut}(N|K')) = G^V \cap \operatorname{Aut}(N|K')$. □

Conjugate valuation rings have conjugate ramification groups. More precisely:

(20.8) $G^V(\sigma B|K) = \sigma \circ G^V \circ \sigma^{-1}$ <u>for any</u> $\sigma \in G$.

<u>Proof</u>: For any $\sigma \in G$, $\rho \in G^V$, and $x \in N^*$ we have $\frac{(\sigma \circ \rho \circ \sigma^{-1})x}{x} - 1 = \sigma(\frac{\rho y}{y} - 1) \in \sigma\mathfrak{M}_B = \mathfrak{M}_{\sigma B}$, where $y = \sigma^{-1}x$; therefore $\sigma \circ G^V \circ \sigma^{-1} \subseteq G^V(\sigma B|K)$. On the other hand, $\sigma^{-1} \circ G^V(\sigma B|K) \circ \sigma \subseteq G^V(\sigma^{-1}(\sigma B)|K) = G^V$. □

Let N' be a normal subextension of $N|K$ with Galois group G'; let $\zeta' = \zeta|N'$, $w' = w|N'$, and Δ' be the value group of w'. We recall that the canonical homomorphism from G onto G' induces surjective homomorphisms $\Theta^Z: G^Z \to G^Z((B \cap N')|K)$ and $\Theta^T: G^T \to G^T((B \cap N')|K)$. In analogy to (19.5) we prove:

(20.9) a) $\Psi_{w',\zeta';K} \circ \Theta^T = \Xi \circ \Psi$, <u>where</u> Ξ <u>is the canonical homo-</u>

morphism from $H_\Omega(\Delta/\Gamma)$ onto $H_\Omega(\Delta'/\Gamma)$.

b) Θ^T induces a homomorphism $\Theta^V\colon G^V \to G^V((B\cap N')|K)$ with kernel $G^V(B|N') = G^V \cap \mathrm{Aut}(N|N')$.

Proof: a) Let $\Psi' = \Psi_{w',\zeta';K}$. For any $\rho \in G^T$ and any non-zero $x \in N'$ we have $(\Psi'(\Theta^T\rho))(w'x+\Gamma) = \zeta(\frac{\rho x}{x}) =$
$= (\Psi\rho)(w'x+\Gamma) = (\Xi(\Psi\rho))(w'x+\Gamma)$, hence $\Psi'\circ\Theta^T = \Xi\circ\Psi$.

b) For any $\sigma \in G^V$ we have $\Psi'(\Theta^T\sigma) = \Xi(\Psi\sigma) = 1_{\Delta'/\Gamma}$, hence $\Theta^T\sigma \in G^V((B\cap N')|K)$. Therefore Θ^T induces a homomorphism $\Theta^V\colon G^V \to G^V((B\cap N')|K)$. Since Θ^T has kernel $G^T\cap \mathrm{Aut}(N|N')$, by (19.5 b), Θ^V has kernel $G^V\cap(G^T\cap\mathrm{Aut}(N|N'))$, which is equal to $G^V(B|N')$, by (20.7). □

In order to prove the surjectivity of the homomorphisms Ψ and Θ^V, we need some facts on ramification fields. The ramification field of B over K is defined as the fixed field $k\,G^V$ of the ramification group $G^V = G^V(B|K)$ and will be denoted by $K^V(B|K)$ (or by K^V if there is no ambiguity). Similarly as in (19.8), we conclude from Infinite Galois Theory:

(20.10) a) $N|K^V$ is a Galois extension with Galois group $\mathrm{Aut}(N|K^V) = G^V$.

b) $K^V|K^T$ is a Galois extension, and the canonical homomorphism from $G^T = \mathrm{Aut}(N|K^T)$ onto $\mathrm{Aut}(K^V|K^T)$ induces a topological isomorphism from G^T/G^V onto $\mathrm{Aut}(K^V|K^T)$.

We denote by Γ^V the value group of $w|K^V$ and by $\mathcal{K}^V$ the residue field of $\zeta|K^V$. The following results, which are proven here for finite Galois extensions $N|K$ such that $K^T(B|K) = K$, will be generalized later to arbitrary normal extensions.

(20.11) Let $N|K$ be a finite Galois extension such that $K^T(B|K) = K$. Then:

a) G^V <u>is the only $\bar{p}$-Sylow group of</u> G (<u>in particular</u>, $G^V = \{\iota_N\}$ <u>if</u> $\bar{p} = 1$) .

b) $e_{B|K^V}$ <u>is a $\bar{p}$-power</u> (<u>in particular</u>, $e_{B|K^V} = 1$ <u>if</u> $\bar{p} = 1$).

c) $\Gamma^V/\Gamma = (\Delta/\Gamma)^{(\bar{p})}$, $e_{(B\cap K^V)|K} = [K^V:K]$, <u>and</u> $f_{(B\cap K^V)|K} = 1$.

d) <u>The homomorphism</u> $\Psi: G^T \to H_\Omega(\Delta/\Gamma)$ <u>is surjective</u>.

<u>Proof</u>: a) By (19.8 a) we have $G^T = G$. By (20.5), G/G^V is isomorphic to a subgroup of $H_\Omega(\Delta/\Gamma)$ and, by (20.2), $H_\Omega(\Delta/\Gamma)$ is isomorphic to $(\Delta/\Gamma)^{(\bar{p})}$; therefore $\bar{p}$ and $(G:G^V)$ are relatively prime. We claim that G^V is a $\bar{p}$-group of G (in particular, $G^V = \{\iota_N\}$ if $\bar{p} = 1$). Otherwise there exists a $\rho \in G^V$ of prime order $q \neq \bar{p}$. Let H be the subgroup of G generated by ρ and L the fixed field of H ; then $H = \mathrm{Aut}(N|L)$, $[N:L] = q$, and $N = L(y)$ for some $y \in N\setminus L$. We set $c = \sum_{i=0}^{q-1} \rho^i y$ and $z = qy - c$. Obviously $c \in L$, $qy \in N\setminus L$, hence $z \neq 0$ and $\sum_{i=0}^{q-1} \rho^i z = 0$. On the other hand, we have $\zeta(\frac{\rho^i z}{z}) = 1$ for any $i \in \mathbb{N}$, since $\rho \in G^V$; therefore $\zeta(z^{-1} \cdot \sum_{i=0}^{q-1} \rho^i z) = q1 \neq 0$, contradicting $\sum_{i=0}^{q-1} \rho^i z = 0$.

b) By a), the order of $G^V = \mathrm{Aut}(N|K^V)$ equals $\bar{p}^m$ for some $m \in \mathbb{N}$. For any $x \in N^*$ we have $\prod_{\sigma\in G^V} \sigma x \in K^V$ and $w(\sigma x) = wx$ for all $\sigma \in G^V$; therefore $\bar{p}^m \cdot wx = \sum_{\sigma\in G^V} w(\sigma x) = x(\prod_{\sigma\in G^V} \sigma x) \in \Gamma^V$, i.e., the order of $wx + \Gamma^V$ is a $\bar{p}$-power. We conclude that Δ/Γ^V is a $\bar{p}$-group and, since it is finite, its order $e_{B|K^V}$ is a $\bar{p}$-power.

c) Let $N' = K^V(B|K)$, $B' = B \cap N'$, $w' = w|N'$, $\zeta' = \zeta|N'$, and $\Psi' = \Psi_{w',\zeta';K}$. By (20.10 b), $N'|K$ is a finite Galois extension and its Galois group G' is isomorphic to G/G^V . By a), $\bar{p}$ and the order of G' are relatively prime. Since $K^T(B'|K) = K^T \cap N' = K$ by (19.10 c), it follows from a) that $G^V(B'|K)$ is a $\bar{p}$-subgroup of G' ; therefore $G^V(B'|K) = \{\iota_{N'}\}$. By (19.7), $G^T(B'|K)$

is the image of $G = G^T$ under the canonical homomorphism from G onto G', hence $G^T(B'|K) = G'$. We conclude from (20.5) that Ψ' is an injective homomorphism from G' into $H_\Omega(\Gamma^V/\Gamma)$, hence $\# G' \leq \# H_\Omega(\Gamma^V/\Gamma)$. On the other hand, $\# H_\Omega(\Gamma^V/\Gamma) = \#(\Gamma^V/\Gamma)^{(\bar{p})} \leq$ $\leq \# \Gamma^V/\Gamma = e_{B'|K} \leq e_{B'|K} \cdot f_{B'|K} \leq [N':K] = \# G'$, by (20.2) and (13.10); therefore the equalities hold. We conclude that $\Psi': G' \to H_\Omega(\Gamma^V/\Gamma)$ is bijective, $f_{B'|K} = 1$, $e_{B'|K} = [N':K]$, and $\Gamma^V/\Gamma = (\Gamma^V/\Gamma)^{(\bar{p})} \subseteq (\Delta/\Gamma)^{(\bar{p})}$. We claim that the inverse inclusion holds, too. In fact, let $\delta \in \Delta$ such that $\bar{p}$ and the order d of $\delta + \Gamma$ are relatively prime. By b) we have $\bar{p}^m \cdot \delta \in \Gamma^V$ for some $m \in \mathbb{N}$ and, choosing $h, k \in \mathbb{Z}$ such that $1 = h \cdot d + k \cdot \bar{p}^m$, we get $\delta = h(d\delta) + k(\bar{p}^m \cdot \delta) \in \Gamma^V$, hence $\delta + \Gamma \in \Gamma^V/\Gamma$.

d) Since Ψ' and $\Theta = \Theta^T$ are surjective, for any $\chi \in H_\Omega(\Delta/\Gamma)$ there is a $\sigma \in G$ such that $\Psi'(\Theta\sigma) = \Xi\chi$, where Ξ is the canonical homomorphism from $H_\Omega(\Delta/\Gamma)$ onto $H_\Omega(\Gamma^V/\Gamma)$. By (20.9 a) we have $\Xi(\Psi\sigma) = \Xi\chi$ and, since Ξ is injective by c) and (20.1 b), we conclude $\Psi\sigma = \chi$. Therefore Ψ is surjective. □

First we generalize statement d) to arbitrary normal extensions, obtaining the following theorem, which is similar to (19.6).

(20.12) THEOREM - <u>The homomorphism</u> $\Psi: G^T \to H_\Omega(\Delta/\Gamma)$ <u>is surjective and induces a topological isomorphism from</u> G^T/G^V <u>onto</u> $H_\Omega(\Delta/\Gamma)$.

<u>Proof</u>: a) We first assume that $N|K$ is finite. By (19.8 a), $N|K^T$ is a finite Galois extension with Galois group $\mathrm{Aut}(N|K^T) = G^T$, hence $G^T(B|K^T) = G^T$ by (19.3) and $K^T(B|K^T) = K^T$ by (19.10 b). By (20.11 d), the homomorphism $\Psi^T = \Psi_{w,\zeta;K^T}: G^T \to H_\Omega(\Delta/\Gamma^T)$ is surjective and, by (20.7), we have $\Psi = I_{K^T} \circ \Psi^T$; therefore it suffices to show that $I_{K^T}: H_\Omega(\Delta/\Gamma^T) \to H_\Omega(\Delta/\Gamma)$ is surjective. In

fact, by (19.12) and (15.8), Γ^T is equal to the value group $\Gamma^\square$ of $w|K_{ins}$ and, by Exercise III-6, $\Gamma^\square/\Gamma$ is a p-group and therefore also a $\bar{p}$-group. We conclude from (20.1 a) that $H_\Omega(\Gamma^T/\Gamma) = \{1_{\Gamma^T/\Gamma}\}$; therefore the image $H_\Omega((\Delta/\Gamma)|(\Gamma^T/\Gamma))$ of I_{K^T} is equal to $H_\Omega(\Delta/\Gamma)$.

b) Let $N|K$ be an arbitrary normal extension. We use similar notations as in the proof of (19.6). For any $\alpha, \beta \in I$, $\alpha \leq \beta$, $\Theta_{\alpha\beta}$ induces a homomorphism $\Theta^T_{\alpha\beta}$ from G^T_β onto G^T_α, by (19.7), and it is easily seen that the topological isomorphism $\Theta: G \to \varprojlim (G_\alpha, \Theta_{\alpha\beta})_I$ maps G^T onto the subgroup $\varprojlim (G^T_\alpha, \Theta^T_{\alpha\beta})_I$ of $\varprojlim (G_\alpha, \Theta_{\alpha\beta})_I$.

Let $\{\Delta_\gamma \mid \gamma \in J\}$ be the set of all subgroups Δ_γ of Δ containing Γ and such that Δ_γ/Γ is finite, where J is ordered such that $\gamma \leq \delta$ if and only if $\Delta_\gamma \subseteq \Delta_\delta$. Let $H_\gamma = H_\Omega(\Delta_\gamma/\Gamma)$ and $\Xi_{\gamma\delta}$ be the canonical homomorphism from H_δ onto H_γ, where $\gamma, \delta \in J$, $\gamma \leq \delta$. Then $(H_\gamma, \Xi_{\gamma\delta})_J$ is an inverse system of finite abelian groups, and the mapping $\chi \longmapsto (\chi|(\Delta_\gamma/\Gamma))_{\gamma\in J}$ is a topological isomorphism from $H_\Omega(\Delta/\Gamma)$ (endowed with the finite topology) onto $\varprojlim (H_\gamma, \Xi_{\gamma\delta})_J$. For any $\alpha \in I$, the value group of $w_\alpha = w|N_\alpha$ is equal to $\Delta_{\bar{\alpha}}$ for a uniquely determined $\bar{\alpha} \in J$, by (13.10). Since $\alpha \mapsto \bar{\alpha}$ is an order preserving mapping from I onto a cofinal subset of J, the canonical mapping $\prod_{\gamma\in J} H_\gamma \to \prod_{\alpha\in I} H_{\bar{\alpha}}$ induces a topological isomorphism from $\varprojlim (H_\gamma, \Xi_{\gamma\delta})_J$ onto $\varprojlim (H_{\bar{\alpha}}, \Xi_{\bar{\alpha}\bar{\beta}})_I$. We conclude that the mapping $\chi \longmapsto (\chi|(\Delta_{\bar{\alpha}}/\Gamma))_{\alpha\in I}$ is a topological isomorphism from $H_\Omega(\Delta/\Gamma)$ onto $\varprojlim (H_{\bar{\alpha}}, \Xi_{\bar{\alpha}\bar{\beta}})_I$.

By (20.7), the family $(\Psi_\alpha)_{\alpha\in I}$ of homomorphisms $\Psi_\alpha = \Psi_{w_\alpha,\zeta_\alpha;K} : G^T_\alpha \to H_{\bar{\alpha}}$ satisfies $\Psi_\alpha \circ \Theta^T_{\alpha\beta} = \Xi_{\bar{\alpha}\bar{\beta}} \circ \Psi_\beta$ for all $\alpha, \beta \in I$, $\alpha \leq \beta$, and therefore defines a homomorphism $\varprojlim (\Psi_\alpha)_I : \varprojlim (G^T_\alpha, \Theta^T_{\alpha\beta})_I \to \varprojlim (H_{\bar{\alpha}}, \Xi_{\bar{\alpha}\bar{\beta}})_I$ such that $\Xi \circ \Psi =$

$\lim\limits_{\leftarrow} (\Psi_\alpha)_I \circ \Theta|G^T$. From the fact that, for all $\alpha \in I$, G^T_α and $H_{\bar{\alpha}}$ are finite and Ψ_α is surjective (as was shown in part (a) of this proof), one concludes that $\lim\limits_{\leftarrow} (\Psi_\alpha)_I$ is surjective (see Bourbaki [4], loc. cit.), hence so is $\Psi = \Xi^{-1} \circ \lim\limits_{\leftarrow} (\Psi_\alpha)_I \circ \Theta|G^T$.

The last statement of the theorem follows from the fact that G^T is compact and $H_\Omega(\Delta/\Gamma)$ is Hausdorff. □

(20.13) COROLLARY - <u>For any normal subextension</u> N' <u>of</u> $N|K$, <u>the homomorphism</u> $\Theta^V : G^V \to G^V((B \cap N')|K)$ <u>is surjective</u>.

<u>Proof</u>: From (20.7), (20.9), (20.12), and (19.5) it follows that the diagram

$$\begin{array}{ccccccc}
 & & & & E & & \\
 & & & & \downarrow & & \\
G^V(B|N') & \xrightarrow{\iota} & G^V & \xrightarrow{\Theta^V} & G^V((B \cap N')|K) & & \\
\downarrow{\scriptstyle\iota} & & \downarrow{\scriptstyle\iota} & & \downarrow{\scriptstyle\iota} & & \\
G^T(B|N') & \xrightarrow{\iota} & G^T & \xrightarrow{\Theta^T} & G^T((B \cap N')|K) & \longrightarrow & E \\
\downarrow{\scriptstyle\Psi_{w,\zeta;N'}} & & \downarrow{\scriptstyle\Psi} & & \downarrow{\scriptstyle\Psi_{w',\zeta';K}} & & \\
H_\Omega(\Delta/\Delta') & \xrightarrow{I_{N'}} & H_\Omega(\Delta/\Gamma) & \xrightarrow{\Xi} & H_\Omega(\Delta'/\Gamma) & & \\
\downarrow & & & & & & \\
E & & & & & &
\end{array}$$

is commutative and has exact rows and columns (where E is the trivial group, Ξ as in (20.9), and the arrows $\xrightarrow{\iota}$ denote identical injections). By diagram chasing one proves that Θ^V is surjective (see Exercise III- 17). □

From (20.12) and (20.10 b) we conclude:

(20.14) COROLLARY - <u>The topological groups</u> $\mathrm{Aut}(K^V|K^T)$ <u>and</u> $H_\Omega(\Delta/\Gamma)$ <u>are canonically isomorphic</u>. <u>In particular</u>,

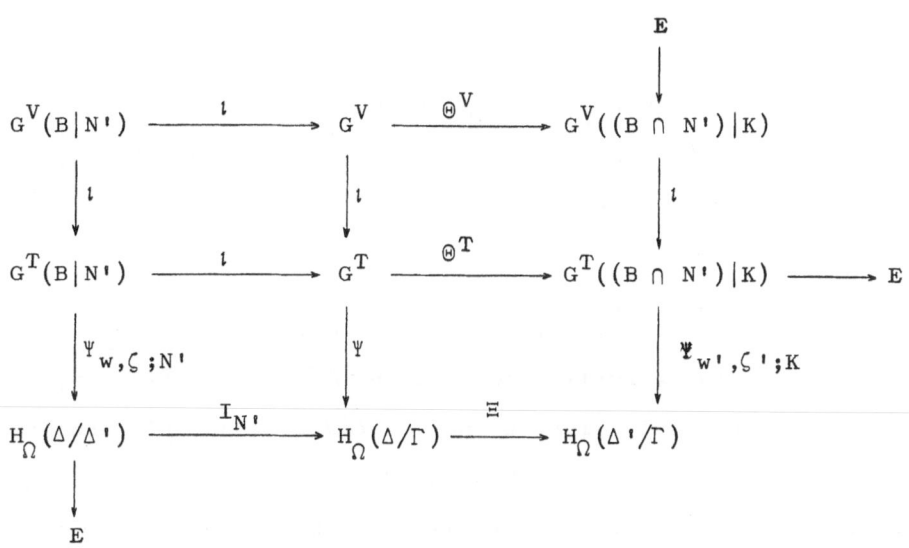

$Aut(K^V|K^T)$ <u>is an abelian group.</u>

The following statements are immediate consequences of (20.7), (20.8), and (20.13).

(20.15) a) $K^V(\sigma B|K) = \sigma K^V$ <u>for any</u> $\sigma \in G$.

b) $K^V(B|K') = K^V \cdot K'$ <u>for any field</u> K' <u>between</u> K <u>and</u> N.

c) $K^V((B \cap N')|K) = K^V \cap N'$ <u>for any normal subextension</u> N' <u>of</u> $N|K$.

The following theorem characterizes the ramification field by a minimal property. It is analogous to (19.11) and generalizes statement (b) of (20.11) to arbitrary normal extensions. We use similar notations as those preceding theorem (19.11).

(20.16) THEOREM - <u>For any field</u> L <u>between</u> K^T <u>and</u> N , <u>we have</u> $K^V \subseteq L$ <u>if and only if</u> Δ/Γ_L <u>is a</u> $\bar{p}$<u>-group</u> (<u>in particular</u>, $\Delta = \Gamma_L$ <u>in the case</u> $\bar{p} = 1$).

<u>Proof</u>: If $K^V \subseteq L$ then $G^V(B|L) = G^V \cap Aut(N|L) = Aut(N|L) \supseteq$ $\supseteq G^T(B|L) \supseteq G^V(B|L)$, by (20.7) and (20.10 a). By (20.12), $H_\Omega(\Delta/\Gamma_L)$ is the trivial group, hence Δ/Γ_L is a $\bar{p}$-group, by (20.1 a). Conversely, let L be a field between K^T and N such that Δ/Γ_L is a $\bar{p}$-group. Then $K^T(B|L) = K^T \cdot L = L$, by (19.10 b), hence $Aut(N|L) = G^T(B|L)$ by (19.8 a). Since $H_\Omega(\Delta/\Gamma_L)$ is the trivial group, by (20.1 a), we have $G^T(B|L) = G^V(B|L)$ by (20.12). We conclude that $L = k(Aut(N|L)) = k(G^V(B|L)) = K^V \cdot L \supseteq K^V$, by (20.15 b). □

Next we generalize statement (c) of (20.11) to arbitrary normal extensions $N|K$. We obtain a theorem which is similar to (19.12).

(20.17) THEOREM - $\Gamma^V/\Gamma^T = (\Delta/\Gamma)^{(\bar{p})}$ <u>and</u> $\mathcal{K}^V = \mathcal{K}^T$.

Proof: Since $K^T(B|K^T) = K^T$ by (19.10 b), we may assume $K = K^T$; then $N|K$ is Galois, $\Gamma = \Gamma^T$, and $\mathcal{K} = \mathcal{K}^T$. Let $\{N_\alpha \mid \alpha \in I\}$ be the set of all finite Galois subextensions of $N|K$; then, for any $\alpha \in I$, we have $K^T((B \cap N_\alpha)|K) = K$, by (19.10 c). We denote by Δ_α (resp. Γ_α^V) the value group of $w|N_\alpha$ (resp. $w|K_\alpha^V$) and by $\mathfrak{n}_\alpha$ (resp. $\mathcal{K}_\alpha^V$) the residue field of $\zeta|N_\alpha$ (resp. $\zeta|K_\alpha^V$), where $K_\alpha^V = K^V((B \cap N_\alpha)|K)$. By (20.11 c) we have $\Gamma_\alpha^V/\Gamma = (\Delta_\alpha/\Gamma)^{(\bar{p})}$ and $\mathcal{K}_\alpha^V = \mathcal{K}$ for all $\alpha \in I$. Since $K^V = \bigcup_{\alpha\in I} K_\alpha^V$ by (20.15 c), we have $\Gamma^V/\Gamma = \bigcup_{\alpha\in I} \Gamma_\alpha^V/\Gamma = \bigcup_{\alpha\in I} (\Delta_\alpha/\Gamma)^{(\bar{p})} = (\bigcup_{\alpha\in I} \Delta_\alpha/\Gamma)^{(\bar{p})} = (\Delta/\Gamma)^{(\bar{p})}$ and $\mathcal{K}^V = \bigcup_{\alpha\in I} \mathcal{K}_\alpha^V = \mathcal{K}$. □

In order to generalize the remaining statement (a) of (20.11) to arbitrary normal extensions $N|K$, we have to extend the notion of a "$\bar{p}$-Sylow group". A subgroup S of the (possibly infinite) Galois group $G = \mathrm{Aut}(N|K)$ will be called a $\bar{p}$-<u>Sylow group</u> of G if, for any $\alpha \in I$, the canonical homomorphism from G onto $G_\alpha = \mathrm{Aut}(N_\alpha|K)$ maps S onto a $\bar{p}$-Sylow group S_α of (the finite group) G_α ; in particular, $S_\alpha = \{\iota_{N_\alpha}\}$ in the case $\bar{p} = 1$. If S is a normal subgroup of G then S is the only $\bar{p}$-Sylow group of G . In fact, for any $\alpha \in I$, S_α is then a normal subgroup of G_α and hence the only $\bar{p}$-Sylow group of G_α , and $\Theta: G \to \varprojlim(G_\alpha, \Theta_{\alpha\beta})_I$ is an isomorphism. In particular, in the case in which $\bar{p} = 1$, $\{\iota_N\}$ is the only $\bar{p}$-Sylow group of G . For more information about generalized Sylow groups see Serre [34], §1.

As a generalization of statement (a) of (20.11), we prove for an arbitrary normal extension $N|K$:

(20.18) THEOREM - G^V <u>is the only</u> $\bar{p}$-<u>Sylow group of</u> G^T (<u>in particular,</u> $G^V = \{\iota_N\}$ <u>in the case</u> $\bar{p} = 1$).

Proof: Since $G^T(B|K^T) = \mathrm{Aut}(N|K^T) = G^T$ and $G^V(B|K^T) = G^V$ by

(19.3), (19.8 a), and (20.7), we may assume $K = K^T$, hence $G = G^T$. For any $\alpha \in I$, $N_\alpha|K$ is a finite Galois extension such that $K^T((B \cap N_\alpha)|K) = K^T \cap N_\alpha = K$, by (19.10 c), and the canonical homomorphism from G onto $G_\alpha = \mathrm{Aut}(N_\alpha|K)$ maps G^V onto $G_\alpha^V = G^V((B \cap N_\alpha)|K)$, by (20.13). By (20.11), G_α^V is a $\bar{p}$-Sylow group of G_α, for any $\alpha \in I$, therefore G^V is a $\bar{p}$-Sylow group of G. Since G^V is a normal subgroup of G^T, by (20.6), G^V is the only p-Sylow group of G. □

In analogy to the considerations at the end of §19, we consider now the set $\mathfrak{K}^{TV}$ of all subextensions of $K^V|K^T$ and the set $\mathfrak{Z}^{TV}$ of all groups between Γ^T and Γ^V. We prove that there is an inclusion-preserving 1-1 correspondence between $\mathfrak{K}^{TV}$ and $\mathfrak{Z}^{TV}$ such that corresponding Galois groups and $\bar{p}$-character groups are canonically isomorphic. For any $L \in \mathfrak{K}^{TV}$ we set $\Psi_L = \Psi_{w|K^V,\zeta|K^V;L}$ and denote by J_L the injective homomorphism $H_\Omega(\pi_L)\colon H_\Omega(\Gamma^V/\Gamma_L) \to H_\Omega(\Gamma^V/\Gamma^T)$, where π_L is the canonical homomorphism from Γ^V/Γ^T onto Γ^V/Γ_L.

(20.19) THEOREM - a) <u>For any</u> $L \in \mathfrak{K}^{TV}$, Ψ_L <u>is an isomorphism from</u> $\mathrm{Aut}(K^V|L)$ <u>onto</u> $H_\Omega(\Gamma^V/\Gamma_L)$, <u>and</u> $J_L \circ \Psi_L$ <u>equals the restriction of</u> Ψ_{K^T} <u>to</u> $\mathrm{Aut}(K^V|L)$.

b) <u>The mapping</u> $\mathfrak{K}^{TV} \to \mathfrak{Z}^{TV}$, <u>defined by</u> $L \mapsto \Gamma_L$, <u>is bijective and inclusion-preserving and so is its inverse</u>.

<u>Proof</u>: a) For any $L \in \mathfrak{K}^{TV}$, $K^V|L$ is a Galois extension and, by (19.7) and (20.13), the canonical homomorphism from $\mathrm{Aut}(N|L)$ onto $\mathrm{Aut}(K^V|L)$ maps $G^T(B|L)$ onto $G^T(B^V|L)$ and $G^V(B|L)$ onto $G^V(B^V|L)$, where $B^V = B \cap K^V$. Since $G^T(B|L) = \mathrm{Aut}(N|L)$ by (19.3) and (19.8 a), and $G^V(B|L) = \mathrm{Aut}(N|K^V)$ by (20.7) and (20.10 a), we conclude that $G^T(B^V|L) = \mathrm{Aut}(K^V|L)$ and $G^V(B^V|L) = \{\iota_{K^V}\}$; therefore Ψ_L is a topological isomorphism from

$\mathrm{Aut}(K^V|L)$ onto $H_\Omega(\Gamma^T/\Gamma_L)$, by (20.12). The last statement follows from (20.7).

b) Let $\mathfrak{G}^{TV}$ (resp. $\mathfrak{H}^{TV}$) be the set of all closed subgroups of $\mathrm{Aut}(K^V|K^T)$ (resp. $H_\Omega(\Gamma^V/\Gamma^T)$) . By a), Ψ_{K^T} defines a bijection $\tilde{\Psi}$ from $\mathfrak{G}^{TV}$ onto $\mathfrak{H}^{TV}$, which is inclusion-preserving, as is its inverse, and which assigns to $\mathrm{Aut}(K^V|L) \in \mathfrak{G}^{TV}$ the element $\Psi_{K^T}(\mathrm{Aut}(K^V|L)) = J_L(\Psi_L(\mathrm{Aut}(K^V|L)) = J_L(H_\Omega(\Gamma^V/\Gamma_L)) =$ $= H_\Omega((\Gamma^V/\Gamma^T) \mid (\Gamma_L/\Gamma^T)) \in \mathfrak{H}^{TV}$, for any $L \in \mathfrak{K}^{TV}$. Therefore the diagram

$$\begin{array}{ccc} \mathfrak{K}^{TV} & \xrightarrow{\ Z\ } & \mathfrak{Z}^{TV} \\ {\scriptstyle \mathcal{G}}\downarrow & & \downarrow{\scriptstyle \mathcal{H}} \\ \mathfrak{G}^{TV} & \xrightarrow[\tilde{\Psi}]{} & \mathfrak{H}^{TV} \end{array}$$

is commutative, where the mappings Z , $\mathcal{G}$, $\mathcal{H}$ are defined by $L \mapsto \Gamma_L$, $L \mapsto \mathrm{Aut}(K^V|L)$, $\Gamma' \mapsto H_\Omega((\Gamma^V/\Gamma^T) \mid (\Gamma'/\Gamma^T))$, respectively. By Infinite Galois Theory, $\mathcal{G}$ is bijective and inclusion-inverting, as is its inverse. The mapping $\mathcal{H}$ is surjective, since $\mathcal{H} \circ Z =$ $= \tilde{\Psi} \circ \mathcal{G}$ is bijective, and is obviously inclusion-inverting. Let Γ_1 , $\Gamma_2 \in \mathfrak{Z}^{TV}$ such that $\mathcal{H}\Gamma_2 \subseteq \mathcal{H}\Gamma_1$; then $((\Gamma_1/\Gamma^T)\cdot(\Gamma_2/\Gamma^T))/(\Gamma_2/\Gamma^T) \cong$ $\cong (\Gamma_1\cdot\Gamma_2)/\Gamma_2$ is a $\bar{p}$-group, by (20.1 c). On the other hand, since Γ^V/Γ^T is $\bar{p}$-free, by (20.17), so is $(\Gamma_1\cdot\Gamma_2)/\Gamma_2$; therefore $\Gamma_1\cdot\Gamma_2 =$ $= \Gamma_2$ and $\Gamma_1 \subseteq \Gamma_2$. We conclude that $\mathcal{H}$ is bijective and its inverse is also inclusion-inverting. Therefore Z is bijective and inclusion-preserving, as is its inverse. □

It is obvious that $\mathfrak{K}^{TV}$ and $\mathfrak{Z}^{TV}$ are lattices, the join (resp. meet) of any two elements being their composition (resp. intersection). It follows that the mapping $L \mapsto \Gamma_L$ in (20.19 b) is a lattice isomorphism from $\mathfrak{K}^{TV}$ onto $\mathfrak{Z}^{TV}$, i.e., for all L , $L' \in$ $\in \mathfrak{K}^{TV}$ we have $\Gamma_{L\cap L'} = \Gamma_L \cap \Gamma_{L'}$ and $\Gamma_{L\cdot L'} = \Gamma_L \cdot \Gamma_{L'}$.

(20.20) COROLLARY - Let $L, M \in \mathfrak{K}^{TV}$ such that $L \subseteq M$. Then Γ_M/Γ_L is $\bar{p}$-free, $e_{(B\cap M)|L} = [M:L]$, and $f_{(B\cap M)|L} = 1$.

Proof: By (20.17) we have $\kappa_M = \kappa_L$, hence $f_{(B\cap M)|L} = 1$, and Γ_M/Γ_L is $\bar{p}$-free. To prove $e_{(B\cap M)|L} = [M:L]$, we first assume that $M|L$ is finite. Let N' be a finite Galois subextension of $K^V|L$ which contains M. Since $K^V((B \cap N')|L) = K^V(B|L) \cap N' = K^V \cdot L \cap N' = N'$ and $K^T((B \cap N')|L) = K^T(B|L) \cap N' = K^T \cdot L \cap N' = L$, by (20.15) and (19.10), we may assume $N = N' = K^V(B|L)$ and $L = K^T(B|L)$. By (13.10) and (20.17), the groups Δ/Γ_L and Δ/Γ_M are finite and $\bar{p}$-free; therefore we conclude from (20.19 a) and (20.2) that $[N:L] = (\Delta:\Gamma_L)$ and $[N:M] = (\Delta:\Gamma_M)$, hence $[M:L] = (\Gamma_M:\Gamma_L) = e_{(B\cap M)|L}$. If $M|L$ is infinite then, for any $n \in \mathbb{N}$, there is a finite subextension M' of $M|L$ such that $[M':L] > n$. Since $(\Gamma_{M'}:\Gamma_L) = [M':L]$ and $\Gamma_{M'} \subseteq \Gamma_{M'}$, the group Γ_M/Γ_L is also infinite. □

The table on the next page gives a survey on the value groups, residue fields, and relative Galois groups of fields between K and N.

Finally, for any finite normal extension $N|K$, we are going to improve the fundamental inequality stated in §17. Let A be any valuation ring of K, $\mathfrak{B}$ the set of all valuation rings of N lying over A, e and f the common ramification index and residue degree, respectively, of all $B \in \mathfrak{B}$ over K, and $\bar{p}$ the characteristic exponent of $A/\mathfrak{M}_A$ (and also of $B/\mathfrak{M}_B$, for all $B \in \mathfrak{B}$).

(20.21) THEOREM - a) There is a $d \in N$ such that $[N:K] = e \cdot f \cdot \#\mathfrak{B} \cdot \bar{p}^d$.

b) The following conditions are equivalent:

TABLE

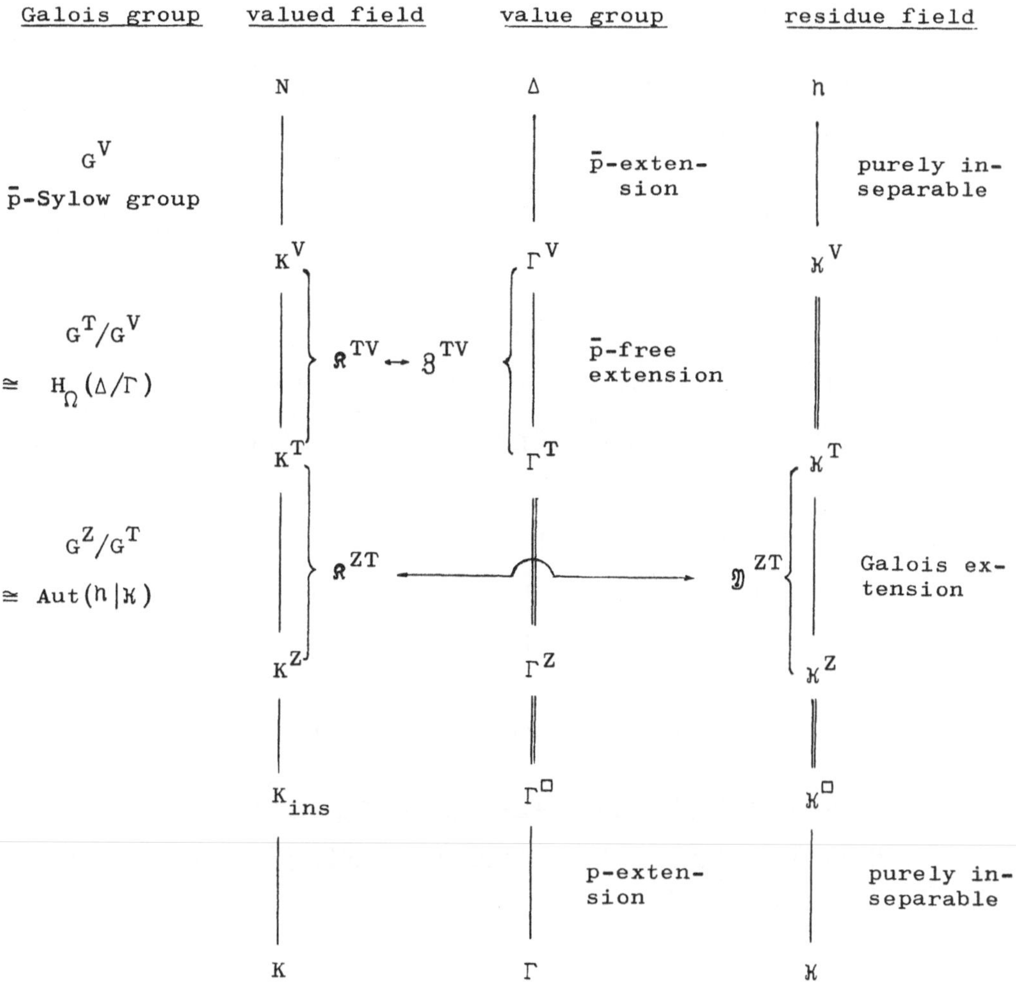

(i) A _is defectless in_ N (i.e., $d = 0$).

(ii) A _is defectless in_ K_{ins} _and_ $B \cap K^V(B|K)$ _is defectless in_ N _for some_ $B \in \mathfrak{B}$.

In this case, (ii) _holds for every_ $B \in \mathfrak{B}$.

Proof: a) Fixing some $B \in \mathfrak{B}$, a Krull valuation w and a place ζ of N corresponding to B , we use the same notation as before. By (15.1 a) and (15.5 b) we have $[K^Z:K_{ins}] = [N:K_{ins}] \cdot [N:K^Z]^{-1} = \# G \cdot (\# G^Z)^{-1} = (G:G^Z) = \# \mathfrak{B}$. By (15.8) we have $\Gamma^Z = \Gamma^{\square}$ and $\kappa^Z = \kappa^{\square}$. By (19.14) and (20.20) we have $[K^T:K^Z] = (\Gamma^T:\Gamma^Z) \cdot [\kappa^T:\kappa^Z]$ and $[K^V:K^T] = (\Gamma^V:\Gamma^T) \cdot [\kappa^V:\kappa^T]$. Since $[N:K^V]$, $(\Delta:\Gamma^V)$, and $[\mathfrak{h}:\kappa^V]$ are $\bar{p}$-powers, by (20.10 a), (20.11), and (19.11), we conclude from (13.10) that $[N:K^V] = (\Delta:\Gamma^V)\cdot[\mathfrak{h}:\kappa^V]\cdot\bar{p}^{d'}$ for some $d' \in \mathbb{N}$. Similarly, by Exercise III- 6 , we have $[K_{ins}:K] = (\Gamma^{\square}:\Gamma) \cdot [\kappa^{\square}:\kappa] \cdot p^{d''}$ for some $d'' \in \mathbb{N}$, where p is the characteristic exponent of K . We may assume that $d'' = 0$ if $p = 1$ and $d' = d'' = 0$ if $\bar{p} = 1$. Therefore, $[N:K] = [N:K^V] \cdot [K^V:K^T] \cdot [K^T:K^Z] \cdot [K^Z:K_{ins}] \cdot [K_{ins}:K] = e\cdot f\cdot \# \mathfrak{B}\cdot\bar{p}^d$, where $d = d' + d''$.

b) Since A is indecomposed in K_{ins} , by (13.8), A is defectless in K_{ins} if and only if $d'' = 0$. Since $B \cap K^V$ is indecomposed in N , by (15.7), $B \cap K^V$ is defectless in N if and only if $d' = 0$. Therefore (ii) holds if and only if $d = 0$, if and only if A is defectless in N . □

As an immediate consequence of Theorem (20.21) we obtain:

(20.22) COROLLARY - <u>Let</u> $N|K$ <u>be a finite Galois extension,</u> A <u>a valuation ring of</u> K , <u>and assume that</u> $K^V(B|K) = N$ <u>for some</u> $B \in \mathfrak{B}$. <u>Then</u> A <u>is defectless in</u> N .

The assumption of Corollary (20.22) holds, in particular, whenever $\bar{p} = 1$. Moreover, in this case we can prove defectlessness even without assuming normality; in fact:

(20.23) COROLLARY - <u>Any valuation ring</u> A <u>of</u> K <u>with</u> Char $A/\mathfrak{M}_A = 0$ <u>is defectless in any finite extension</u> L <u>of</u> K .

<u>Proof</u>: Since $p = \bar{p} = 1$, L is separable over K , hence is contained in some finite Galois extension N of K . Let B be any valuation ring of N which lies over A . We have $G^V(B|K) =$

$= \{\iota_N\}$ by (20.11 a), hence $K^V(B|K) = N$, by (20.10 a). By (20.22) and (18.1), A is defectless in L. □

§21 Higher ramification groups

Let $N|K$ be a normal extension with Galois group G and B a valuation ring of N. We recall that the ramification group G^V of B over K consists of all $\sigma \in G$ such that $\frac{\sigma x}{x} - 1 \in \mathfrak{M}_B$ for all $x \in N^*$. Setting $x^{[\sigma]} = \frac{\sigma x}{x} - 1$ we may write $G^V =$ $= \{\sigma \in G \mid x^{[\sigma]} \in \mathfrak{M}_B$ for all $x \in N^*\}$.

We are going to define a chain of subgroups of G^V, the so-called higher ramification groups, by replacing $\mathfrak{M}_B$ by arbitrary proper ideals of B. We shall use the following statements:

(21.1) a) <u>For any</u> $x, y \in N^*$ <u>and any</u> $\sigma \in G$ <u>we have</u>

$$(x\cdot y)^{[\sigma]} = x^{[\sigma]} + y^{[\sigma]} + x^{[\sigma]} \cdot y^{[\sigma]} .$$

b) <u>For any</u> $x \in N^*$, $\sigma \in G$, <u>and</u> $\tau \in G^V$ <u>there is a</u> $u \in 1 + \mathfrak{M}_B$ <u>such that</u> $x^{[\tau\circ\sigma]} = x^{[\tau]} + x^{[\sigma]} \cdot u$.

<u>Proof</u>: a) Follows from $(x\cdot y)^{[\sigma]} + 1 = \frac{\sigma(x\cdot y)}{x\cdot y} = \frac{\sigma x}{x} \cdot \frac{\sigma y}{y} =$ $= (x^{[\sigma]} + 1)\cdot(y^{[\sigma]} + 1) = x^{[\sigma]} \cdot y^{[\sigma]} + x^{[\sigma]} + y^{[\sigma]} + 1$.

b) If $x^{[\sigma]} = 0$, then $\sigma x = x$, hence $x^{[\tau\circ\sigma]} = x^{[\tau]}$. Assume $x^{[\sigma]} \neq 0$. Then $x^{[\tau\circ\sigma]} + 1 = \frac{\tau(\sigma x)}{x} = \frac{\tau x}{x} \cdot \tau(\frac{\sigma x}{x}) =$ $= (x^{[\tau]} + 1)\cdot(\tau x^{[\sigma]} + 1)$ and $\tau x^{[\sigma]} = x^{[\sigma]} \cdot \frac{\tau(x^{[\sigma]})}{x^{[\sigma]}} =$ $= x^{[\sigma]} \cdot ((x^{[\sigma]})^{[\tau]} + 1)$, hence $x^{[\tau\circ\sigma]} + 1 =$ $= (x^{[\tau]} + 1)\cdot(x^{[\sigma]} \cdot ((x^{[\sigma]})^{[\tau]} + 1) + 1) = x^{[\tau]} + x^{[\sigma]} \cdot u + 1$, where $u = (x^{[\tau]} + 1)\cdot((x^{[\sigma]})^{[\tau]} + 1) \in 1 + \mathfrak{M}_B$. □

Let $\mathfrak{J}$ be the set of all proper ideals of B . For any $\mathfrak{A} \in \mathfrak{J}$ we set $G^V(\mathfrak{A}) = \{\sigma \in G \mid x^{[\sigma]} \in \mathfrak{A}$ for all $x \in N^*\}$. Obviously $G^V(\mathfrak{M}_B) = G^V$ contains $G^V(\mathfrak{A})$ for any $\mathfrak{A} \in \mathfrak{J}$, and $G^V((0))$ is the trivial group $E = \{\iota_N\}$. Moreover, $G^V(\bigcap_{\mathfrak{A}\in\mathcal{J}} \mathfrak{A}) = \bigcap_{\mathfrak{A}\in\mathcal{J}} G^V(\mathfrak{A})$ for any non-empty subset $\mathcal{J}$ of $\mathfrak{J}$.

(21.2) THEOREM - <u>For any</u> $\mathfrak{A} \in \mathfrak{J}$, $G^V(\mathfrak{A})$ <u>is a normal subgroup of</u> $G^Z(B|K)$.

<u>Proof</u>: Let $\rho, \sigma \in G^V(\mathfrak{A})$. Since $\rho^{-1} \in G^V$, for any $x \in N^*$ there exist $u_1, u_2 \in U_B$ such that $0 = x^{[\rho^{-1}\circ\rho]} = x^{[\rho^{-1}]} + x^{[\rho]}.u_1$ and $x^{[\rho^{-1}\circ\sigma]} = x^{[\rho^{-1}]} + x^{[\sigma]}.u_2 = -x^{[\rho]}.u_1 + x^{[\sigma]}.u_2$, by (21.1 b); hence $\rho^{-1}\circ\sigma \in G^V(\mathfrak{A})$. Therefore $G^V(\mathfrak{A})$ is a subgroup of G . For any $\tau \in G^Z$, $\sigma \in G^V(\mathfrak{A})$, and $x \in N^*$ we have $x^{[\tau\circ\sigma\circ\tau^{-1}]} = \tau((\tau^{-1}x)^{[\sigma]}) \in \tau\mathfrak{A} \subseteq \mathfrak{A}$ (cf Exercise III- 8), hence $\tau\circ\sigma\circ\tau^{-1} \in G^V(\mathfrak{A})$. Therefore $G^V(\mathfrak{A})$ is a normal subgroup of G^Z . □

$G^V(\mathfrak{A})$ is called the <u>higher ramification group of</u> B <u>over</u> K <u>corresponding to</u> $\mathfrak{A}$. The set $\mathcal{G}^V = \{G^V(\mathfrak{A}) \mid \mathfrak{A} \in \mathfrak{J}\}$ of all higher ramification groups of B over K has the following properties:

(21.3) a) <u>An inclusion-preserving mapping from</u> $\mathfrak{J}$ <u>onto</u> $\mathcal{G}^V$ <u>is given by</u> $\mathfrak{A} \mapsto G^V(\mathfrak{A})$, <u>and</u> $\mathcal{G}^V$ <u>is totally ordered by inclusion</u>.

b) <u>For any</u> $H \in \mathcal{G}^V$ <u>there is a smallest ideal</u> $\mathfrak{A}_H \in \mathfrak{J}$ <u>such that</u> $H = G^V(\mathfrak{A}_H)$.

c) <u>If</u> $\mathcal{G}^V \setminus \{E\}$ <u>has a minimal element, then</u> $G^V(\mathfrak{A}) = E$ <u>for some non-zero ideal</u> $\mathfrak{A} \in \mathfrak{J}$.

<u>Proof</u>: a) The first statement is obvious. The second one follows from (6.3).

b) Since $H \in \mathcal{G}^V$, the set $\mathcal{J} = \{\mathfrak{B} \in \mathfrak{J} \mid G^V(\mathfrak{B}) = H\}$ is not empty. The ideal $\mathfrak{A}_H = \bigcap_{\mathfrak{B}\in\mathcal{J}} \mathfrak{B}$ has the desired property.

c) Let H be a minimal element of $\mathfrak{G}^V \setminus \{E\}$, and let $\mathfrak{A}_H \in \mathfrak{J}$ as defined in b). Then $\mathfrak{A}_H \neq (0)$, there is a non-zero element $x \in \mathfrak{A}$ such that $x \cdot B \subset \mathfrak{A}_H$, and $G^V(x \cdot B) \subset G^V(\mathfrak{A}_H) = H$. By the minimal property of H we have $G^V(x \cdot B) = E$. □

Note that the hypothesis in c) is satisfied whenever G^V is finite.

The ideal $\mathfrak{A}_H \in \mathfrak{J}$ determined in (21.3 b) is called the ramification ideal of B corresponding to H. It is obvious that an inclusion-preserving 1-1 correspondence between the set $\mathfrak{J}^V$ of all ramification ideals of B and the set $\mathfrak{G}^V$ is given by

$$\mathfrak{A} \mapsto G^V(\mathfrak{A}) \quad (\mathfrak{A} \in \mathfrak{J}^V) \quad \text{and} \quad H \mapsto \mathfrak{A}_H \quad (H \in \mathfrak{G}^V)$$

(cf Exercise III- 20).

In the case of a discrete valuation ring B of N, the proper non-zero ideals of B are in 1-1 correspondence with the positive integers (cf (4.5)) and therefore may be replaced by them. In particular, for any $r \in \mathbb{N} \setminus \{0\}$, $G^V(\mathfrak{M}_B^r)$ is called the r-th ramification group of B over K and is denoted by G_r^V. For any $H \in \mathfrak{G}^V \setminus \{E\}$, the number $r_H = \max\{r \mid H = G_r^V\}$ satisfies $\mathfrak{M}_B^{r_H} = \mathfrak{A}_H$ and is called the ramification number corresponding to H. The ramification numbers, in natural order, form a (finite or infinite) sequence $r_1, r_2, \ldots$; therefore the higher ramification groups can be written in a chain $G^V = G_1^V = \ldots = G_{r_1}^V \supset G_{r_1+1}^V = \ldots = G_{r_2}^V \supset G_{r_2+1}^V \ldots G_{r_i}^V \supset G_{r_{i+1}}^V \ldots$ This chain is stationary if and only if the sequence $r_1, r_2, \ldots$ is finite, if and only if $G_r^V = E$ for some $r > 0$ (cf (21.3 c)).

Considering again an arbitrary valuation ring B of N, we recall that $G^Z(B|K)/G^T(B|K)$ is canonically isomorphic to $\mathrm{Aut}(\mathfrak{n}|\mathfrak{K})$ and that $G^T(B|K)/G^V(B|K)$ is canonically isomorphic to

$H_\Omega(\Delta/\Gamma)$, as was shown in §19 and §20 by means of the surjective homomorphisms Φ and Ψ , respectively. Similarly, we are going to show that, for any proper non-zero principal ideal $\mathfrak{A}$ of B , the factor group $G^V(\mathfrak{A})/G^V(\mathfrak{A}\cdot\mathfrak{M}_B)$ is an abelian group of exponent $\bar{p}$. (Note that if $\bar{p} = 1$ then $G^V = E$, so we may assume $\bar{p} \neq 1$.) For this purpose, we shall indicate a homomorphism from $G^V(\mathfrak{A})$, with kernel $G^V(\mathfrak{A}\cdot\mathfrak{M}_B)$, into the abelian group $\mathrm{Hom}(N^*/K^*, \mathfrak{h}^+)$ of all homomorphisms of the factor group N^*/K^* into the additive group $\mathfrak{h}^+$ of the residue field $\mathfrak{h}$. However, this homomorphism will depend on the choice of a generating element of $\mathfrak{A}$ and will be not surjective, in general. The restriction to principal ideals becomes natural in view of the following proposition (which would fit better in §6):

(21.4) Let $\mathfrak{A}$ be a non-zero ideal of any valuation ring B . The following conditions are equivalent:

(i) $\mathfrak{A}$ is a principal ideal of B .

(ii) $\mathfrak{A}\cdot\mathfrak{M}_B$ is strictly contained in $\mathfrak{A}$.

In this case, $\mathfrak{A} = s\cdot B$ for any $s \in \mathfrak{A}\setminus\mathfrak{A}\cdot\mathfrak{M}_B$, and $\mathfrak{A}\cdot\mathfrak{M}_B = s\cdot\mathfrak{M}_B$ is the largest ideal of B which is strictly contained in $\mathfrak{A}$.

Proof: (i) ⇒ (ii): If $\mathfrak{A} = r\cdot B \neq (0)$ then $\mathfrak{A}\cdot\mathfrak{M}_B = r\cdot\mathfrak{M}_B \neq r\cdot B$.

(ii) ⇒ (i): Let $s \in \mathfrak{A}\setminus\mathfrak{A}\cdot\mathfrak{M}_B$. For any $a \in \mathfrak{A}$ we have $\frac{a}{s} \in B$, since otherwise $\frac{s}{a} \in \mathfrak{M}_B$ and $s = a\cdot\frac{s}{a} \in \mathfrak{A}\cdot\mathfrak{M}_B$, contrary to the hypothesis. Therefore $a = s\cdot\frac{a}{s} \in s\cdot B$ for any $a \in \mathfrak{A}$, hence $\mathfrak{A} = s\cdot B$. Let $\mathfrak{S}$ be the set of all ideals of B which are strictly contained in $\mathfrak{A}$. Obviously $\mathfrak{A}\cdot\mathfrak{M}_B$ is a maximal element of $\mathfrak{S}$ and, since by (6.3) $\mathfrak{S}$ is totally-ordered, it is even the largest element of $\mathfrak{S}$. □

(21.5) THEOREM - Let B be a valuation ring of N and $\mathfrak{A} = s\cdot B$ a proper non-zero principal ideal of B . Then:

a) For any $\sigma \in G^V(\mathfrak{A})$ there is one and only one $\sigma_s \in \operatorname{Hom}(N^*/K^*, \mathfrak{h}^+)$ such that $\sigma_s(x \cdot K^*) = \zeta(\frac{x^{[\sigma]}}{s})$ for all $x \in N^*$.

b) The mapping $\sigma \mapsto \sigma_s$ $(\sigma \in G^V(\mathfrak{A}))$ is a homomorphism $\chi_s: G^V(\mathfrak{A}) \to \operatorname{Hom}(N^*/K^*, \mathfrak{h}^+)$ with kernel $G^V(\mathfrak{A} \cdot \mathfrak{M}_B)$.

Proof: a) Let $\sigma \in G^V(\mathfrak{A})$. For any $x \in N^*$ we have $\frac{(x \cdot y)^{[\sigma]}}{s} = \frac{x^{[\sigma]}}{s} + \frac{y^{[\sigma]}}{s} + \frac{x^{[\sigma]}}{s} \cdot \frac{y^{[\sigma]}}{s} \cdot s$ by (21.1 a), and $\zeta(\frac{(x \cdot y)^{[\sigma]}}{s}) = \zeta(\frac{x^{[\sigma]}}{s}) + \zeta(\frac{y^{[\sigma]}}{s})$, since $\zeta_s = 0$. Moreover, $\zeta(\frac{a^{[\sigma]}}{s}) = 0$ for all $a \in K^*$. Therefore $x \mapsto \zeta(\frac{x^{[\sigma]}}{s})$ $(x \in N^*)$ is a homomorphism from N^* into $\mathfrak{h}^+$ whose kernel contains K^*, hence induces a homomorphism $\sigma_s: N^*/K^* \to \mathfrak{h}^+$. Obviously, σ_s is the only element of $\operatorname{Hom}(N^*/K^*, \mathfrak{h}^+)$ with the desired property.

b) By (21.1 b), for any $\sigma, \tau \in G^V(\mathfrak{A})$ and any $x \in N^*$ there is a $u \in 1 + \mathfrak{M}_B$ such that $\frac{x^{[\tau \circ \sigma]}}{s} = \frac{x^{[\tau]}}{s} + \frac{x^{[\sigma]}}{s} \cdot u$, hence $(\tau \cdot \sigma)_s(x \cdot K^*) = \zeta(\frac{x^{[\tau \circ \sigma]}}{s}) = \zeta(\frac{x^{[\tau]}}{s}) + \zeta(\frac{x^{[\sigma]}}{s}) \cdot \zeta u = \tau_s(x \cdot K^*) + \sigma_s(x \cdot K^*)$; therefore the mapping $\sigma \mapsto \sigma_s$ $(\sigma \in G^V(\mathfrak{A}))$ is a homomorphism χ_s from $G^V(\mathfrak{A})$ into $\operatorname{Hom}(N^*/K^*, \mathfrak{h}^+)$. Finally, we have $\sigma_s = 0$ if and only if $\zeta(\frac{x^{[\sigma]}}{s}) = 0$ for all $x \in N^*$, if and only if $x^{[\sigma]} \in s \cdot \mathfrak{M}_B = \mathfrak{A} \cdot \mathfrak{M}_B$ for all $x \in N^*$, if and only if $\sigma \in G^V(\mathfrak{A} \cdot \mathfrak{M}_B)$. Therefore $G^V(\mathfrak{A} \cdot \mathfrak{M}_B)$ is the kernel of χ_s. □

From Theorem (21.5) and the fact that $\operatorname{Hom}(N^*/K^*, \mathfrak{h}^+)$ is an abelian group of exponent $\bar{p}$, we get as an immediate consequence:

(21.6) COROLLARY - For any proper non-zero principal ideal $\mathfrak{A}$ of B, $G^V(\mathfrak{A})/G^V(\mathfrak{A} \cdot \mathfrak{M}_B)$ is an abelian group of exponent $\bar{p}$. If it is finite, it is a direct sum of a finite number of cyclic groups of order $\bar{p}$.

In the case of a discrete valuation ring B of N , any proper non-zero ideal $\mathfrak{A}$ of B is principal and equals $\mathfrak{M}_B^r$ for some $r \in \mathbb{N} \setminus \{0\}$, hence $G^V(\mathfrak{A}) = G_r^V$ and $G^V(\mathfrak{A} \cdot \mathfrak{M}_B) = G_{r+1}^V$. Therefore in the strictly descending (finite or infinite) chain of higher ramification groups

$$G^V = G_{r_1}^V \supset G_{r_2}^V \supset \ldots \supset G_{r_i}^V \supset G_{r_{i+1}}^V \supset \ldots ,$$

which are all normal in G^V (and even in $G^Z(B|K)$, by (21.2)), the corresponding factor groups $G_{r_i}^V / G_{r_{i+1}}^V$ are abelian groups of exponent $\bar{p}$.

For non-discrete valuation rings B of N , Corollary (21.6) is less useful; in fact, B has non-principal ideals $\mathfrak{C}$, and the corollary does not give any information about the factor groups of $G^V(\mathfrak{C})$.

For a more profound theory of higher ramification groups in the case of a discrete valuation ring see Serre [33], Chap.IV.

§22 Unramified and tamely ramified extensions

In §15, §19, and §20 we characterized decomposition fields, inertia fields, and ramification fields by minimal properties. In this section we are going to characterize them by maximal properties, assuming a condition of defectlessness. For this purpose, we first have to extend the notion "defectless" to infinite field extensions.

Let A be a valuation ring of K and $L|K$ an arbitrary algebraic extension. We say that A is <u>defectless in</u> L if A is defectless (in the usual sense) in any finite subextension of $L|K$. This definition coincides with the usual one in the case of finite extensions, because of (18.1). It is easy to extend (18.1) to infinite extensions $L|K$:

(22.1) Let L' be a finite subextension of the algebraic extension $L|K$. For any valuation ring A of K the following conditions are equivalent:

(i) A is defectless in L.

(ii) A is defectless in L' and any valuation ring of L' which lies over A is defectless in L.

However, we do not know whether (22.1) holds also for infinite subextensions L' of $L|K$.

Let $N|K$ be a Galois extension, B a valuation ring of N, and $A = B \cap K$. As in previous sections we set $K^Z = K^Z(B|K)$, $K^T = K^T(B|K)$, and $K^V = K^V(B|K)$. In (20.22) we proved that A is defectless in N whenever $N|K$ is finite and $K^V = N$. More generally, we prove now for arbitrary Galois extensions $N|K$:

(22.2) For any field K'' between K and K^V such that $K^V|K''$ is Galois, $B \cap K''$ is defectless in K^V.

In particular, for any field K'' between K^Z and K^V, $B \cap K''$ is defectless and indecomposed in K^V.

Proof: Let $K \subseteq K'' \subseteq K^V$ such that $K^V|K''$ is Galois. Then any finite subextension L'' of $K^V|K''$ is contained in some finite Galois subextension N'' of $K^V|K''$. By (20.15) we have $K^V((B \cap N'')|K'') = K^V(B|K'') \cap N'' = K^V{\cdot}K'' \cap N'' = N''$. We conclude from (20.22) that $B \cap K''$ is defectless in N'', hence also defectless in L'', by (18.1). Therefore $B \cap K''$ is defectless in K^V.

The second statement follows from the fact that $K^V|K^Z$ is

Galois, by (21.2), and from (15.7). □

Let (L,B) be an extension of the valued field (K,A), w a Krull valuation and ζ a place of L, both corresponding to B, Δ (resp. Γ) the value group of w (resp. $w|K$), $\mathfrak{h}$ (resp. $\mathfrak{K}$) the residue field of ζ (resp. $\zeta|K$), and $\bar{p}$ the characteristic exponent of $\mathfrak{h}$. We recall that (L,B) is said to be <u>immediate over</u> (K,A) if $\Delta = \Gamma$ and $\mathfrak{h} = \mathfrak{K}$. We say that (L,B) is <u>unramified</u> (resp. <u>tamely ramified</u>) <u>over</u> (K,A) if $\Delta = \Gamma$ (resp. Δ/Γ is $\bar{p}$-free) and $\mathfrak{h}|\mathfrak{K}$ is separable. It is obvious that these definitions do not depend on the choice of w and ζ, that the implications

immediate ⇒ unramified ⇒ tamely ramified

hold, and that the following transitivity is satisfied:

(22.3) <u>For any field</u> L' <u>between</u> K <u>and</u> L <u>the following conditions are equivalent:</u>

(i) (L,B) <u>is immediate</u> (<u>resp. unramified, resp. tamely ramified</u>) <u>over</u> (K,A).

(ii) (L,B) <u>is immediate</u> (<u>resp. unramified, resp. tamely ramified</u>) <u>over</u> (L', B ∩ L') <u>and</u> (L', B ∩ L') <u>is immediate</u> (<u>resp. unramified, resp. tamely ramified</u>) <u>over</u> (K,A).

Moreover, from the fact that L is the union of its finite subextensions, one concludes easily:

(22.4) (L,B) <u>is immediate over</u> (K,A) <u>if and only if</u> (L', B ∩ L') <u>is immediate over</u> (K,A) <u>for any finite subextension</u> L' <u>of</u> L|K.

<u>The same is true for "unramified" and "tamely ramified" instead of "immediate"</u>.

We say that A is <u>immediate</u> (resp. <u>unramified</u>, resp.

tamely ramified) in L if A is indecomposed in L and (L,B) is immediate (resp. unramified, resp. tamely ramified) over (K,A), where $B = I_L(A)$ is the unique valuation ring of L which lies over A. From (13.6), (22.3), and (22.4) we conclude:

(22.5) a) Let $K \subseteq L' \subseteq L$. The valuation ring A is immediate in L if and only if A is immediate in L' and $I_{L'}(A)$ is immediate in L.

b) A is immediate in L if and only if A is immediate in any finite subextension of L.

The same statements hold with "unramified" and "tamely ramified" instead of "immediate".

Note that A is defectless and immediate in L if and only if $L = K$. Assuming that A is defectless in L and $L|K$ is finite, the notions "unramified in L" and "tamely ramified in L" can be described by means of the ramification index and residue degree. In fact, we get as an immediate consequence of the fundamental inequality (17.5):

(22.6) For any finite extension $L|K$ the following conditions are equivalent:

(i) A is defectless and indecomposed in L.

(ii) $e_{B|K} \cdot f_{B|K} = [L:K]$ for some valuation ring B of L lying over A.

In this case, $B = I_L(A)$; moreover, A is unramified (resp. tamely ramified) in L if and only if $e_{B|K} = 1$ (resp. $(\bar{p}, e_{B|K}) = 1$) and $\mathfrak{h}|\mathfrak{K}$ is separable.

Let $N|K$ be again a Galois extension, B a valuation ring of N, and $A = B \cap K$. Assuming defectlessness, we are going to show that K^Z (resp. K^T, resp. K^V) is the largest subextension

L of $N|K$ such that $(L, B \cap L)$ is immediate (resp. unramified, resp. tamely ramified) over (K,A) . More precisely:

(22.7) THEOREM - Let L be a subextension of $N|K$. If

(i) $L \subseteq K^Z$ (resp. $\subseteq K^T$, resp. $\subseteq K^V$),

then

(ii) $(L, B \cap L)$ is immediate (resp. unramified, resp. tamely ramified) over (K,A) .

If $B \cap K^Z$ (resp. $B \cap K^T$, resp. $B \cap K^V$) is defectless in N , then (i) $\Leftrightarrow$ (ii).

Proof: (i) $\Rightarrow$ (ii) follows from (15.8) (resp. (19.12), resp. (20.17)) and (22.3).

(ii) $\Rightarrow$ (i): Assume that $K \subseteq L \subseteq N$ and $(L, B \cap L)$ is immediate over (K,A) , and let $L^Z = K^Z(B|L)$; then $L^Z \supseteq K^Z$ by (15.6 b). By (15.8), $(L^Z, B \cap L^Z)$ is immediate over $(L, B \cap L)$, hence also over (K,A) and over $(K^Z, B \cap K^Z)$. Since $B \cap K^Z$ is indecomposed and defectless in N , we have $[K":K^Z] = e".f"$ for any finite subextension $K"$ of $L^Z|K^Z$, where $e" = e_{(B\cap K")|K^Z}$ and $f" = f_{(B\cap K")|K^Z}$. Since $(L^Z, B \cap L^Z)$ is immediate over $(K^Z, B \cap K^Z)$, we have $e" = f" = 1$, hence $K" = K^Z$; therefore $K^Z = L^Z \supseteq L$.

Similarly, replacing "immediate" by "unramified" (resp. "tamely ramified") and the superscript ^{Z} by ^{T} (resp. ^{V}) , we conclude that $(L^T, B \cap L^T)$ (resp. $(L^V, B \cap L^V)$) is unramified (resp. tamely ramified) over $(K^T, B \cap K^T)$ (resp. $(K^V, B \cap K^V)$) and that it suffices to prove $e" = f" = 1$ for any finite subextension $K"$ of $L^T|K^T$ (resp. $L^V|K^V$).

Since $(L^T, B \cap L^T)$ is unramified over $(K^T, B \cap K^T)$, we have $e" = 1$ and the residue field $\mathcal{K}"$ of $\zeta|K"$ is separable over $\mathcal{K}^T$. On the other hand, $\mathcal{K}"|\mathcal{K}^T$ is purely inseparable, by (19.11), hence $\mathcal{K}" = \mathcal{K}^T$, $f" = 1$.

Since $(L^V, B \cap L^V)$ is tamely ramified over $(K^V, B \cap K^V)$, Γ''/Γ^V is $\bar{p}$-free and $\mathcal{K}''|\mathcal{K}^T$ is separable, where Γ'' is the value group of $w|K''$ and $\mathcal{K}''$ the residue field of $\zeta|K''$. On the other hand, Γ''/Γ^V is a $\bar{p}$-group by (20.16) and $\mathcal{K}''|\mathcal{K}^V$ is purely inseparable by (19.11) ; therefore $\Gamma'' = \Gamma^V$ and $\mathcal{K}'' = \mathcal{K}^V$, hence $e'' = f'' = 1$. □

As a consequence of this theorem and of the characterization of K^Z, K^T, and K^V by minimal properties (cf (15.7), (19.11), and (20.16)) we get:

(22.8) COROLLARY - a) _If_ $B \cap K^Z$ _is defectless in_ N _then_ K^Z _is the only field_ L _between_ K _and_ N _such that_ $\Gamma_L = \Gamma$, $\mathcal{K}_L = \mathcal{K}$, _and_ $B \cap L$ _is indecomposed in_ N.

b) _If_ $B \cap K^T$ _is defectless in_ N _then_ K^T _is the only field_ L _between_ K^Z _and_ N _such that_ $\Gamma_L = \Gamma$ _and_ $\mathcal{K}_L$ _is the separable closure of_ $\mathcal{K}$ _in_ $\mathfrak{n}$.

c) _If_ $B \cap K^V$ _is defectless in_ N _then_ K^V _is the only field_ L _between_ K^T _and_ N _such that_ $\Gamma_L/\Gamma = (\Delta/\Gamma)^{(\bar{p})}$ _and_ $\mathfrak{n}|\mathcal{K}$ _is separable._

Note that $B \cap K^Z$, $B \cap K^T$, and (trivially) $B \cap K^V$ are defectless in N whenever $N = K^V$ and, in particular, whenever $\bar{p} = 1$, as follows from (22.2). Furthermore, by (18.7), $B \cap K^Z$ (resp. $B \cap K^T$, resp. $B \cap K^V$) is defectless in N if it is a discrete valuation ring; this occurs in particular, whenever B is discrete. If A is discrete, we can conclude only that $B \cap K^Z$ and $B \cap K^T$ are discrete, by (15.8) and (19.12). Finally, if A is defectless in N and $K^Z|K$ is finite then $B \cap K^Z$ is defectless in N, by (22.1), and similar statements hold for $B \cap K^T$ and $B \cap K^V$.

Considering only subfields L of N which contain K^Z, we get as an immediate consequence of (22.7):

(22.9) COROLLARY - a) If $B \cap K^T$ is defectless in N, then K^T is the largest field between K^Z and N in which $B \cap K^Z$ is unramified.

b) If $B \cap K^V$ is defectless in N, then K^V is the largest field between K^Z and N in which $B \cap K^Z$ is tamely ramified.

Moreover, with the same hypothesis as in (22.9), we can prove that "unramified" (resp. "tamely ramified") implies "defectless":

(22.10) COROLLARY - Let L be a finite subextension of $N|K^Z$.

a) Assume that $B \cap K^T$ is defectless in N. If $B \cap K^Z$ is unramified in L then $e_{(B\cap L)|K^Z} = 1$ and $f_{(B\cap L)|K^Z} = [L:K^Z]$.

b) Assume that $B \cap K^V$ is defectless in N. If $B \cap K^Z$ is tamely ramified in L then $(\bar{p}, e_{(B\cap L)|K^Z}) = 1$ and $e_{(B\cap L)|K^Z} \cdot f_{(B\cap L)|K^Z} = [L:K^Z]$.

Proof: Since $(K^Z, B \cap K^Z)$ is immediate over (K,A), $(L, B \cap L)$ is unramified (resp. tamely ramified) over (K,A), by (22.3); therefore $L \subseteq K^T$ (resp. $\subseteq K^V$) by (22.7). By (22.2), $B \cap K^Z$ is defectless in K^V, hence defectless in L. Therefore the asserted equalities follow from (22.6) applied on the extension $L|K^Z$. □

CHAPTER IV

Fields with Prescribed Valuations

§23 Introduction and notation

Given an algebraic number field K, it is known that any valuation ring A of K, $A \neq K$, is discrete (by 11.5) and (13.15)) and is defectless in any finite extension L of K (by (18.7)). It is natural to ask whether the equality $\sum_{i=1}^{r} e_{B_i|K} \cdot f_{B_i|K} = [L:K]$ is the only relationship between the number r of valuation rings $B_1,\ldots,B_r$ of L lying over A, the ramification indices $e_{B_1|K},\ldots,e_{B_r|K}$, and the residue degrees $f_{B_1|K},\ldots,f_{B_r|K}$; or, in other words, whether for any prescription of positive integers r, $e_1,\ldots,e_r$, $f_1,\ldots,f_r$ with $\sum_{i=1}^{r} e_i \cdot f_i =$ $= n$ there is a finite extension L of K of degree n such that there exist exactly r valuation rings $B_1,\ldots,B_r$ of L lying over A and $e_{B_i|K} = e_i$, $f_{B_i|K} = f_i$ for $i = 1,\ldots,r$.

This question was answered affirmatively by Hasse [15] in 1925. He proved that even for finitely many discrete valuation rings $A^1,\ldots,A^k$ of K and prescriptions r^j, $e_1^j,\ldots,e_{r^j}^j$, $f_1^j,\ldots,f_{r^j}^j$ such that $\sum_{i=1}^{r^j} e_i^j \cdot f_i^j = n$ $(j = 1,\ldots,k)$ there is a field extension L of K of degree n with the desired properties.

In 1959, Krull generalized Hasse's result to rank one valuation rings $A^1,\ldots,A^k$ of an arbitrary field K, prescribing even the extensions Δ_i^j of the value group Γ^j instead of the indices e_i^j and the extensions $\mathfrak{L}_i^j$ of the residue field $\varkappa^j$ instead of the degress f_i^j (cf Krull [22] and Endler [6], Chapter IV).

We shall obtain, in §27, Krull's results as a consequence of a more general theory, in which completions are prescribed instead of value groups and residue fields, and also archimedean valuations are admitted. This theory will be exposed in §24 through §26; see also Endler [9]. As to generalizations to Krull valuations of higher rank, cf. Ribenboim [28], [29], Endler [7], and Hill [18].

To motivate the prescription of completions, we recall Theorem (2.12), which states that, for any valued field (K, φ) and any finite separable extension L of K, there are only finitely many valuations $\psi_1,\dots,\psi_r$ of L which extend φ and $\sum_{i=1}^{r} [\hat{L}_i:\hat{K}] = [L:K]$, where $(\hat{K},\hat{\varphi})$, $(\hat{L}_1,\hat{\varphi}),\dots,(\hat{L}_r,\hat{\varphi})$ is a completion of (K,φ), $(L,\psi_1),\dots,(L,\psi_r)$, respectively. Now it is natural to ask whether the number r of extensions of φ and their completions $\hat{L}_1,\dots,\hat{L}_r$ can be prescribed arbitrarily. Considering primitive elements of the extensions $L|K$, $\hat{L}_1|\hat{K},\dots,\hat{L}_r|\hat{K}$, it is convenient to make the following definitions (using the same notation as in (2.12)): A φ-<u>prescription of degree</u> n <u>and length</u> r is any r-tuple $\Upsilon = (\hat{x}_1,\dots,\hat{x}_r)$ of elements $\hat{x}_i \in \Omega$ which are pairwise non-conjugate over $\hat{K}$ [30] and satisfy $\sum_{i=1}^{r} [\hat{K}(\hat{x}_i):\hat{K}] = n$. A <u>solution</u> of Υ is any element y (in a field extension of K) which has the following properties:

a) y is separable over K and $[K(y):K] \leq n$.

b) There are at least r valuations $\psi_1,\dots,\psi_r$ of $K(y)$ which extend φ.

c) For any $i \in \{1,\dots,r\}$ there is a completion $\mu_i:(K(y),\psi_i) \to (\hat{K}(\hat{x}_i),\hat{\varphi})$ such that $\mu_i|K = \iota_K$.

We hasten to show that, in this case, the following holds:

---

[30] This condition can be avoided (cf Endler [9]), but it simplifies the following considerations and is natural in view of the fact that the elements $\hat{y}_1,\dots,\hat{y}_r$ in Theorem (2.12) are non-conjugate over $\hat{K}$ by construction.

(23.1) a') $[K(y):K] = n$.

b') $\psi_1,\ldots,\psi_r$ <u>are the only valuations of</u> $K(y)$ <u>which extend</u> φ .

c') $\hat{K}(\mu_i y) = \hat{K}(\hat{x}_i)$ <u>for any</u> $i \in \{1,\ldots,r\}$.

<u>Proof</u>: For any $i \in \{1,\ldots,r\}$, $\mu_i y \in \Omega$ is a root of $P_{y|K}$ and therefore a root of some irreducible polynomial $\hat{P}_i \in \hat{K}[X]$ which divides $P_{y|K}$; moreover, $\omega \circ \mu_i$ coincides with ψ_i . Since $\mu_i : (K(y),\psi_i) \to (\hat{K}(\mu_i y),\hat{\varphi})$ is a completion, by (2.12), we have $\hat{K}(\mu_i y) = \hat{K}(\hat{x}_i)$, by (2.4). Since $[K(y):K] \le n = \sum_{i=1}^{r} [\hat{K}(\hat{x}_i):\hat{K}] = \sum_{i=1}^{r} [\hat{K}(\mu_i y):\hat{K}]$, we conclude from (2.12) that a') and b') hold. □

For technical reasons, we shall consider solutions y of $\mathfrak{X} = (\hat{x}_1,\ldots,\hat{x}_r)$ which satisfy the equations $\mu_1 y = \hat{x}_1,\ldots,\mu_r y = \hat{x}_r$ approximately. More precisely, a solution y of $\mathfrak{X}$ will be called an ϵ-<u>solution of</u> $\mathfrak{X}$ if, for any $i \in \{1,\ldots,r\}$, μ_i can be chosen such that $\omega(\mu_i y - \hat{x}_i) \le \epsilon$. Note that, in Theorem (2.12), y is a 0-solution of the φ-prescription $(\hat{y}_1,\ldots,\hat{y}_r)$. Of course, an arbitrary φ-prescription cannot be expected to have a 0-solution.

It is convenient to introduce the following terminology. For any positive integer r and any r-tupel $Z = (\hat{z}_1,\ldots,\hat{z}_r) \in \Omega \times\ldots\times \Omega$ (r times) we set $\omega Z = \max\{\omega\hat{z}_1,\ldots,\omega\hat{z}_r\}$. We denote by $\mathfrak{T}_n$ the set of all φ-prescriptions of degree n and arbitrary length. For any $\mathfrak{X} \in \mathfrak{T}_n$ we denote by $r(\mathfrak{X})$ the length of $\mathfrak{X}$ and by $\hat{x}_i$ the i-th component of $\mathfrak{X}$, i.e., $\mathfrak{X} = (\hat{x}_1,\ldots,\hat{x}_{r(\mathfrak{X})})$. Note that any $\mathfrak{X} \in \mathfrak{T}_n$ with $r(\mathfrak{X}) = n$ is an element of $\hat{K} \times\ldots\times \hat{K}$ (n times). The set of all $\mathfrak{X} \in \mathfrak{T}_n$ with $r(\mathfrak{X}) = 1$ will be of special interest and will be denoted by $\mathfrak{D}_n$. For any $\epsilon \ge 0$ let $\mathfrak{B}_\epsilon\mathfrak{X}$ be the set of all $\mathfrak{Y} \in \mathfrak{T}_n$ such that $r(\mathfrak{Y}) = r(\mathfrak{X})$, $\hat{K}(\hat{y}_i) = \hat{K}(\hat{x}_i)$ for all $i \in \{1,\ldots,r(\mathfrak{X})\}$ and $\omega(\mathfrak{Y} - \mathfrak{X}) \le \epsilon$, where $\mathfrak{Y} - \mathfrak{X}$ denotes the $r(\mathfrak{X})$-tuple $(\hat{y}_1-\hat{x}_1,\ldots,\hat{y}_{r(\mathfrak{X})} - \hat{x}_{r(\mathfrak{X})})$. The set $\{\mathfrak{B}_\epsilon\mathfrak{X} \mid \epsilon > 0\}$ will turn out to be a fundamental system of neighborhoods of $\mathfrak{x}$ with respect to a topology induced by φ (cf §24).

Moreover, we denote by $\mathfrak{m}_n$ the set of all monic polynomials in $K[X]$ of degree n, by $\mathfrak{J}_n$ the set of all irreducible polynomials $F \in \mathfrak{m}_n$, and by $\mathfrak{S}_n$ the set of all separable polynomials $F \in \mathfrak{m}_n$. (F is called separable if F and its derivative F' are relatively prime or, equivalently, if F has only simple roots.) Similarly $\hat{\mathfrak{m}}_n$, $\hat{\mathfrak{J}}_n$, and $\hat{\mathfrak{S}}_n$ are defined with respect to the polynomial ring $\hat{K}[X]$. Finally, we denote by ρ the mapping

$$\mathfrak{x} \mapsto \prod_{i=1}^{r(\mathfrak{x})} P_{\hat{x}_i|\hat{K}} \qquad (\mathfrak{x} \in \mathfrak{X}_n)$$

and prove:

(23.2) ρ <u>maps</u> $\mathfrak{X}_n$ <u>onto</u> $\hat{\mathfrak{S}}_n$ <u>and maps</u> $\mathfrak{D}_n$ <u>onto</u> $\hat{\mathfrak{J}}_n \cap \hat{\mathfrak{S}}_n$.

<u>Proof</u>: For any $\mathfrak{x} \in \mathfrak{X}_n$, the elements $\hat{x}_1,\ldots,\hat{x}_{r(\mathfrak{x})}$ are separable and pairwise non-conjugate over $\hat{K}$; therefore $\rho\mathfrak{x} \in \hat{\mathfrak{S}}_n$. On the other hand, any $\hat{F} \in \hat{\mathfrak{S}}_n$ is a product $\hat{P}_1 \cdot \ldots \cdot \hat{P}_r$ of finitely many distinct polynomials $\hat{P}_1,\ldots,\hat{P}_r \in \hat{\mathfrak{J}}_n \cap \hat{\mathfrak{S}}_n$. For any $i \in \{1,\ldots,r\}$, we have $\hat{P}_i = P_{\hat{y}_i|\hat{K}}$ for some $\hat{y}_i \in \Omega$, the elements $\hat{y}_1,\ldots,\hat{y}_r$ are pairwise non-conjugate over $\hat{K}$, and the equality $\sum_{i=1}^{r} [\hat{K}(\hat{y}_i):\hat{K}] = \sum_{i=1}^{r} \deg \hat{P}_i = \deg P = n$ holds. Therefore $\mathfrak{y} = (\hat{y}_1,\ldots,\hat{y}_r) \in \mathfrak{X}_n$ and $\rho\mathfrak{y} = F$. Obviously, $\mathfrak{x} \in \mathfrak{D}_n$ if and only if $\rho\mathfrak{x} \in \hat{\mathfrak{J}}_n \cap \hat{\mathfrak{S}}_n$. □

Note that the mapping $\rho: \mathfrak{X}_n \to \hat{\mathfrak{S}}_n$ is not bijective, in general. In fact, we have $\rho\mathfrak{x} = \rho\mathfrak{y}$ if and only if $r(\mathfrak{x}) = r(\mathfrak{y})$ and $\hat{x}_i$ is $\hat{K}$-conjugate to $\hat{y}_{\pi i}$ $(i = 1,\ldots,r(\mathfrak{x}))$ for some permutation π of $\{1,\ldots,r(\mathfrak{x})\}$.

By means of the mapping ρ, it is easy to characterize the minimal polynomials of ϵ-solutions of a given $\mathfrak{x} \in \mathfrak{X}_n$:

(23.3) Let $\mathfrak{X} \in \mathfrak{X}_n$, $F \in \mathfrak{M}_n$, and $\epsilon \geq 0$. The following conditions are equivalent:

(i) $F \in \mathcal{J}_n \cap \rho(\mathfrak{B}_\epsilon \mathfrak{X})$.

(ii) Some root of F is an ϵ-solution of $\mathfrak{X}$.

In this case, any root y of F is an ϵ-solution of $\mathfrak{X}$, and $F = P_{y|K} \in \mathcal{J}_n \cap \mathcal{S}_n$.

Proof: We set $r = r(\mathfrak{X})$. (i) $\Rightarrow$ (ii): Since $F \in \mathcal{J}_n$, we have $F = P_{y|K}$ and $[K(y):K] = n$ for any root y of F . Since $F = \rho\mathfrak{Y}$ for some $\mathfrak{Y} \in \mathfrak{B}_\epsilon \mathfrak{X}$, we have $F \in \hat{\mathcal{S}}_n$ by (23.2); therefore $F \in \mathcal{J}_n \cap \hat{\mathcal{S}}_n = \mathcal{J}_n \cap \mathcal{S}_n$, and y is separable over K . Moreover, we have $r(\mathfrak{Y}) = r$, $\hat{K}(\hat{y}_i) = \hat{K}(\hat{x}_i)$ for all $i \in \{1,\ldots,r\}$, and $\omega(\mathfrak{Y} - \mathfrak{X}) \leq \epsilon$. By (2.12), there exist r valuations $\psi_1,\ldots,\psi_r$ of $K(y)$ extending φ and completions $\lambda_i: (K(y),\psi_i) \to (\hat{K}(\hat{y}_i),\hat{\varphi})$ such that $\lambda_i|K = \iota_K$ and $\lambda_i y = \hat{y}_i$, for all $i \in \{1,\ldots,r\}$. Therefore y is an ϵ-solution of $\mathfrak{X}$.

(ii) $\Rightarrow$ (i): Let y be an ϵ-solution of $\mathfrak{X}$ such that $F(y) = 0$. We conclude from (23.1) that $F = P_{y|K} \in \mathcal{J}_n \cap \mathcal{S}_n$, that there are exactly r valuations $\psi_1,\ldots,\psi_r$ of $K(y)$ extending φ and that, for any $i \in \{1,\ldots,r\}$, there is a completion $\mu_i: (K(y),\psi_i) \to (\hat{K}(\hat{x}_i),\hat{\varphi})$ such that $\mu_i| K = \iota_K$, $\hat{K}(\mu_i y) = \hat{K}(\hat{x}_i)$, and $\omega(\mu_i y - \hat{x}_i) \leq \epsilon$. The elements $\mu_1 y,\ldots,\mu_r y \in \Omega$ are pairwise non-conjugate over $\hat{K}$, since otherwise there would exist a $\hat{K}$-automorphism σ of Ω such that $\mu_i = \sigma \circ \mu_j$ for some pair $i,j \in \{1,\ldots,r\}$, $i \neq j$, hence $\psi_i = \omega \circ \mu_i = \omega \circ \sigma \circ \mu_j = \omega \circ \mu_j = \psi_j$, a contradiction. Therefore $(\mu_1 y,\ldots,\mu_r y) \in \mathfrak{X}_n$ and even $\in \mathfrak{B}_\epsilon \mathfrak{X}$. The minimal polynomials $P_{\mu_1 y|\hat{K}},\ldots,P_{\mu_r y|\hat{K}}$ divide $P_{y|K} = F$ and so does their product. Since this product has degree $\sum_{i=1}^{r} [\hat{K}(\mu_i y):\hat{K}] = \sum_{i=1}^{r} [\hat{K}(\hat{x}_i):\hat{K}] = n$, it is equal to F ; therefore $F = \rho(\mu_1 y,\ldots,\mu_r y) \in \rho(\mathfrak{B}_\epsilon \mathfrak{X})$. □

If we consider a finite number of valuations $\varphi^1,\dots,\varphi^k$ of K, the reference to φ^j will be indicated by a superscript ^{j}. For example, $(\hat{K}^j, \hat{\varphi}^j)$ denotes a completion of (K,φ^j), Ω^j a separable closure of $\hat{K}^j$, $\mathfrak{T}_n^j$ the set of all φ^j-prescriptions of degree n, $\hat{\mathfrak{m}}_n^j$ the set of all monic polynomials in $\hat{K}^j[X]$ of degree n, ρ^j the mapping from $\mathfrak{T}_n^j$ onto $\hat{\mathfrak{S}}_n^j$ indicated in (23.2), etc. Moreover, we set $\overline{\mathfrak{m}}_n = \hat{\mathfrak{m}}_n^1 \times\dots\times \hat{\mathfrak{m}}_n^k$, $\overline{\mathfrak{S}}_n = \hat{\mathfrak{S}}_n^1 \times\dots\times \hat{\mathfrak{S}}_n^k$, $\overline{\mathfrak{T}}_n = \mathfrak{T}_n^1 \times\dots\times \mathfrak{T}_n^k$, $\overline{\mathfrak{B}}_\varepsilon\overline{\mathfrak{T}} = \mathfrak{B}_\varepsilon^1\mathfrak{T}^1 \times\dots\times \mathfrak{B}_\varepsilon^k\mathfrak{T}^k$ where $\mathfrak{T} = (\mathfrak{T}^1,\dots,\mathfrak{T}^k) \in \overline{\mathfrak{T}}_n$. Any $\mathfrak{T} = (\mathfrak{T}^1,\dots,\mathfrak{T}^k) \in \overline{\mathfrak{T}}_n$ will be called a $(\varphi^1,\dots,\varphi^k)$-*prescription* of degree n; note that $\mathfrak{T}^1,\dots,\mathfrak{T}^k$ may have different lengths. We conclude from (23.2) that

(23.4) $\overline{\mathfrak{T}} \mapsto (\rho^1\mathfrak{T}^1,\dots,\rho^k\mathfrak{T}^k)$ $(\overline{\mathfrak{T}} \in \overline{\mathfrak{T}}_n)$ *is a mapping* $\overline{\rho}$ *from* $\overline{\mathfrak{T}}_n$ *onto* $\overline{\mathfrak{S}}_n$.

Let $\overline{\mathfrak{T}} \in \overline{\mathfrak{T}}_n$. A common ε-solution of $\mathfrak{T}^1,\dots,\mathfrak{T}^k$ will be called an ε-*solution* of the $(\varphi^1,\dots,\varphi^k)$-prescription $\overline{\mathfrak{T}}$. Denoting by $\nabla: \mathfrak{m}_n \to \overline{\mathfrak{m}}_n$ the diagonal mapping $F \mapsto (F,\dots,F)$ $(F \in \mathfrak{m}_n)$, we get as an immediate consequence of (23.3):

(23.5) *Let* $\overline{\mathfrak{T}} \in \overline{\mathfrak{T}}_n$, $F \in \mathfrak{m}_n$, *and* $\varepsilon \ge 0$. *The following conditions are equivalent:*

(i) $F \in \mathfrak{J}_n$ *and* $\nabla F \in \overline{\rho}(\overline{\mathfrak{B}}_\varepsilon\overline{\mathfrak{T}})$.

(ii) *Some root of* F *is an* ε-*solution of* $\overline{\mathfrak{T}}$.

In this case, any root y *of* F *is an* ε-*solution of* $\overline{\mathfrak{T}}$, *and* $F = P_{y|K} \in \mathfrak{J}_n \cap \mathfrak{S}_n$.

In §25 we shall characterize those $\overline{\mathfrak{T}} \in \overline{\mathfrak{T}}_n$ which have an ε-solution for any $\varepsilon > 0$. We shall see, in particular, that $\overline{\mathfrak{T}}$ has this property whenever at least one of its components $\mathfrak{T}^1,\dots,\mathfrak{T}^k$ has length 1. We denote by $\overline{\mathfrak{D}}_n$ the set of all $\overline{\mathfrak{T}} \in \overline{\mathfrak{T}}_n$ such that $r(\mathfrak{T}^j) = 1$ for some $j \in \{1,\dots,k\}$; i.e., $\overline{\mathfrak{D}}_n = \bigcup_{j=1}^{k} \mathfrak{D}_n^{(j)}$ where

$$\mathfrak{D}_n^{(j)} = \mathfrak{T}_n^1 \times\ldots\times \mathfrak{T}_n^{j-1} \times \mathfrak{D}_n^j \times \mathfrak{T}_n^{j+1} \times\ldots\times \mathfrak{T}_n^k .$$

§24 Topological preliminaries

At the end of §16 we proved Krasner's lemma for henselian valuation rings and showed that it holds also for valued fields (K,φ) such that φ has only one extension to the algebraic closure $ac(K)$ of K. Replacing $ac(K)$ by the separable closure $sc(K)$, we obtain as a consequence:

(24.1) Let (K,φ) be a valued field such that there is only one valuation ψ of $sc(K)$ which extends φ. Then for any $x \in sc(K)$ there is an $\varepsilon_x > 0$ such that, for any $y \in sc(K)$, $\psi(y-x) < \varepsilon_x$ implies $K(x) \subseteq K(y)$.

Let now (K,φ) be an arbitrary valued field, $(\hat{K},\hat{\varphi})$ a completion of (K,φ), Ω a separable closure of $\hat{K}$, and ω the unique extension of $\hat{\varphi}$ to Ω (cf (2.8)). We get as an immediate consequence of (24.1):

(24.2) For any $\hat{x} \in \Omega$ there is an $\varepsilon_{\hat{x}} > 0$ such that, for any $\hat{y} \in \Omega$, $\omega(\hat{y}-\hat{x}) < \varepsilon_{\hat{x}}$ implies $\hat{K}(\hat{x}) \subseteq \hat{K}(\hat{y})$.

We define in $\mathfrak{T}_n$ a Hausdorff topology in the following way. Let Ω be endowed with the topology defined by ω, $\Omega_r = \Omega \times\ldots\times \Omega$ (r times) with the product topology, $\bigcup_{r=1}^{n} \Omega_r$ with the sum topology, and consider $\mathfrak{T}_n$ as a subspace of $\bigcup_{r=1}^{n} \Omega_r$. Using (24.2) we show:

(24.3) For any $\mathfrak{X} \in \mathfrak{T}_n$, $\{\mathfrak{B}_\varepsilon \mathfrak{X} \mid \varepsilon > 0\}$ is a fundamental system of neighborhoods of $\mathfrak{X}$ in $\mathfrak{T}_n$.

Proof: For any $\epsilon > 0$ let $\mathfrak{B}'_\epsilon\mathfrak{X}$ be the set of all $\mathfrak{Y} \in \mathfrak{X}_n$ such that $r(\mathfrak{Y}) = r(\mathfrak{X})$ and $\omega(\mathfrak{Y}-\mathfrak{X}) \le \epsilon$. Obviously $\{\mathfrak{B}'_\epsilon\mathfrak{X} \mid \epsilon > 0\}$ is a fundamental system of neighborhoods of $\mathfrak{X}$ in $\mathfrak{X}_n$, and $\mathfrak{B}_\epsilon\mathfrak{X} \subseteq \mathfrak{B}'_\epsilon\mathfrak{X}$ for any $\epsilon \ge 0$. We claim that $\mathfrak{B}_\epsilon\mathfrak{X} = \mathfrak{B}'_\epsilon\mathfrak{X}$ for sufficiently small $\epsilon > 0$. In fact, let $0 < \epsilon < \min\{\epsilon_{\hat{x}_i} \mid i = 1,\dots,r(\mathfrak{X})\}$ and $\mathfrak{Y} \in \mathfrak{B}'_\epsilon\mathfrak{X}$. By (24.2) we have $\hat{K}(\hat{x}_i) \subseteq \hat{K}(\hat{y}_i)$ for all $i \in \{1,\dots,r(\mathfrak{X})\}$, and even the equalities hold since $\sum_{i=1}^{r(\mathfrak{X})}[\hat{K}(\hat{x}_i):\hat{K}] = n = \sum_{i=1}^{r(\mathfrak{X})}[\hat{K}(\hat{y}_i):\hat{K}]$; therefore $\mathfrak{Y} \in \mathfrak{B}_\epsilon\mathfrak{X}$. We conclude that $\{\mathfrak{B}_\epsilon\mathfrak{X} \mid \epsilon > 0\}$ is also a fundamental system of neighborhoods of $\mathfrak{X}$ in $\mathfrak{X}_n$. □

In the set $\mathfrak{m}_n$ of all monic polynomials in $K[X]$ of degree n, a Hausdorff topology is defined in the following way. We extend φ to the polynomial ring $K[X]$ by setting

$$\varphi(a_m\cdot X^m + \dots + a_1\cdot X + a_o) = \max\{\varphi a_o,\dots,\varphi a_m\} .$$[31]

This extension defines a Hausdorff topology in $K[X]$, and $\mathfrak{m}_n$ is considered as a subspace of $K[X]$. For any $\epsilon \ge 0$ and $F \in \mathfrak{m}_n$, let $\mathfrak{U}_\epsilon(F)$ be the set of all $G \in \mathfrak{m}_n$ such that $\varphi(G-F) \ge \epsilon$. It is obvious that $\{\mathfrak{U}_\epsilon(F) \mid \epsilon > 0\}$ is a fundamental system of neighborhoods of F in $\mathfrak{m}_n$. Similarly, a Hausdorff topology is defined in $\hat{\mathfrak{m}}_n$, by means of the valuation $\hat{\varphi}$, and $\{\hat{\mathfrak{U}}_\epsilon(\hat{F}) \mid \epsilon > 0\}$ is a fundamental system of neighborhoods of $\hat{F}$ in $\hat{\mathfrak{m}}_n$.

The following theorem asserts the continuity of polynomial roots in the case of separable polynomials. It can be generalized to arbitrary polynomials (see Exercise IV-1).

(24.4) THEOREM - Let K be a separably closed field and φ a valuation of K. For any $F \in \mathfrak{S}_n$ and any $\epsilon > 0$ there is a $\delta > 0$ such that, for any $G \in \mathfrak{U}_\delta(F)$, a 1-1 correspondence between

[31] Note that the extension of φ to $K[X]$ does not satisfy condition V_2 of §1, unless φ is non-archimedean. Therefore φ extends to a valuation of the field $K(X)$ only if φ is non-archimedean.

<u>the set</u> R_F <u>of all roots of</u> F <u>and the set</u> R_G <u>of all roots of</u> G <u>is defined by the relation</u>

$$\{(x,y) \in R_F \times R_G \mid \varphi(y-x) \le \epsilon\} .$$

<u>In particular,</u> $\mathfrak{U}_\delta(F) \subseteq \mathcal{S}_n$.

<u>Proof</u>: We set $F = X^n + a_1 \cdot X^{n-1} + \ldots + a_n$, $\rho = 2n \cdot \varphi F$ and claim that $\varphi x < \rho$ for any root $x \in R_F$. In fact, otherwise we would have $(\varphi x)^i \ge \rho^i \ge \rho$ for all $i \ge 1$, hence $\varphi(F(x) \cdot x^{-n}) =$ $= \varphi(1 + a_1 \cdot x^{-1} + \ldots + a_n \cdot x^{-n}) \ge \varphi 1 - \varphi(a_1 \cdot x^{-1} + \ldots + a_n \cdot x^{-n}) \ge$ $\ge 1 - (\varphi a_1 \cdot (\varphi x)^{-1} + \ldots + \varphi a_n \cdot (\varphi x)^{-n}) \ge 1 - \varphi F \cdot ((\varphi x)^{-1} + \ldots + (\varphi x)^{-n}) \ge$ $\ge 1 - \varphi F \cdot n \cdot \rho^{-1} = \frac{1}{2} > 0$, which is impossible. Let $\delta = \epsilon^n \cdot (n \cdot \rho^{n-1})^{-1}$ and $G = X^n + b_1 \cdot X^{n-1} + \ldots + b_n = \prod_{i=1}^{n} (X - y_i) \in \mathfrak{U}_\delta(F)$. For any $x \in R_F$ we have $\prod_{i=1}^{n} \varphi(x-y_i) = \varphi(\prod_{i=1}^{n} (x-y_i)) = \varphi(G(x)) = \varphi(G(x)-F(x)) =$ $= \varphi(\sum_{i=1}^{n} (b_i - a_i) \cdot x^{n-i}) \le \sum_{i=1}^{n} \varphi(b_i - a_i) \cdot (\varphi x)^{n-i} \le \delta \cdot (\rho^{n-1} + \ldots + \rho^1 + 1) \le$ $\le \delta \cdot n \cdot \rho^{n-1} = \epsilon^n$, hence $\varphi(x-y_i) \le \epsilon$ for some $i \in \{1,\ldots,n\}$. Therefore there is a mapping $\theta: R_F \to R_G$ such that $\varphi(\theta x - x) \le \epsilon$ for all $x \in R_F$. Assuming, without loss of generality, that $\epsilon < \frac{1}{2} \cdot \min\{\varphi(x-x') \mid x,x' \in R_F , x \ne x'\}$, we get $\varphi(\theta x - \theta x') \ge$ $\ge \varphi(x-x') - \varphi(\theta x - x) - \varphi(\theta x' - x') > 0$ for all pairs $x, x' \in R_F$ such that $x \ne x'$; therefore θ is injective. Since $\# R_G \le n = \# R_F$, θ is even bijective, hence $\# R_G = n$, $G \in \mathcal{S}_n$. □

Let (K,φ) be again an arbitrary valued field. We use (24.4) and (24.1) for proving:

(24.5) COROLLARY - a) $\mathcal{S}_n$ <u>is open in</u> $\mathfrak{M}_n$.

b) <u>If</u> φ <u>has only one extension to</u> $sc(K)$ <u>then</u> $\mathcal{J}_n \cap \mathcal{S}_n$ <u>is open in</u> $\mathfrak{M}_n$.

<u>Proof</u>: φ extends to a valuation φ' of $sc(K)$. Let $\mathfrak{M}_n'$, $\mathcal{S}_n'$, $\mathfrak{U}_\epsilon'$

be defined with respect to $(sc(K), \varphi')$.

a) Let $F \in \mathcal{S}_n$. Since $F \in \mathcal{S}'_n$, there is a $\delta > 0$ such that $\mathcal{U}'_\delta(F) \subseteq \mathcal{S}'_n$, by (24.4), hence $\mathcal{U}_\delta(F) = \mathcal{U}'_\delta(F) \cap K[X] \subseteq \mathcal{S}'_n \cap$ $\cap K[X] = \mathcal{S}_n$.

b) Let $F \in \mathcal{J}_n \cap \mathcal{S}_n$ and $x \in sc(K)$ be a root of F . Since $F \in \mathcal{S}'_n$, we conclude from (24.4) that there is a $\delta > 0$ such that, for any $G \in \mathcal{U}_\delta(F) \subseteq \mathcal{U}'_\delta(F)$, there exists a root $y \in sc(K)$ of G with $\varphi'(y-x) < \varepsilon_x$; therefore $K(x) \subseteq K(y)$ by (24.1). Since $n = [K(x):K] \leq [K(y):K] = \deg P_{y|K} \leq \deg G \leq n$, we have $K(x) =$ $= K(y)$ and $G = P_{y|K} \in \mathcal{J}_n \cap \mathcal{S}_n$. □

We get as an immediate consequence of (24.5):

(24.6) COROLLARY - $\hat{\mathcal{S}}_n$ <u>and</u> $\hat{\mathcal{J}}_n \cap \hat{\mathcal{S}}_n$ <u>are open in</u> $\hat{\mathfrak{m}}_n$.

In this context we show also:

(24.7) THEOREM - <u>If</u> φ <u>is a non-trivial valuation of</u> K <u>then</u> $\mathcal{S}_n$ <u>is dense in</u> $\mathfrak{m}_n$ <u>and</u> $\hat{\mathcal{S}}_n$ <u>is dense in</u> $\hat{\mathfrak{m}}_n$.

<u>Proof</u>: Let $F = X^n + a_1 \cdot X^{n-1} + \ldots + a_n \in \mathfrak{m}_n$ and let $Y_1, \ldots, Y_n$ be indeterminates over $K[X]$. Since $a_1 + Y_1, \ldots, a_n + Y_n$ are algebraically independent over K , the polynomial $F_Y = X^n +$ $+ (a_1+Y_1) \cdot X^{n-1} + \ldots + (a_n+Y_n) \in K(Y_1, \ldots, Y_n)[X]$ is separable; therefore the resultant $Res(F_Y, F'_Y)$ is non-zero, where F'_Y is the derivative of F_Y (with respect to X). Since φ is non-trivial, for any $\varepsilon > 0$ the set $\{c \in K \mid \varphi c < \varepsilon\}$ is infinite, hence contains elements $c_1, \ldots, c_n$ such that $Res(F_c, F'_c) \neq 0$, where $F_c = X^n + (a_1+c_1) \cdot X^{n-1} + \ldots + (a_n+c_n) \in K[X]$; obviously $F_c \in \mathcal{U}_\varepsilon(F) \cap$ $\mathcal{S}_n$. Therefore $\mathcal{S}_n$ is dense in $\mathfrak{m}_n$. Since φ is non-trivial, so is $\hat{\varphi}$. Therefore the same reasoning shows that $\hat{\mathcal{S}}_n$ is dense in $\hat{\mathfrak{m}}_n$. □

We note that (24.7) holds also for any non-trivial Krull valuation v instead of φ [32]; in fact, it suffices to replace $\{c \in K \mid \varphi c < \epsilon\}$ by $\{c \in K \mid vc > \epsilon\}$, where ϵ is any element of the value group of v . We want to mention, without proof, that (24.4) and (24.5) hold for Krull valuations too (cf Exercise IV-5).

The following theorem, concerning the mapping $\mathfrak{P}$ from $\mathfrak{X}_n$ ondo $\hat{\mathfrak{S}}_n$, is crucial for the following section.

(24.8) THEOREM - <u>The mapping</u> $\mathfrak{P}: \mathfrak{X}_n \to \hat{\mathfrak{S}}_n$ <u>is continuous and open.</u>

<u>Proof</u>: a) Let $\mathfrak{x} \in \mathfrak{X}_n$, $r = r(\mathfrak{x})$, $\hat{L}_i = \hat{K}(\hat{x}_i)$, and E_i be the set of all $\hat{K}$-monomorphisms $\sigma: \hat{L}_i \to \Omega$ $(i = 1,\ldots,r)$. Let $\delta > 0$ and $\mathfrak{y} \in \mathfrak{B}_\delta\mathfrak{x}$; then $r(\mathfrak{y}) = r$, $\hat{K}(\hat{y}_i) = \hat{L}_i$, and $\omega(\sigma\hat{y}_i - \sigma\hat{x}_i) = (\omega \circ \sigma)(\hat{y}_i - \hat{x}_i) = \omega(\hat{y}_i - \hat{x}_i) \le \delta$ for all $\sigma \in E_i$, $i = 1,\ldots,r$. Since the coefficients of a polynomial are continuous functions of its roots, for any $\epsilon > 0$ there is a $\delta > 0$ such that $\mathfrak{y} \in \mathfrak{B}_\delta\mathfrak{x}$ implies $\hat{\varphi}(\mathfrak{P}\mathfrak{y} - \mathfrak{P}\mathfrak{x}) = \hat{\varphi}(\prod_{i=1}^{r} \prod_{\sigma\in E_i} (X-\sigma\hat{y}_i) - \prod_{i=1}^{r} \prod_{\sigma\in E_i} (X-\sigma\hat{x}_i)) \le \epsilon$, i.e., $\mathfrak{P}\mathfrak{y} \in \hat{\mathfrak{U}}_\epsilon(\mathfrak{P}\mathfrak{x})$. Therefore $\mathfrak{P}$ is continuous.

b) To prove that $\mathfrak{P}$ is open we have to show that, for any $\mathfrak{x} \in \mathfrak{X}_n$ and $\epsilon > 0$, there is a $\delta > 0$ such that $\hat{\mathfrak{U}}_\delta(\mathfrak{P}\mathfrak{x}) \subseteq \mathfrak{P}(\mathfrak{B}_\epsilon\mathfrak{x})$. Let $r = r(\mathfrak{x})$, $Z_\mathfrak{x}$ the set of all roots of $\mathfrak{P}\mathfrak{x}$ in Ω , $\epsilon_\mathfrak{x} = \min\{\omega(\hat{z}-\hat{z}') \mid \hat{z}, \hat{z}' \in Z_\mathfrak{x} , \hat{z} \neq \hat{z}'\}$, and choose $\epsilon > 0$ such that $\epsilon < \min\{\epsilon_{\hat{x}_1},\ldots,\epsilon_{\hat{x}_r}\}$ and $2\epsilon < \epsilon_\mathfrak{x}$. By (24.4) there is a $\delta > 0$ such that any $\hat{G} \in \hat{\mathfrak{U}}_\delta(\mathfrak{P}\mathfrak{x})$ has roots $\hat{y}_1,\ldots,\hat{y}_r \in \Omega$ with $\omega(\hat{y}_i - \hat{x}_i) \le \epsilon$. These are non-conjugate over $\hat{K}$; in fact, otherwise there would exist $i,j \in \{1,\ldots,r\}$, $i \neq j$, and a $\hat{K}$-automorphism σ

---

[32] Of course, the Krull valuation v of K has to be extended to $K[X]$ by setting $v(\sum_{i=0}^{m} a_i \cdot X^i) = \min\{va_0,\ldots,va_m\}$.

of Ω such that $\sigma\hat{y}_i = \hat{y}_j$, hence we would have $\omega(\hat{x}_j - \sigma\hat{x}_i) \leq$ $\leq \omega(\hat{x}_j-\hat{y}_j) + \omega(\sigma\hat{y}_i - \sigma\hat{x}_i) = \omega(\hat{x}_j-\hat{y}_j) + \omega(\hat{y}_i-\hat{x}_i) \leq 2\varepsilon < \varepsilon_{\mathfrak{X}}$, hence $\hat{x}_j = \sigma\hat{x}_i$, which is impossible since $\hat{x}_1,\dots,\hat{x}_r$ are pairwise non-conjugate over $\hat{K}$. Therefore $\hat{G}$ is a multiple of $\prod_{i=1}^{r} P_{\hat{y}_i|\hat{K}}$. For all $i \in \{1,\dots,r\}$ we have $\omega(\hat{y}_i-\hat{x}_i) \leq \varepsilon < \varepsilon_{\hat{x}_i}$, hence $\hat{K}(\hat{x}_i) \subseteq$ $\subseteq \hat{K}(\hat{y}_i)$ by (24.2), hence $n = \sum_{i=1}^{r} [\hat{K}(\hat{x}_i):\hat{K}] \leq \sum_{i=1}^{r} [\hat{K}(\hat{y}_i):\hat{K}] =$ $= \deg(\prod_{i=1}^{r} P_{\hat{y}_i|\hat{K}}) \leq \deg \hat{G} = n$. Therefore $\mathfrak{y} = (\hat{y}_1,\dots,\hat{y}_r) \in \mathfrak{B}_\varepsilon\mathfrak{X} \subseteq \mathfrak{X}_n$ and $\hat{G} = \rho\mathfrak{y}$. □

Considering finitely many valuations $\varphi^1,\dots,\varphi^k$ of K , we endow $\overline{\mathfrak{X}}_n = \mathfrak{X}_n^1 \times\dots\times \mathfrak{X}_n^k$ with the product of the topologies of $\mathfrak{X}_n^1,\dots,\mathfrak{X}_n^k$ and endow $\overline{\mathfrak{m}}_n = \hat{\mathfrak{m}}_n^1 \times\dots\times \hat{\mathfrak{m}}_n^k$ with the product of the topologies of $\hat{\mathfrak{m}}_n^1,\dots,\hat{\mathfrak{m}}_n^k$. Setting $\overline{\mathfrak{U}}_\varepsilon(\hat{F}_1,\dots,\hat{F}_k) = \hat{\mathfrak{U}}_\varepsilon^1(\hat{F}_1) \times\dots\times \hat{\mathfrak{U}}_\varepsilon^k(\hat{F}_k)$ for all $\varepsilon \geq 0$ and $(\hat{F}_1,\dots,\hat{F}_k) \in \overline{\mathfrak{m}}_n$, it is obvious that $\{\overline{\mathfrak{U}}_\varepsilon(\hat{F}_1,\dots,\hat{F}_k) \mid \varepsilon > 0\}$ is a fundamental system of neighborhoods of $(\hat{F}_1,\dots,\hat{F}_k)$ in $\overline{\mathfrak{m}}_n$. Moreover, it follows from (24.3) that $\{\overline{\mathfrak{B}}_\varepsilon\overline{\mathfrak{X}} \mid \varepsilon > 0\}$ is a fundamental system of neighborhoods of $\overline{\mathfrak{X}}$ in $\overline{\mathfrak{X}}_n$. Finally, we get as immediate consequences of (24.6), (24.7), and (24.8):

(24.9) $\overline{\mathfrak{S}}_n$ <u>and</u> $\mathfrak{J}_n \cap \overline{\mathfrak{S}}_n$ <u>are open in</u> $\overline{\mathfrak{m}}_n$.

(24.10) <u>If</u> $\varphi^1,\dots,\varphi^k$ <u>are non-trivial then</u> $\overline{\mathfrak{S}}_n$ <u>is dense in</u> $\overline{\mathfrak{m}}_n$.

(24.11) <u>The mapping</u> $\overline{\rho}: \overline{\mathfrak{X}}_n \to \overline{\mathfrak{S}}_n$ <u>is continuous and open.</u>

§25 *Solvable* $(\varphi^1,\dots,\varphi^k)$-*prescriptions*

Let $\varphi^1,\dots,\varphi^k$ be pairwise non-equivalent non-trivial valuations of K.[33] Given $\epsilon > 0$ and a $(\varphi^1,\dots,\varphi^k)$-prescription $\bar{\mathfrak{X}} \in \bar{\mathfrak{T}}_n$, we show that each root of any irreducible monic polynomial $F \in K[X]$ such that ∇F is sufficiently proximate to $(\rho^1\mathfrak{X}^1,\dots,\rho^k\mathfrak{X}^k)$ is an ϵ-solution of $\bar{\mathfrak{X}}$. If $\bar{\mathfrak{X}} \in \bar{\mathfrak{D}}_n$ then the hypothesis of irreducibility is irrelevant.

(25.1) a) *Let* $\bar{\mathfrak{X}} \in \bar{\mathfrak{T}}_n$. *For any* $\epsilon > 0$ *there is a* $\delta > 0$ *such that, for any* $F \in \mathcal{J}_n$, $\nabla F \in \bar{\mathfrak{U}}_\delta(\bar{\rho}\bar{\mathfrak{X}})$ *implies that each root of* F *is an* ϵ-*solution of* $\bar{\mathfrak{X}}$.

b) *Let* $\bar{\mathfrak{X}} \in \bar{\mathfrak{D}}_n$. *For any* $\epsilon > 0$ *there is a* $\delta > 0$ *such that, for any* $F \in \mathfrak{M}_n$, $\nabla F \in \bar{\mathfrak{U}}_\delta(\bar{\rho}\bar{\mathfrak{X}})$ *implies that each root of* F *is an* ϵ-*solution of* $\bar{\mathfrak{X}}$.

Proof: a) For any $\epsilon > 0$ there is a $\delta > 0$ such that $\bar{\mathfrak{U}}_\delta(\bar{\rho}\bar{\mathfrak{X}}) \subseteq \bar{\rho}(\mathfrak{B}_\epsilon\bar{\mathfrak{X}})$, by (24.11). Let $F \in \mathcal{J}_n$ such that $\nabla F \in \bar{\mathfrak{U}}_\delta(\bar{\rho}\bar{\mathfrak{X}})$. By (23.5), each root of F is an ϵ-solution of $\bar{\mathfrak{X}}$.

b) We may assume $\bar{\mathfrak{X}} = (\mathfrak{X}^1,\dots,\mathfrak{X}^k) \in \mathfrak{D}_n^1 \times \mathfrak{T}_n^2 \times\dots\times \mathfrak{T}_n^k$, hence $\rho^1\mathfrak{X}^1 \in \hat{\mathcal{J}}_n^1 \cap \hat{\mathfrak{S}}_n^1$, by (23.2). For sufficiently small $\delta' > 0$ we have $\hat{\mathfrak{U}}^1_{\delta'}(\rho^1\mathfrak{X}^1) \subseteq \hat{\mathcal{J}}_n^1 \cap \hat{\mathfrak{S}}_n^1$, by (24.6), hence $\hat{\mathfrak{U}}^1_{\delta'}(\rho^1\mathfrak{X}^1) \cap \mathfrak{M}_n \subseteq \mathcal{J}_n$. Let $\delta'' = \min\{\delta,\delta'\}$; if $F \in \mathfrak{M}_n$ such that $\nabla F \in \bar{\mathfrak{U}}_{\delta''}(\bar{\rho}\bar{\mathfrak{X}})$, then $F \in \mathcal{J}_n$ and each root of F is an ϵ-solution of $\bar{\mathfrak{X}}$, by a). □

A $(\varphi^1,\dots,\varphi^k)$-prescription $\bar{\mathfrak{X}} \in \bar{\mathfrak{T}}_n$ will be called *solvable* if for any $\epsilon > 0$ there exists an ϵ-solution of $\bar{\mathfrak{X}}$.[34] We show that the solvable $(\varphi^1,\dots,\varphi^k)$-prescriptions of degree n are

---

[33] The hypothesis that $\varphi^1,\dots,\varphi^k$ are pairwise non-equivalent will not be used before (25.4).

[34] A student of mine, Bastos [2], has proved that $\bar{\mathfrak{X}}$ is solvable whenever it has a solution (that is, an ϵ-solution for some $\epsilon > 0$).

exactly those $\bar{\mathfrak{X}} \in \bar{\mathfrak{X}}_n$ for which $\bar{\mathcal{P}}\bar{\mathfrak{X}}$ is adherent to the diagonal of $\mathcal{J}_n$ (or of $\mathcal{J}_n \cap \mathcal{S}_n$) . In fact, denoting by $\mathcal{C}\ell(\ )$ the topological closure in $\bar{\mathfrak{m}}_n$ we prove:

(25.2) THEOREM - <u>For any</u> $\bar{\mathfrak{X}} \in \bar{\mathfrak{X}}_n$ <u>the following conditions are equivalent</u>:

(i) $\bar{\mathfrak{X}}$ <u>is solvable</u>.

(ii) $\bar{\mathcal{P}}\bar{\mathfrak{X}} \in \mathcal{C}\ell(\nabla(\mathcal{J}_n \cap \mathcal{S}_n))$.

(iii) $\bar{\mathcal{P}}\bar{\mathfrak{X}} \in \mathcal{C}\ell(\nabla\mathcal{J}_n)$.

<u>If</u> $\bar{\mathfrak{X}} \in \bar{\mathfrak{D}}_n$ <u>then these conditions are also equivalent to</u>

(iv) $\bar{\mathcal{P}}\bar{\mathfrak{X}} \in \mathcal{C}\ell(\nabla\mathfrak{m}_n)$.

<u>Proof</u>: (i) ⇒ (ii): For any $\epsilon > 0$ there is a $\delta > 0$ such that $\bar{\mathcal{P}}(\bar{\mathfrak{B}}_\delta\bar{\mathfrak{X}}) \subseteq \bar{\mathfrak{U}}_\epsilon(\bar{\mathcal{P}}\bar{\mathfrak{X}})$, by (24.11). By (i), there is a δ-solution y of $\bar{\mathfrak{X}}$ and, by (23.5), we have $P_{y|K} \in \mathcal{J}_n \cap \mathcal{S}_n$ and $\nabla P_{y|K} \in \bar{\mathcal{P}}(\bar{\mathfrak{B}}_\delta\bar{\mathfrak{X}})$. Therefore $\bar{\mathfrak{U}}_\epsilon(\bar{\mathcal{P}}\bar{\mathfrak{X}}) \cap \nabla(\mathcal{J}_n \cap \mathcal{S}_n) \neq \emptyset$ for all $\epsilon > 0$ hence (ii) holds.

(ii) ⇒ (iii) and (iii) ⇒ (iv) are obvious.

(iii) ⇒ (i): For any $\delta > 0$ there is an $F_\delta \in \mathcal{J}_n$ such that $\nabla F_\delta \in \bar{\mathfrak{U}}_\delta(\bar{\mathcal{P}}\bar{\mathfrak{X}})$, by (iii). Given any $\epsilon > 0$, let $\delta > 0$ be chosen as in (25.1 a); then any root of F_δ is an ϵ-solution of $\bar{\mathfrak{X}}$.

(iv) ⇒ (i), in case $\bar{\mathfrak{X}} \in \bar{\mathfrak{D}}_n$, is proven similarly. It suffices to replace $\mathcal{J}_n$ by $\mathfrak{m}_n$ and replace (25.1 a) by (25.1 b). □

Let $\bar{\mathfrak{S}}_n$ be the set of all solvable $\bar{\mathfrak{X}} \in \bar{\mathfrak{X}}_n$. Using the fact that $\bar{\mathcal{S}}_n$ is dense in $\bar{\mathfrak{m}}_n$ and equals $\bar{\mathcal{P}}\bar{\mathfrak{X}}_n$ (cf (24.10) and (23.4)), we get as an immediate consequence of (25.2):

(25.3) COROLLARY - <u>The following conditions are equivalent</u>:

(i) $\bar{\mathfrak{S}}_n = \bar{\mathfrak{X}}_n$ (<u>i.e., all</u> $\bar{\mathfrak{X}} \in \bar{\mathfrak{X}}_n$ <u>are solvable</u>).

(ii) $\nabla(\mathcal{J}_n \cap \mathcal{S}_n)$ <u>is dense in</u> $\bar{\mathcal{S}}_n$.

(iii) $\nabla(\mathcal{J}_n \cap \mathcal{S}_n)$ <u>is dense in</u> $\bar{\mathfrak{m}}_n$.

(iv) $\nabla\mathcal{J}_n$ <u>is dense in</u> $\bar{\mathfrak{m}}_n$.

We claim that they are also equivalent to

(v) $\nabla\mathcal{J}_n$ <u>is dense in</u> $\nabla\mathfrak{m}_n$.

In fact, this is an immediate consequence of the following generalization of the approximation theorem (1.8)-(1.9).

(25.4) $\nabla\mathfrak{m}_n$ <u>is dense in</u> $\bar{\mathfrak{m}}_n$.

<u>Proof</u>: Let $(\hat{F}_1,\ldots,\hat{F}_k) \in \bar{\mathfrak{m}}_n$ and $\varepsilon > 0$. For any $j \in \{1,\ldots,k\}$ there is an $F_j \in \mathfrak{m}_n$ such that $\hat{\varphi}^j(F_j-\hat{F}_j) \le \frac{\varepsilon}{2}$, since K is dense in $\hat{K}^j$ (with respect to $T_{\hat{\varphi}^j}$). By (1.8) there is an $F \in \mathfrak{m}_n$ such that $\varphi^j(F-F_j) \le \frac{\varepsilon}{2}$ for all $j \in \{1,\ldots,k\}$. Therefore $\hat{\varphi}^j(F-\hat{F}_j) \le \varepsilon$ for all $j \in \{1,\ldots,k\}$, hence $\nabla F \in \nabla\mathfrak{m}_n \cap \bar{\mathfrak{U}}_\varepsilon(\hat{F}_1,\ldots,\hat{F}_k)$.

□

The equivalence of (i) and (v) can be reformulated in the following way.

(25.5) THEOREM - <u>We have</u> $\bar{\mathfrak{S}}_n = \bar{\mathfrak{T}}_n$ <u>if and only if, for any</u> $F \in \mathfrak{m}_n$ <u>and any</u> $\varepsilon > 0$, <u>there is a</u> $P \in \mathcal{J}_n$ <u>such that</u> $\varphi^j(P-F) \le \varepsilon$ <u>for all</u> $j \in \{1,\ldots,k\}$.

Note that the equality $\bar{\mathfrak{S}}_n = \bar{\mathfrak{T}}_n$ is not always true, not even in the case $k = 1$. For example, if K is φ-complete then no $\mathfrak{X} \in \mathfrak{T}_n$ with $r(\mathfrak{X}) > 1$ can be solvable, because of (2.8). On the other hand, $(\varphi^1,\ldots,\varphi^k)$-prescriptions $\bar{\mathfrak{X}} = (\mathfrak{X}^1,\ldots,\mathfrak{X}^k)$ such that $\min\{r(\mathfrak{X}^1),\ldots,r(\mathfrak{X}^k)\} = 1$ are always solvable; in fact, (25.4) and the last statement of (25.2) yield:

(25.6) THEOREM - <u>We have always</u> $\bar{\mathfrak{D}}_n \subseteq \bar{\mathfrak{S}}_n$.

This theorem can be used to get an interesting sufficient

condition for the equality $\overline{\mathfrak{S}}_n = \overline{\mathfrak{T}}_n$:

(25.7) COROLLARY - We have $\overline{\mathfrak{S}}_n = \overline{\mathfrak{T}}_n$ whenever there is a non-trivial valuation φ^o of K which is non-equivalent to $\varphi^1,\ldots,\varphi^k$ and such that $\mathfrak{D}_n^o \neq \emptyset$.

Proof: Let $\mathfrak{T}^o \in \mathfrak{D}_n^o$. For any $\overline{\mathfrak{T}} = (\mathfrak{T}^1,\ldots,\mathfrak{T}^k) \in \overline{\mathfrak{T}}_n$ and any $\varepsilon > 0$, the (k+1)-tuple $(\mathfrak{T}^o, \mathfrak{T}^1,\ldots,\mathfrak{T}^k)$ is a $(\varphi^o, \varphi^1,\ldots,\varphi^k)$-prescription of degree n and, by (25.6), has an ε-solution, since $r(\mathfrak{T}^o) = 1$. This is also an ε-solution of $\mathfrak{T}$. □

In the next section, we shall characterize those non-trivial valuations of K for which $\mathfrak{D}_n \neq \emptyset$. In particular, we shall see that any discrete valuation has this property for all $n \geq 1$.

§26 Henselian and antihenselian valuations

Let φ be a non-trivial valuation of K and n a positive integer. We call φ n-henselian (resp. n-antihenselian) if, for any separable extension L of K of degree n , the number of valuations of L which extend φ is $r = 1$ (resp. $r > 1$) . It is obvious that any (resp. no) valuation is 1-henselian (resp. 1-antihenselian). Note that φ may be n-henselian and n-antihenselian at the same time; in fact, this occurs if and only if K has no separable extension of degree n . Using the notation introduced in §23, we prove:

(26.1) The following conditions are equivalent:

(i) φ is n-antihenselian.

(ii) $\mathfrak{D}_n = \emptyset$.

(iii) $\hat{K}$ has no separable extension of degree n .

Proof: (i) ⇒ (ii): Assume $\mathfrak{D}_n \neq \emptyset$, say, $\mathfrak{X} \in \mathfrak{D}_n$. By (25.6), $\mathfrak{X}$ is solvable. Let y be a solution of $\mathfrak{X}$; then $K(y)$ is a separable extension of K of degree n and has only one valuation which extends φ . Therefore φ is not n-antihenselian.

(ii) ⇒ (iii): Assume that $\hat{K}$ has a separable extension $\hat{L}$ of degree n . Then $L = K(\hat{x})$ for some $\hat{x} \in \Omega$, and obviously $(\hat{x}) \in \mathfrak{D}_n$.

(iii) ⇒ (i): Assume that φ is not n-antihenselian and let $K(y)$ be a separable extension of K which has only one valuation extending φ . Then $P_{y|K}$ is separable and irreducible in $\hat{K}[X]$, by (2.12), and therefore has a root $\hat{y} \in \Omega$ such that $[\hat{K}(\hat{y}):\hat{K}] = n$, contradicting (iii). □

Also n-henselian valuations can be characterized by means of $\mathfrak{D}_n$; in fact:

(26.2) The following conditions are equivalent:

(i) φ is n-henselian.

(ii) Any $\mathfrak{X} \in \mathfrak{T}_n$ which has a solution is in $\mathfrak{D}_n$.

(iii) $\mathfrak{S}_n = \mathfrak{D}_n$.

In this case, any non-trivial valuation φ^o of K which is non-equivalent to φ is n-antihenselian, unless $n = 1$.

Proof: (i) ⇒ (ii): Let y be a solution of $\mathfrak{X} \in \mathfrak{T}_n$. Then $K(y)$ is a separable extension of K of degree n , and there is only one valuation of $K(y)$ which extends φ . Therefore $r(\mathfrak{X}) = 1$, $\mathfrak{X} \in \mathfrak{D}_n$.

(ii) ⇒ (iii): Obviously $\mathfrak{S}_n \subseteq \mathfrak{D}_n$, by (ii). The equality follows from (25.6).

(iii) ⇒ (i): Assume that φ is not n-henselian. Then there is a finite separable extension $K(y)$ of K of degree n which has $r > 1$ valuations extending φ . By (2.12), y is a 0-solution of some $\mathfrak{Y} = (\hat{y}_1,\ldots,\hat{y}_r) \in \mathfrak{T}_n$, hence $\mathfrak{Y} \in \mathfrak{S}_n$, but $\mathfrak{Y} \notin \mathfrak{D}_n$.

To prove the last statement, we assume that φ is n-henselian and $n > 1$, and we choose an n-tuple $Z = (\hat{z}_1,\ldots,\hat{z}_n)$ of pairwise distinct elements $\hat{z}_1,\ldots,\hat{z}_n \in \hat{K}$; obviously $Z \in \mathfrak{T}_n \setminus \mathfrak{D}_n$. Suppose that $\mathfrak{D}_n^o \neq \emptyset$, say, $\mathfrak{X}^o \in \mathfrak{D}_n^o$. Then $(\mathfrak{X}^o, Z)$ is a solvable (φ^o,φ)-prescription of degree n , by (25.6), hence $Z \in \mathfrak{S}_n$, in contradiction to (iii). Therefore $\mathfrak{D}_n^o = \emptyset$, that is, φ^o is n-antihenselian, by (26.1). □

Let $\varphi^1,\ldots,\varphi^k$ be pairwise non-equivalent non-trivial valuations of K .

(26.3) THEOREM - If φ^1 is n-henselian then $\bar{\mathfrak{S}}_n = \bar{\mathfrak{D}}_n = \mathfrak{D}_n^1 \times \mathfrak{T}_n^2 \times \ldots \times \mathfrak{T}_n^k$.

Proof: For any $\bar{\mathfrak{X}} = (\mathfrak{X}^1,\ldots,\mathfrak{X}^k) \in \bar{\mathfrak{S}}_n$ we have $\mathfrak{X}^1 \in \mathfrak{S}_n^1 = \mathfrak{D}_n^1$, by (26.2), hence $\bar{\mathfrak{S}}_n \subseteq \mathfrak{D}_n^1 \times \mathfrak{T}_n^2 \times\ldots\times \mathfrak{T}_n^k \subseteq \bar{\mathfrak{D}}_n$. The equalities hold because of (25.6). □

By (26.3) and (25.7), we have determined the set $\bar{\mathfrak{S}}_n$ of all solvable $(\varphi^1,\ldots,\varphi^k)$-prescriptions of degree n in two important cases:

(26.4) COROLLARY - a) $\bar{\mathfrak{S}}_n = \bar{\mathfrak{D}}_n$ whenever one of the valuations $\varphi^1,\ldots,\varphi^k$ is n-henselian.

b) $\bar{\mathfrak{S}}_n = \bar{\mathfrak{T}}_n$ whenever there exists a non-trivial valuation φ^o of K which is non-equivalent to $\varphi^1,\ldots,\varphi^k$ and is not n-antihenselian.

We shall prove in §28 the existence of fields with k valuations $\varphi^1,\ldots,\varphi^k$ none of which is n-henselian and such that any non-trivial valuation of K which is non-equivalent to $\varphi^1,\dots,\varphi^k$ is n-antihenselian. It would be interesting to determine $\bar{\mathfrak{S}}_n$ in this case, too.

A non-trivial valuation φ of K is called henselian (resp. antihenselian) if it is n-henselian (resp. n-antihenselian) for any integer $n > 1$. We get as an immediate consequence of (26.2):

(26.5) If φ is henselian then any non-trivial valuation of K which is non-equivalent to φ is antihenselian.

It is clear that any non-trivial valuation of a separably closed field is henselian and antihenselian. On the other hand, if K has a valuation which is henselian and antihenselian then K is separably closed. Therefore (26.5) yields:

(26.6) Any field K which is not separably closed has at most one henselian valuation (up to equivalence).

The following proposition gives several characterizations of antihenselian valuations. It shows, in particular, that such a valuation has exactly n extensions to any finite separable field extension of degree n .

(26.7) The following conditions are equivalent:

(i) φ is antihenselian.

(ii) $\hat{K}$ is separably closed, i.e., $\hat{K} = \Omega$.

(iii) For any $n \geq 1$ we have $\mathfrak{T}_n \subseteq \hat{K} \times \dots \times \hat{K}$ (n times).

(iv) For any $n \geq 1$ we have $\mathfrak{S}_n \subseteq \hat{K} \times \dots \times \hat{K}$ (n times).

(v) Any finite separable extension L of K has exactly [L:K] valuations which extend φ .

Proof: (i) ⇒ (ii) follows from (26.1).

(iii) ⇒ (iv) and (v) ⇒ (i) are obvious.

(ii) ⇒ (iii): For any $n \geq 1$ and any $\mathfrak{X} = (\hat{x}_1,\ldots,\hat{x}_r) \in \mathfrak{X}_n$ we have $\hat{x}_1,\ldots,\hat{x}_r \in \Omega = \hat{K}$ and $n = \sum_{i=1}^{r} [\hat{K}(\hat{x}_i):\hat{K}] = r$; therefore $\mathfrak{X} \in \hat{K} \times \ldots \times \hat{K}$ (n times).

(iv) ⇒ (v): Suppose there is a finite separable extension $L = K(y)$ of K of degree n which has only $r < n$ valuations extending φ . By (2.12), y is a 0-solution of some $\mathfrak{Y} = (\hat{y}_1,\ldots,\hat{y}_r) \in \mathfrak{X}_n$, hence $\mathfrak{Y} \in \mathfrak{S}_n$, contradicting (iv). □

Let φ be an <u>archimedean</u> valuation of K . We recall that the completion $(\hat{K},\hat{\varphi})$ of (K,φ) is isomorphic to $(\mathbb{C},|\ |_{\mathbb{C}}^{\rho})$ or to $(\mathbb{R},|\ |_{\mathbb{R}}^{\rho})$ for some $\rho > 0$ (cf (2.11)). In the first case we have $\Omega = \hat{K}$ and φ is called complex-archimedean; in the second case we have $[\Omega:\hat{K}] = 2$ and φ is called real-archimedean. In either case we have Char $K = 0$, hence $sc(K) = ac(K)$. We conclude from (26.1):

(26.8) <u>Let</u> φ <u>be an archimedean valuation of</u> K . <u>Then</u>:

a) φ <u>is</u> n-<u>antihenselian for any</u> $n \geq 3$.

b) φ <u>is antihenselian if and only if</u> φ <u>is complex-archimedean</u>.

It is clear that no complex-archimedean valuation of K is henselian, unless $K = ac(K)$. For real-archimedean valuations we prove:

(26.9) <u>Let</u> φ <u>be a real-archimedean valuation of</u> K . <u>Then</u> $K \neq ac(K)$, <u>and the following conditions are equivalent</u>:

(i) φ <u>is henselian</u>.

(ii) $ac(K)$ <u>is a finite extension of</u> K .

(iii) $ac(K) = K(\sqrt{-1})$.

(iv) K <u>is real-closed</u>.

<u>Proof</u>: Since φ is not antihenselian, by (26.8 b), K is not separably closed, hence $K \neq ac(K)$. The equivalences

(ii) ⇔ (iii) ⇔ (iv) follow from Artin-Schreier's theory (cf Jacobson [19], Chap. VI, §2 and §11).

(i) ⇒ (ii): For all $n \geq 3$, φ is n-antihenselian, by (26.8 a), and n-henselian. Therefore K has no finite separable extension of degree $n \geq 3$, hence $ac(K) = sc(K)$ is a finite extension of K .

(iii) ⇒ (i): None of the valuations of $ac(K)$ which extend φ can be real-archimedean. Therefore their number equals $\frac{1}{2}\cdot[ac(K):K] = 1$, by (2.13 b). □

Let K be any real-closed field which can be imbedded in $\mathbb{R}$. We conclude from (26.5), (26.8 b) and (26.9) that K has exactly one real-archimedean valuation, up to equivalence, which is henselian, whereas any other (non-archimedean or complex-archimedean) valuation of K is antihenselian, i.e. has two extensions to the algebraic closure $ac(K) = K(\sqrt{-1})$.

The meaning of "n-antihenselian" and "antihenselian" for non-archimedean valuations will be discussed at the end of the next section. Here we want to mention only the fact that any discrete valuation is non-n-antihenselian for any $n \geq 1$. In fact, since $\hat{\varphi}$ is discrete, the field $\hat{K}$ admits a separable Eisenstein extension $\hat{L}$ of degree n (i.e., $\hat{L} = \hat{K}(\hat{y})$ for some root $\hat{y} \in \Omega$ of some separable eisensteinian polynomial $\hat{F} \in \hat{K}[X]$ of degree n ; cf Exercise III-13).

§27 Prescription of value groups and residue fields

In this section, we study the possibility of prescribing value groups and residue fields, instead of completions, in the case of non-archimedean valuations. The following theorems are essentially due to Krull (see Krull [22] and Endler [6]) but they are proven here in a different way.

We first consider the special case in which for one non-archimedean non-trivial valuation φ only one extension is prescribed. Here it is convenient to replace φ by the corresponding exponential valuation v. More generally, the following theorem is proven even for Krull valuations v of arbitrary rank.

(27.1) THEOREM - Let A be a valuation ring of K, $A \neq K$, and let v (resp. π) be a Krull valuation (resp. place) of K corresponding to A with value group Γ (resp. residue field $\mathcal{K}$). For any $n \geq 1$, any totally ordered group $\Delta \supseteq \Gamma$ and any field $\mathcal{L} \supseteq \mathcal{K}$ such that $(\Delta:\Gamma)\cdot[\mathcal{L}:\mathcal{K}] = n$ there exist a field L, a valuation ring B of L, a Krull valuation w and a place ζ of K, both corresponding to B, with the following properties:

a) $L|K$ is a separable extension of degree n.

b) B is the only valuation ring of L which lies over A.

c) w lies over v and has Δ as value group.

d) ζ lies over π and has $\mathcal{L}$ as residue field.

Proof: Setting $e = (\Delta:\Gamma)$ and $f = [\mathcal{L}:\mathcal{K}]$ we have $n = e\cdot f$. We assume first that Δ/Γ is cyclic and $\mathcal{L}|\mathcal{K}$ is a simple extension. Let $\delta \in \Delta$ be a representative of a generator of Δ/Γ such that $\delta > 0$. Since $e\cdot\delta \in \Gamma$ we have $e\cdot\delta = vd$ for some $d \in K$.

Let $F \in \mathfrak{m}_f \cap A[X]$ be a representative of the minimal polynomial $P_{\alpha|\mathcal{K}} \in \mathcal{K}[X]$ of some primitive element α of $\mathfrak{L}|\mathcal{K}$. We set $G = F^e - d$. Since $G \in \mathfrak{m}_n$, there is an $H \in \mathfrak{S}_n$ such that $v(G-H) > vd$, by (24.7). Let y be a root of H, $L = K(y)$, $\mathfrak{B}$ the set of all valuation rings of L which lie over A, $B_o \in \mathfrak{B}$, and w_o (resp. ζ_o) a Krull valuation (resp. place) of L, corresponding to B_o, which lies over v (resp. π). Since F is in $A[X]$, so are G and H, hence $y \in I_L(A) = \bigcap_{B\in\mathfrak{B}} B \subseteq B_o$, by (13.3 b). Since $F(y)^e = d + G(y) = d + ((G-H)(y))$ and $w_o((G-H)(y)) \geq w_o(G-H) > w_o d$, we have $e \cdot w_o(F(y)) = w_o(F(y)^e) = w_o d = e \cdot \delta$; therefore there exists a (necessarily injective and order preserving) homomorphism g from Δ into the value group of w_o such that $g|\Gamma = \iota_\Gamma$ and $g\delta = w_o(F(y))$. Since $w_o(F(y)) > 0$, we have $P_{\alpha|\mathcal{K}}(\zeta_o y) = \zeta_o(F(y)) = 0$; therefore there exists a $\mathcal{K}$-momomorphism h from $\mathfrak{L}$ into the residue field of ζ_o such that $h\alpha = \zeta_o y$. We conclude that $[L:K] = \deg P_{y|K} \leq \deg H = n = (\Delta:\Gamma)\cdot[\mathfrak{L}:\mathcal{K}] \leq e_{B_o|K}\cdot f_{B_o|K} \leq \sum_{B\in\mathfrak{B}} e_{B|K}\cdot f_{B|K}$, and from the fundamental inequality (17.5) it follows that the equality signs hold. Therefore the separable extension $L|K$ has degree n, $\mathfrak{B} = \{B_o\}$, g and h are isomorphisms, and the Krull valuation $w = g^{-1} \circ w_o$ and the place $\zeta = h^{-1} \circ \zeta_o$ of L have the desired properties.

In the general case, since Δ/Γ and $\mathfrak{L}|\mathcal{K}$ are finitely generated, there exist finite chains $\Gamma = \Delta_o \subseteq \ldots \subseteq \Delta_s = \Delta$ and $\mathcal{K} = \mathfrak{L}_o \subseteq \ldots \subseteq \mathfrak{L}_s = \mathfrak{L}$ of intermediary totally ordered groups Δ_j and intermediary fields $\mathfrak{L}_j$ such that, for any $j \in \{1,\ldots,s\}$, Δ_j/Δ_{j-1} is cyclic and $\mathfrak{L}_j|\mathfrak{L}_{j-1}$ a simple extension. Therefore, starting with (K,A,v,π), we successively construct quadruples (L_j,B_j,w_j,ζ_j) $(j = 0,1,\ldots,s)$ such that, for any $j \in \{1,\ldots,s\}$, L_j is a separable extension of L_{j-1} of degree $(\Delta_j:\Delta_{j-1})\cdot[\mathfrak{L}_j:\mathfrak{L}_{j-1}]$, B_j is the only valuation ring of L_j lying over B_{j-1}, and w_j (resp. ζ_j) is a Krull valuation (resp. place) of L_j which lies over w_{j-1} (resp. ζ_{j-1}) and has value group Δ_j (resp. residue field $\mathcal{L}_j$). It is obvious that L_s , B_s , w_s , ζ_s have the desired properties. □

We consider again a non-archimedean non-trivial valuation φ of K . We denote by A_φ its valuation ring (of rank 1), by Γ_φ its value group (contained in the multiplicative group $\mathbb{R}_+^*$ of all positive real numbers), and by $\mathcal{K}_\varphi$ its residue field. We recall that, for any finite extension L of K , the valuations $\psi_1,\ldots,\psi_r$ of L which extend φ are in 1-1 correspondence with the valuation rings $B_1,\ldots,B_r$ of L which lie over A_φ and that the fundamental inequality $\sum_{i=1}^{r} e_i \cdot f_i \leq [L:K]$ holds, where $e_i = e_{B_i|K} = (\Gamma_{\psi_i}:\Gamma_\varphi)$ and $f_i = f_{B_i|K} = [\mathcal{K}_{\psi_i}:\mathcal{K}_\varphi]$ [35], for $i = 1,\ldots,r$. If the equality $\sum_{i=1}^{r} e_i \cdot f_i = [L:K]$ holds we say that φ , or A_φ , is <u>defectless</u> in L .

It is natural to ask whether the number r of extensions of φ and their value groups and residue fields can be prescribed arbitrarily. For this purpose, it is convenient to make the following definitions. A <u>Krull</u> φ<u>-prescription</u> of degree n and length r is any 2r-tuple $\xi = (\Delta_1,\ldots,\Delta_r, \mathcal{L}_1,\ldots,\mathcal{L}_r)$ consisting of r totally ordered groups $\Delta_i \supseteq \Gamma_\varphi$ and r fields $\mathcal{L}_i \supseteq \mathcal{K}_\varphi$ such that $\sum_{i=1}^{r} (\Delta_i:\Gamma_\varphi)\cdot[\mathcal{L}_i:\mathcal{K}_\varphi] = n$. By a <u>solution</u> of ξ we mean any element y (in a field extension of K) with the following properties:

a) y is separable over K and $[K(y):K] = n$.

b) There are exactly r valuations $\psi_1,\ldots,\psi_r$ of $K(y)$ which extend φ .

c) For any $i \in \{1,\ldots,r\}$, Δ_i is Γ_φ-isomorphic to Γ_{ψ_i} [36] and

---

[35] We identify $\mathcal{K}_\varphi$ with its canonical image in $\mathcal{K}_{\psi_i}$, as was explained in §3.

[36] We say that Δ is Γ-isomorphic to Δ' if there exists an isomorphism (of ordered groups) $\Delta \to \Delta'$ which induces the identity ι_Γ of Γ .

$\mathfrak{L}_i$ is $\mathcal{K}_\varphi$-isomorphic to $\mathcal{K}_{\psi_i}$.

In the special case $r = 1$, Theorem (27.1) yields:

(27.2) COROLLARY - Any Krull φ-prescription $(\Delta,\mathfrak{L})$ of length 1 has a solution.

Proof: Since $(\Delta:\Gamma_\varphi) < \infty$ and Γ_φ is a subgroup of the divisible group $\mathbb{R}_+^*$, we may assume that Δ is a subgroup of $\mathbb{R}_+^*$, too. By (3.5), $v = -\log\varphi$ is an exponential valuation of K corresponding to A_φ with value group $\Gamma = \log\Gamma_\varphi$, and $\Delta' = \log\Delta$ is a totally ordered group containing Γ such that $(\Delta':\Gamma) = (\Delta:\Gamma_\varphi)$. Setting $A = A_\varphi$, $\pi = \pi_A$, $\mathcal{K} = \mathcal{K}_\varphi$ and replacing Δ by Δ', we construct L, B, w, ζ according to (27.1). Then $\psi = e^{-w}$ is the only valuation of L which lies over φ, we have $A_\psi = B$, $\Gamma_\psi = \Delta$, and ζ induces a $\mathcal{K}_\varphi$-isomorphism from $\mathcal{K}_\psi$ onto $\mathfrak{L}$. Moreover, $L|K$ has a primitive element y, and y is a solution of $(\Delta,\mathfrak{L})$. □

For Krull φ-prescriptions ξ of arbitrary length $r \geq 1$, the question of existence of a solution will be reduced to the analogous question for φ-prescriptions $\mathfrak{X}$, which we considered in §25. We first note that any Krull φ-prescriptions $\xi = (\Delta_1,\ldots,\Delta_r,\mathfrak{L}_1,\ldots,\mathfrak{L}_r)$ of length r yields r Krull φ-prescriptions $\xi_i = (\Delta_i,\mathfrak{L}_i)$ of length 1 $(i = 1,\ldots,r)$, which can also be considered as Krull $\hat{\varphi}$-prescriptions, since $\Gamma_{\hat{\varphi}} = \Gamma_\varphi$ and $\mathcal{K}_{\hat{\varphi}} = \mathcal{K}_\varphi$ (cf (3.12)). Let $\mathfrak{X} = (\hat{x}_1,\ldots,\hat{x}_r) \in \mathfrak{X}_n$ be a φ-prescription (in the sense of §23) of degree n and length r. We say that $\mathfrak{X}$ is defectless if, for any $i \in \{1,\ldots,r\}$, $\hat{\varphi}$ is defectless in $\hat{K}(\hat{x}_i)$. We say that ξ is derived from $\mathfrak{X}$ if, for any $i \in \{1,\ldots,r\}$, $\hat{x}_i$ is a solution of the Krull $\hat{\varphi}$-prescription $(\Delta_i,\mathfrak{L}_i)$.

It is obvious that Krull φ-prescriptions ξ can be derived only from defectless φ-prescriptions $\mathfrak{X}$. On the other hand, we show that, for any defectless φ-prescription $\mathfrak{X}$, there is one, and essentially only one, Krull φ-prescription ξ which is derived from $\mathfrak{X}$, and that any solution of $\mathfrak{X}$ is also a solution of ξ.

(27.3) Let $\mathfrak{X} = (\hat{x}_1,\dots,\hat{x}_r) \in \mathfrak{X}_n$ be defectless and, for any $i \in \{1,\dots,r\}$, let $\hat{\varphi}_i$ be the unique valuation of $\hat{K}(\hat{x}_i)$ which extends $\hat{\varphi}$. Then:

a) $(\Gamma_{\hat{\varphi}_1},\dots,\Gamma_{\hat{\varphi}_r}, \mathcal{K}_{\hat{\varphi}_1},\dots,\mathcal{K}_{\hat{\varphi}_r})$ is a Krull φ-prescription which is derived from $\mathfrak{X}$.

b) Any Krull φ-prescription $\xi = (\Delta_1,\dots,\Delta_r,\mathcal{L}_1,\dots,\mathcal{L}_r)$ which is derived from $\mathfrak{X}$ has degree n and, for any $i \in \{1,\dots,r\}$, Δ_i is Γ_φ-isomorphic to $\Gamma_{\hat{\varphi}_i}$ and $\mathcal{L}_i$ is $\mathcal{K}_\varphi$-isomorphic to $\mathcal{K}_{\hat{\varphi}_i}$.

c) Any solution of $\mathfrak{X}$ is a solution of ξ.

Proof: a) is obvious.

b) If $\xi = (\Delta_1,\dots,\Delta_r, \mathcal{L}_1,\dots,\mathcal{L}_r)$ is derived from $\mathfrak{X}$ then, for any $i \in \{1,\dots,r\}$, $\hat{x}_i$ is a solution of the Krull $\hat{\varphi}$-prescription $(\Delta_i,\mathcal{L}_i)$; therefore Δ_i is Γ_φ-isomorphic to $\Gamma_{\hat{\varphi}_i}$ and $\mathcal{L}_i$ is $\mathcal{K}_\varphi$-isomorphic to $\mathcal{K}_{\hat{\varphi}_i}$. Moreover, ξ has degree $\sum_{i=1}^{r} (\Delta_i:\Gamma_\varphi)\cdot[\mathcal{L}_i:\mathcal{K}_\varphi] = \sum_{i=1}^{r} (\Gamma_{\hat{\varphi}_i}:\Gamma_\varphi)\cdot[\mathcal{K}_{\hat{\varphi}_i}:\mathcal{K}_\varphi] = \sum_{i=1}^{r} [\hat{K}(\hat{x}_i):\hat{K}] = n$, since $\mathfrak{X}$ is defectless.

c) Let ξ be derived from $\mathfrak{X}$ and let y be a solution of $\mathfrak{X}$. By (23.1) we have $[K(y):K] = n$, φ has exactly r extensions $\psi_1,\dots,\psi_r$ to $K(y)$ and, for any $i \in \{1,\dots,r\}$, there is a completion $\mu_i: (K(y),\psi_i) \to (\hat{K}(\hat{x}_i),\hat{\varphi}_i)$ such that $\mu_i|K = \iota_K$. From (3.12) we conclude that, for any $i \in \{1,\dots,r\}$, μ_i induces a Γ_φ-isomorphism from Γ_{ψ_i} onto $\Gamma_{\hat{\varphi}_i}$ and a $\mathcal{K}_\varphi$-isomorphism from $\mathcal{K}_{\psi_i}$ onto $\mathcal{K}_{\hat{\varphi}_i}$; therefore, by b), Γ_{ψ_i} is Γ_φ-isomorphic to Δ_i and $\mathcal{K}_{\psi_i}$ is $\mathcal{K}_\varphi$-isomorphic to $\mathcal{L}_i$. Hence y is a solution of ξ. □

Conversely, we prove by means of (27.2):

(27.4) Any Krull φ-prescription ξ of degree n is derived from some φ-prescription $\mathfrak{X} \in \mathfrak{X}_n$.

Proof: Let $\xi = (\Delta_1,\dots,\Delta_r , \mathfrak{L}_1,\dots,\mathfrak{L}_r)$. For any $i \in \{1,\dots,r\}$, the Krull $\hat{\varphi}$-prescription $(\Delta_i,\mathfrak{L}_i)$ has a solution $\hat{y}_i \in \Omega$, by (27.2), and for any $\hat{a}_i \in \hat{K}$, $\hat{y}_i+\hat{a}_i \in \Omega$ is a solution, too. The elements $\hat{a}_1,\dots,\hat{a}_r \in \hat{K}$ can be chosen such that $\hat{y}_1+\hat{a}_1,\dots,\hat{y}_r+\hat{a}_r$ are pairwise non-conjugate over $\hat{K}$. Therefore $\mathfrak{X} = (\hat{y}_1+\hat{a}_1,\dots,\hat{y}_r+\hat{a}_r)$ is a φ-prescription, ξ is derived from $\mathfrak{X}$ and, by (27.3), $\mathfrak{X} \in \mathfrak{X}_n$. □

Note that a Krull φ-prescription ξ may be derived from different φ-prescriptions $\mathfrak{X}$, $\mathfrak{Y}$, even in the case of length 1. For example, let φ be the p-adic valuation of $K = \mathbb{Q}$ (for some prime number p), let $n \in \mathbb{N}$ such that $n > 1$ and $(p,n) = 1$, let $\hat{x} \in \Omega$ be an n-th root of p and $\hat{y} \in \Omega$ be an n-th root of $\hat{u}\cdot p$, where $\hat{u} \in \hat{K} = \hat{\mathbb{Q}}_p$ is a primitive $(p-1)$ root of unity (cf the end of §5). The Krull φ-prescription $(\Gamma_\varphi^{\frac{1}{n}} ,\mathfrak{K}_\varphi)$ is derived from both $(\hat{x})$ and $(\hat{y})$, but $\hat{K}(\hat{x})$ is not $\hat{K}$-isomorphic to $\hat{K}(\hat{y})$ (see Hasse [16], §16).

Let $\varphi^1,\dots,\varphi^k$ be pairwise non-equivalent non-archimedean non-trivial valuations of K . By a Krull $(\varphi^1,\dots,\varphi^k)$-prescription of degree n we mean any k-tuple $\bar{\xi} = (\xi^1,\dots,\xi^k)$, where ξ^j is a Krull φ^j-prescription of degree n $(j = 1,\dots,k)$. Note that $\xi^1,\dots,\xi^k$ may have different lengths. Any common solution of $\xi^1,\dots,\xi^k$ is called a solution of $\bar{\xi}$.

Let T_n be the set of all Krull $(\varphi^1,\dots,\varphi^k)$-prescriptions of degree n and Σ_n the set of those which have a solution. We prove the following analogue of Theorem (25.5):

(27.5) THEOREM - We have $\Sigma_n = T_n$ if for any $F \in \mathfrak{M}_n$ and any $\epsilon > 0$ there is a $P \in \mathcal{J}_n$ such that $\varphi^j(P-F) \leq \epsilon$ for all

$j \in \{1,\dots,k\}$.

Proof: Let $\bar{\xi} = (\xi^1,\dots,\xi^k) \in T_n$. For any $j \in \{1,\dots,k\}$, ξ^j is derived from some $\mathfrak{x}^j \in \mathfrak{X}_n^j$, by (27.4), and we have $\bar{\mathfrak{x}} = (\mathfrak{x}^1,\dots,\mathfrak{x}^k) \in \bar{\mathfrak{S}}_n$, by (25.5); therefore $\mathfrak{x}^1,\dots,\mathfrak{x}^k$ have a common solution y . By (27.3 c), y is a solution of $\bar{\xi}$, hence $\bar{\xi} \in \Sigma_n$. □

Let Ξ_n be the set of those $\bar{\xi} = (\xi^1,\dots,\xi^k) \in T_n$ such that at least one of the components $\xi^1,\dots,\xi^k$ has length 1. Using the same argument as in the proof of (27.5), we conclude from (25.6) and (25.7) (or (26.4 b)):

(27.6) THEOREM - We have always $\Xi_n \subseteq \Sigma_n$.

(27.7) THEOREM - We have $\Sigma_n = T_n$ whenever there exists a non-trivial valuation φ^o of K which is non-equivalent to $\varphi^1,\dots,\varphi^k$ and is not n-antihenselian.

The analogue of (26.4 a) is also true:

(27.8) THEOREM - $\Sigma_n = \Xi_n$ whenever one of the valuations $\varphi^1,\dots,\varphi^k$ is n-henselian.

Proof: We may assume that φ^1 is n-henselian. Let $\bar{\xi} = (\xi^1,\dots,\xi^k) \in \Sigma_n$ and let y be a solution of $\bar{\xi}$. Then $K(y)$ is a separable extension of K of degree n , and there is only one valuation of $K(y)$ which extends φ^1 . Since y is a solution of the Krull φ^1-prescription ξ^1 , ξ^1 must have length 1. Therefore $\bar{\xi} \in \Xi_n$. □

Note that the valuation φ^o in (27.7) need not be non-archimedean. In fact, if $n = 2$ then any real-archimedean valuation of K has the desired property. Therefore, if the field K can be imbedded in $\mathbb{R}$ then all Krull $(\varphi^1,\dots,\varphi^k)$-prescriptions of degree 2 have solutions.

For arbitrary n, any discrete valuation is non-n-antihenselian, as was proven at the end of §26. Therefore, if K has infinitely many non-equivalent discrete valuations then all Krull $(\varphi^1,\dots,\varphi^k)$-prescriptions of any degree n have solutions (and so have all $\bar{\mathfrak{X}} \in \bar{\mathfrak{X}}_n$). This occurs, in particular, if K is an algebraic number field or an algebraic function field.

Besides the discrete valuations, there exist also many other non-archimedean valuations which are non-n-antihenselian. In fact, we show that any non-archimedean non-trivial valuation φ of K is non-n-antihenselian, unless it is n-saturated in the following sense: φ is called n-saturated if $(\Delta:\Gamma_\varphi)\cdot[\mathfrak{L}:\mathcal{K}_\varphi] \neq n$ for any totally ordered group $\Delta \supseteq \Gamma_\varphi$ and any field $\mathfrak{L} \supseteq \mathcal{K}_\varphi$. The notions "n-antihenselian" and "n-saturated" are tightly related to each other; in fact:

(27.9) Let φ be a non-archimedean non-trivial valuation of K, and let $n > 1$.

a) If φ is n-antihenselian then φ is n-saturated.

b) If φ is n-saturated and is defectless in any separable extension of K of degree n, then φ is n-antihenselian.

Proof: a) If φ is not n-saturated then $(\Delta:\Gamma_\varphi)\cdot[\mathfrak{L}:\mathcal{K}_\varphi] = n$ for some $\Delta \supseteq \Gamma_\varphi$ and some $\mathfrak{L} \supseteq \mathcal{K}_\varphi$, and $(\Delta,\mathfrak{L})$ can be considered as a Krull $\hat{\varphi}$-prescription of degree n and length 1. It has a solution $\hat{y}$, by (27.2), and since $\hat{K}(\hat{y})$ is a separable extension of $\hat{K}$ of degree n, φ is not n-antihenselian, by (26.1).

b) Suppose that φ is not n-antihenselian; then it has exactly one extension ψ to some separable extension L of K of degree n. Since φ is defectless in L, we have $(\Gamma_\psi:\Gamma_\varphi)\cdot[\mathcal{K}_\psi:\mathcal{K}_\varphi] = n$, hence φ is not n-saturated. ☐

From (27.9) we get another proof of the fact that any discrete valuation is non-n-antihenselian:

(27.10) Any discrete valuation φ is non-n-saturaded and non-n-antihenselian for any $n \geq 1$.

Proof: Since Γ_φ is isomorphic to the additive group $\mathbb{Z}$, we have $(\Gamma_\varphi^{\frac{1}{n}}:\Gamma_\varphi)\cdot[\mathcal{K}_\varphi:\mathcal{K}_\varphi] = (\frac{1}{n}\mathbb{Z}:\mathbb{Z}) = n$, hence φ is not n-saturated. By (27.9 a), φ is not n-antihenselian. □

From (27.9 a) and (26.2) we conclude:

(27.11) a) If K has no separable extension of degree n then any non-archimedean non-trivial valuation of K is n-saturated.

b) If K has an n-henselian non-trivial valuation φ then any non-archimedean non-trivial valuation of K which is non-equivalent to φ is n-saturated, unless $n = 1$.

A non-archimedean non-trivial valuation φ is called saturated if it is n-saturated for any $n > 1$. It is obvious that this occurs if and only if Γ_φ is a divisble group (i.e., $\Gamma_\varphi^m = \Gamma_\varphi$ for all $m \geq 1$) and $\mathcal{K}_\varphi$ is algebraically closed. The following statement is an immediate consequence of (27.11).

(27.12) a) If K is separably closed then all non-archimedean non-trivial valuations of K are saturated.

b) If K has a henselian non-trivial valuation φ then any non-archimedean non-trivial valuation of K which is non-equivalent to φ is saturated.

Note that the valuation φ in (27.12) and (27.11) need not be non-archimedean. For example, if K is a real-closed subfield of $\mathbb{R}$ then the restriction of the absolute value $|\ |_{\mathbb{R}}$ to

K is henselian, by (26.9); therefore all non-archimedean non-trivial valuations of K are saturated, by (27.12 b). Moreover, we conclude from (27.12 b) and (2.8) that if φ is non-trivial and (K,φ) is complete then any non-archimedean non-trivial valuation of K which is non-equivalent to φ is saturated and, in particular, non-discrete.

Finally, we prove as an analogue of (27.9):

(27.13) For any non-archimedean non-trivial valuation φ of K the following conditions are equivalent:

(i) φ is antihenselian.

(ii) φ is saturated and is defectless in any finite separable extension of K .

Proof: (i) ⇒ (ii): By (27.9 a) φ is saturated. Any finite separable extension L of K has exactly $n = [L:K]$ valuations $\psi_1,\dots,\psi_n$ which extend φ , by (26.7), hence $n \le \sum_{i=1}^{n} (\Gamma_{\psi_i}:\Gamma_\varphi)\cdot[\mathcal{K}_{\psi_i}:\mathcal{K}_\varphi] \le n$; therefore φ is defectless in L .

(ii) ⇒ (i) follows immediately from (27.9 b). □

For a generalization of (27.13) to Krull valuations see (17.15) and Exercise IV-11. Note that the condition of defectlessness is irrelevant whenever the residue field has characteristic zero (cf (20.23)).

§28 The case of infinite field extensions

We recall that there exist fields with exactly one discrete valuation (up to equivalence); for example, $\hat{\varphi}_p$ is the only discrete valuation of the field of p-adic numbers $\hat{\mathbb{Q}}_p$ (cf §5 and §27). The question, posed by Ribenboim, whether for any positive integer k there exist fields L with exactly k discrete valuations $\psi^1,\dots,\psi^k$ (up to equivalence) was answered affirmatively by Krull [23]. Ribenboim [31] dealt with the additional question whether the residue fields of these valuations can be prescribed arbitrarily. He showed that this is not always possible, but it is possible under the hypothesis that the prescribed fields are algebraic (not necessarily finite) extensions of residue fields $\mathcal{K}_{\varphi^1},\dots,\mathcal{K}_{\varphi^k}$, respectively, where $\varphi^1,\dots,\varphi^k$ are non-equivalent discrete valuations of some field K [37]; in this case, L can be obtained as a separable extension of K and $\psi^1,\dots,\psi^k$ as extensions of $\varphi^1,\dots,\varphi^k$ to L, respectively.

More generally, let $\varphi^1,\dots,\varphi^k$ be pairwise non-equivalent non-archimedean non-trivial (but not necessarily discrete) valuations of K. Prescribing any 2k-tuple $(\Delta^1,\dots,\Delta^k, \mathfrak{L}^1,\dots,\mathfrak{L}^k)$ such that, for each $j \in \{1,\dots,k\}$, Δ^j is a <u>torsion extension</u> of Γ_{φ^j} (i.e., a totally ordered group containing Γ_{φ^j} such that $\Delta^j/\Gamma_{\varphi^j}$ is a torsion group) and $\mathfrak{L}^j$ is an algebraic extension of $\mathcal{K}_{\varphi^j}$, we ask whether there exists a separable extension L of K with the following property: For any $j \in \{1,\dots,k\}$ there is a valuation ψ^j of L extending φ^j whose value group is Γ_{φ^j}-isomorphic to Δ^j and whose residue field is $\mathcal{K}_{\varphi^j}$-isomorphic to $\mathfrak{L}^j$. One should note that we are interested only in one of the (possibly infinitely many) valuations of L which extend φ^j.

The following theorem shows that the answer is affirmative and that L can be chosen such that all non-trivial valuations of

---

[37] This hypothesis holds, in particular, whenever the prescribed fields have the same characteristic (cf Ribenboim [31]).

L which are non-equivalent to $\psi^1,\dots,\psi^k$ are antihenselian and therefore non-discrete.

(28.1) THEOREM - <u>For any</u> $j \in \{1,\dots,k\}$ <u>let</u> Δ^j <u>be a torsion extension of</u> Γ_{φ^j} <u>and</u> $\mathfrak{L}^j$ <u>be an algebraic extension of</u> $\mathcal{K}_{\varphi^j}$. <u>There exist a separable extension</u> L <u>of</u> K <u>and valuations</u> $\psi^1,\dots,\psi^k$ <u>of</u> L <u>with the following properties</u>:

a) <u>For any</u> $j \in \{1,\dots,k\}$, ψ^j <u>extends</u> φ^j , Γ_{ψ^j} <u>is</u> Γ_{φ^j}<u>-isomorphic to</u> Δ^j , <u>and</u> $\mathcal{K}_{\psi^j}$ <u>is</u> $\mathcal{K}_{\varphi^j}$<u>-isomorphic to</u> $\mathfrak{L}^j$.

b) <u>Any non-trivial valuation</u> ψ <u>of</u> L <u>which is non-equivalent to</u> $\psi^1,\dots,\psi^k$ <u>is antihenselian.</u>

<u>Proof</u>: Let $\mathfrak{S}$ be the set of all (3k+1)-tuples $\Phi_\iota = (K_\iota, \varphi_\iota^1,\dots,\varphi_\iota^k, \gamma_\iota^1,\dots,\gamma_\iota^k, \varkappa_\iota^1,\dots,\varkappa_\iota^k)$ such that $K \subseteq K_\iota \subseteq sc(K)$, φ_ι^j is a valuation of K_ι which extends φ^j , $\gamma_\iota^j : \Gamma_{\varphi_\iota^j} \to \Delta^j$ is a Γ_{φ^j}-monomorphism and $\varkappa_\iota^j : \mathcal{K}_{\varphi_\iota^j} \to \mathfrak{L}^j$ is a $\mathcal{K}_{\varphi^j}$-monomorphism, for any $j \in \{1,\dots,k\}$. The set $\mathfrak{S}$ is non-empty and is inductively ordered with respect to its natural ordering, as can be checked easily. By Zorn's lemma, $\mathfrak{S}$ has a maximal element $\Psi = (L, \psi^1,\dots,\psi^k, \delta^1,\dots,\delta^k, \lambda_1,\dots,\lambda^k)$.

a) We claim that $\delta^j : \Gamma_{\psi^j} \to \Delta^j$ and $\lambda^j : \mathcal{K}_{\psi^j} \to \mathfrak{L}^j$ are isomorphisms, for any $j \in \{1,\dots,k\}$. Otherwise we may assume that δ^1 or λ^1 is not surjective. In the first case, there is a totally ordered group $\Delta_o^1 \supseteq \Gamma_{\psi^1}$ such that $(\Delta_o^1 : \Gamma_{\psi^1}) < \infty$, say $= m$, and δ^1 extends to a Γ_{φ^1}-monomorphism $\Delta_o^1 \to \Delta^1$. Let $\xi^1 = (\Delta_o^1, \mathcal{K}_{\psi^1})$ and, for any $j \in \{2,\dots,k\}$, let ξ^j be the 2m-tuple $(\Gamma_{\psi^j},\dots,\Gamma_{\psi^j}, \mathcal{K}_{\psi^j},\dots,\mathcal{K}_{\psi^j})$. By (27.6) the Krull $(\psi^1,\dots,\psi^k)$-prescription $\xi = (\xi^1,\dots,\xi^k)$ has a solution $y \in sc(K)$. It is obvious that, by means of extensions χ^j of ψ^j to $L(y)$ $(j = 1,\dots,k)$, we can construct an element of $\mathfrak{S}$ which is strictly larger than Ψ , contradicting the maximality of ψ . If λ^1 is not surjective, one argues similarly.

b) Suppose that there is a non-trivial valuation ψ^o of L which is non-equivalent to $\psi^1,\dots,\psi^k$ and is not antihenselian; then ψ^o is non-n-antihenselian for some $n > 1$. For any $j \in \{1,\dots,k\}$ let ξ^j be the 2n-tuple $(\Gamma_{\psi^j},\dots,\Gamma_{\psi^j}, \mathcal{K}_{\psi^j},\dots,\mathcal{K}_{\psi^j})$; then $\overline{\xi} = (\xi^1,\dots,\xi^k)$ is a Krull $(\psi^1,\dots,\psi^k)$-prescription of degree n and has a solution $y \in sc(K)$, by (27.7). It is obvious that, by means of extensions χ^j of ψ^j to $L(y)$ $(j = 1,\dots,k)$, we can construct an element of $\mathcal{S}$ which is strictly larger than Ψ , contradicting the maximality of Ψ . □

Prescribing, in particular, the 2k-tuple $(\Gamma_{\varphi^1},\dots,\Gamma_{\varphi^k}, \mathcal{K}_{\varphi^1},\dots,\mathcal{K}_{\varphi^k})$ of improper extensions, we obtain:

(28.2) COROLLARY - <u>There exist a separable extension</u> L <u>of</u> K <u>and valuations</u> $\psi^1,\dots,\psi^k$ <u>of</u> L <u>with the following properties:</u>

a) (L,ψ^j) <u>is an immediate extension of</u> (K,φ^j) <u>for any</u> $j \in \{1,\dots,k\}$.

b) <u>Any non-trivial valuation</u> ψ <u>of</u> L <u>which is non-equivalent to</u> $\psi^1,\dots,\psi^k$ <u>is antihenselian.</u>

c) <u>If</u> φ^1 <u>is not n-saturated then</u> $\psi^2,\dots,\psi^k$ <u>are not n-henselian.</u>

<u>Proof</u>: a) and b) are immediate consequences of (28.1).

c) Assume that $k > 1$ and that one of the valuations $\psi^2,\dots,\psi^k$ is n-henselian. Then ψ^1 is n-antihenselian, by (26.2), and therefore n-saturated, by (27.9 a). Since (L,ψ^1) is an immediate extension of (K,φ^1) , the valuation φ^1 is n-saturated, too. □

Note that statement b) of Theorem (28.1) and of Corollary

(28.2) refers to non-archimedean as well as to archimedean valuations ψ .

If $\varphi^1,\dots,\varphi^k$ are discrete valuations of K then the valuations $\psi^1,\dots,\psi^k$ are the only discrete valuations of L (up to equivalence), whereas any other non-archimedean non-trivial valuation of L is saturated and defectless and any archimedean valuation of L is complex-archimedean (cf (27.13) and (26.8 b)). Moreover, if $k \geq 2$ then $\varphi^1,\dots,\varphi^k$ are non-n-henselian for any $n > 1$, and therefore L admits separable extensions of any degree n .

Assuming the existence of a non-m-antihenselian non-trivial valuation φ^o of K which is non-equivalent to $\varphi^1,\dots,\varphi^k$, we can generalize Theorem (28.1) by prescribing, for any $j \in \{1,\dots,k\}$, the value groups $\Delta_1^j,\dots,\Delta_{r_j}^j$ and residue fields $\mathfrak{L}_1^j,\dots,\mathfrak{L}_{r_j}^j$ for r_j extensions $\psi_1^j,\dots,\psi_{r_j}^j$ of φ^j , where $r_j \leq m$.

(28.3) COROLLARY - _For any_ $j \in \{1,\dots,k\}$ _let_ $\Delta_1^j,\dots,\Delta_{r_j}^j$ _be torsion extensions of_ Γ_{φ^j} _and_ $\mathfrak{L}_1^j,\dots,\mathfrak{L}_{r_j}^j$ _be algebraic extensions of_ $\mathcal{K}_{\varphi^j}$. _Assume that_ K _has a non-trivial valuation_ φ^o _which is non-equivalent to_ $\varphi^1,\dots,\varphi^k$ _and is non-m-antihenselian for some_ $m \geq \max\{r_1,\dots,r_k\}$. _Then there exist a separable extension_ L _of_ K _and valuations_ ψ_i^j _of_ L $(i = 1,\dots,r_j;\ j = 1,\dots,k)$ _with the following properties:_

a) _For any_ $j \in \{1,\dots,k\}$ _and any_ $i \in \{1,\dots,r_j\}$, ψ_i^j _extends_ φ^j , $\Gamma_{\psi_i^j}$ _is_ Γ_{φ^j}_-isomorphic to_ Δ_i^j , _and_ $\mathcal{K}_{\psi_i^j}$ _is_ $\mathcal{K}_{\varphi^j}$_-isomorphic to_ $\mathfrak{L}_i^j$.

b) _Any non-trivial valuation of_ L _which is non-equivalent to all valuations_ ψ_i^j $(i = 1,\dots,r_j;\ j = 1,\dots,k)$ _is antihenselian._

Proof: For any $j \in \{1,\dots,k\}$ let ξ^j be the 2m-tuple $(\Gamma_{\varphi^j},\dots,\Gamma_{\varphi^j} , \mathcal{K}_{\varphi^j},\dots,\mathcal{K}_{\varphi^j})$; then $\bar{\xi} = (\xi^1,\dots,\xi^k)$ is a

Krull $(\varphi^1,\ldots,\varphi^k)$-prescription of degree m and has a solution $z \in sc(K)$, by (27.7). For any $j \in \{1,\ldots,k\}$, $K(z)$ has m valuations $\varphi_1^j,\ldots,\varphi_m^j$ which extend φ^j , and each of them has Γ_{φ^j} as value group and $\mathcal{K}_{\varphi^j}$ as residue field. Applying (28.1) on $K(z)$ and the valuations φ_i^j of $K(z)$ $(i = 1,\ldots,r_j ; j = 1,\ldots,k)$, we obtain a field L and valuations ψ_i^j of L $(i = 1,\ldots,r_j;j=1,\ldots,k)$ with the desired properties. □

We mention without proof that, in the hypothesis of Corollary (28.3), the non-m-antihenselian valuation φ^o may be replaced by two non-equivalent non-antihenselian valuations (for example, two non-equivalent real-archimedean valuations). Note also that the hypothesis of (28.3) is satisfied whenever K has infinitely many non-equivalent discrete valuations.

Finally, applying (28.3) on the field $\mathbb{Q}$ and replacing non-archimedean valuations by the corresponding exponential valuations, we obtain:

(28.4) COROLLARY - Let $\Delta_1,\ldots,\Delta_s$ be subgroups of the additive group of $\mathbb{Q}$ which contain $\mathbb{Z}$ and let $\mathcal{L}_1,\ldots,\mathcal{L}_s$ be absolutely algebraic fields of prime characteristics. There exist an algebraic extension L of $\mathbb{Q}$ and exponential valuations $w_1,\ldots,w_s$ of L with the following properties:

a) For any $h \in \{1,\ldots,s\}$, $w_h|\mathbb{Q}$ is normalized, Δ_h is the value group of w_h , and $\mathcal{L}_h$ is isomorphic to the residue field of w_h .

b) Any non-trivial exponential valuation of L which is non-equivalent to $w_1,\ldots,w_s$ has the additive group $\mathbb{Q}$ as value group, has an algebraically closed absolutely algebraic field of prime characteristic as residue field, and has $[M:L]$ extensions to any finite field extension M of L .

c) L <u>has no real-archimedean valuation</u>.

<u>Proof</u>: Let $p_1,\ldots,p_k$ be the distinct prime numbers occurring as characteristics of the fields $\mathfrak{L}_1,\ldots,\mathfrak{L}_s$. Renumber these fields in the form $\mathfrak{L}_i^j$, where $j = 1,\ldots,k$, and $i = 1,\ldots,r_j$, and $r_1 +\ldots+ r_k = s$, so that Char $\mathfrak{L}_i^j = p_j$. For any $j \in \{1,\ldots,k\}$ let φ^j be the non-archimedean valuation of $\mathbb{Q}$ which corresponds by (3.5) to the normalized p_j-adic valuation v_{p_j} of $\mathbb{Q}$. For any $i \in \{1,\ldots,r_j\}$, the isomorphic image of Δ_i^j under the exponential function is a torsion extension of Γ_{φ^j} , and $\mathfrak{L}_i^j$ is isomorphic to an algebraic extension of $\mathcal{K}_{\varphi^j}$. Let L and ψ_i^j be constructed as in (28.3). Then the exponential valuation w_i^j of L which corresponds to ψ_i^j by (3.5) has obviously the property stated in a), for any $i \in \{1,\ldots,r_j\}$ and $j \in \{1,\ldots,k\}$. Let w be any non-trivial exponential valuation of L . By §4 and (13.11), the value group of w is a non-trivial subgroup of the additive group of $\mathbb{Q}$, and its residue field is an absolutely algebraic field of prime characteristic. If w is non-equivalent to w_i^j , for all $i \in \{1,\ldots,r_j\}$ and $j \in \{1,\ldots,k\}$, then the corresponding non-archimedean valuation ψ of L is non-trivial and non-equivalent to all ψ_i^j $(i = 1,\ldots,r_j \,;\, j = 1,\ldots,k)$ and therefore is anti-henselian, by (28.3 b). The statement b) therefore follows from (27.13). Any archimedean valuation of L is antihenselian, by (28.3 b), hence is complex-archimedean, by (26.8 b). □

In particular, assigning to each prime number p a non-negative integer r_p such that $r_p = 0$ for almost all p , there exists an algebraic extension L of $\mathbb{Q}$ such that, for any p , L has exactly r_p discrete exponential valuations $w_{p,i}$ which extend v_p , $(L,w_{p,i})$ is an immediate extension of $(\mathbb{Q},v_p)$ $(i = 1,\ldots,r_p)$, and all non-trivial exponential valuations of L which are non-equivalent to all $w_{p,i}$ are saturated.

Exercises

Chapter I

I-1) Using the notation of (1.16), let α be any mapping from $\mathcal{P} \cup \{\infty\}$ into $\mathbb{R}$ such that $\prod_{p \in \mathcal{P} \cup \{\infty\}} (\varphi_p x)^{\alpha(p)} = 1$ for all non-zero $x \in \mathbb{Q}$. Prove that α is constant.

I-2) What is the relation between the approximation theorem (1.8) and the Chinese remainder theorem for the ring $\mathbb{Z}$?

I-3) Let φ be a non-archimedean valuation of K and let $x_n \in K$ for any $n \in \mathbb{N}$. Prove:

a) $(x_n)_{n \in \mathbb{N}}$ is φ-Cauchy if and only if $(x_{n+1} - x_n)_{n \in \mathbb{N}}$ φ-converges to zero.

b) If $(\sum_{i=0}^{n} \varphi x_i)_{n \in \mathbb{N}}$ is convergent (in $\mathbb{R}$) then $(\sum_{i=0}^{n} x_i)_{n \in \mathbb{N}}$ is φ-Cauchy.

Which of these statements are true for archimedean valuations?

I-4) Let v be an exponential valuation of K and let $L = K(x)$, where x is transcendental over K. Prove that there exists exactly one exponential valuation w of L such that $w(\sum_{i=0}^{n} a_i \cdot x^i) = \min\{v\, a_o, \ldots, v\, a_n\}$ for all $a_o, \ldots, a_n \in K$, $n \in \mathbb{N}$. Is w the only exponential valuation of L which extends v ?

I-5) Let K and v be as in I-4). Prove that for any field extension L of K there is an exponential valuation of L which extends v. (Hint: Extend v first to a maximal pure transcendental subextension L_o of $L|K$.)

I-6) For some prime numbers p and some rational numbers r, indicate the p-adic expansion $r = \sum_{i=-\infty}^{\infty} a_i \cdot p^i$ (with $a_i \in \{0, 1, \ldots, p-1\}$ for all $i \in \mathbb{Z}$).

I-7) Let p , q be distinct prime numbers. Prove that $(q^n)_{n\in\mathbb{N}}$ has a convergent subsequence in $\hat{\mathbb{Q}}_p$. Does the sequence itself converge in $\hat{\mathbb{Q}}_p$?

I-8) Using the notation of (4.4), let $y = \frac{a\cdot z + b}{c\cdot z + d}$, where a , b , c , $d \in K_o$ and $a\cdot d - b\cdot c \neq 0$. For any $P \in \mathfrak{J} \cup \{\infty\}$ determine an element $\pi P \in \mathfrak{J} \cup \{\infty\}$ such that $v_{z;P} = v_{y;\pi P}$, and show that the mapping $P \mapsto \pi P$ is a permutation of $\mathfrak{J} \cup \{\infty\}$.

I-9) Try to generalize (4.4) and Exercise I-8) to pure transcendental extensions $K_o(z_1,\ldots,z_n)$ of K of finite transcendence degree $n \geq 1$.

I-10) Let A be a valuation ring of K . Prove or disprove the following statements:

a) Any A-submodule M of K such that $M \neq K$ is a fractionary ideal of A .

b) Let M_1 , M_2 be A-submodules of K such that $A \subseteq M_1 \subset M_2$. There is some ideal $\mathfrak{A}$ of A such that $\mathfrak{A}\cdot M_1 \subseteq A$ but $\mathfrak{A}\cdot M_2 \nsubseteq A$.

I-11) Let $F \in \mathbb{Z}[X_1,\ldots,X_n]$ and let p be any prime number. Prove the equivalence of the following conditions:

(i) For any $k \in \mathbb{N}$ there is an n-tuple $(x_1^{(k)},\ldots,x_n^{(k)})$ of elements of $\mathbb{Z}$ such that $F(x_1^{(k)},\ldots,x_n^{(k)}) \equiv 0 \mod p^k$.

(ii) There is an n-tuple $(\hat{x}_1,\ldots,\hat{x}_n)$ of p-adic integers (i.e. elements of the valuation ring of the p-adic valuation $\hat{v}_p$ of $\hat{\mathbb{Q}}_p$) such that $F(\hat{x}_1,\ldots,\hat{x}_n) = 0$.

Chapter II

II-1) Let $\mathfrak{A}$ be a set of valuation rings of the field K such that $K \notin \mathfrak{A}$ and $\#\mathfrak{A} > 1$, and let $\mathfrak{A}' = \{A_1 \cdot A_2 \mid A_1, A_2 \in \mathfrak{A}\}$.

a) Prove that $\mathfrak{A}'$ is equal to the set of all finite products $A_1 \cdot \ldots \cdot A_n$ $(n \geq 1)$ of valuation rings $A_1, \ldots, A_n \in \mathfrak{A}$.

b) Which relations hold between the following conditions?

(i) The valuation rings of $\mathfrak{A}$ are pairwise independent.

(ii) The valuation rings of $\mathfrak{A}$ are pairwise incomparable (with respect to inclusion).

(iii) $\mathfrak{A}' = \mathfrak{A} \cup \{K\}$.

II-2) Let A be a valuation ring of the field K and R be an arbitrary subring of K (containing the unit element of K).

Prove:

a) $A \cdot R = \{a \cdot r \mid a \in A, r \in R\}$.

b) A subset S of K is a proper ideal of $A \cdot R$ if and only if S is a proper ideal of A and an R-submodule of K.

c) The prime ideals of $A \cdot R$ coincide with the prime ideals of A which are contained in $\mathfrak{M}_{A \cdot R}$.

d) $\mathfrak{M}_A$ is an R-submodule of K if and only if $R \subseteq A$.

e) The following conditions are equivalent:

(i) $A \cdot R = K$.

(ii) (0) is the only proper ideal of A which is an R-submodule of K.

(iii) (0) is the only prime ideal of A which is an R-submodule of K.

II-3) Let v be a Krull valuation of K with value group Γ.

Prove that the isolated subgroups Φ of Γ are in 1-1 correspondence with the prime ideals $\mathfrak{P}$ of the valuation ring A of v by $\mathfrak{P} = \{x \in A \mid vx \notin \Phi\}$ and

$\Phi = \{\gamma \in \Gamma \mid \gamma = vx$ or $= vx^{-1}$ for some $x \in A \setminus \mathfrak{P}\}$.

II-4) Let $\pi: \tilde{K} \to \tilde{L}$ be a place. Prove that $\mathrm{Char}\, L \neq \mathrm{Char}\, K$ if and only if there is a prime number p such that $(\mathbb{Q}, \mathbb{Z}_{p\cdot\mathbb{Z}})$ can be imbedded in (K, A_π) (i.e., $\mu\, \mathbb{Z}_{p\cdot\mathbb{Z}} = A_\pi \cap \mu\mathbb{Q}$ for some monomorphism $\mu: \mathbb{Q} \to K$). In this case, $\mathrm{Char}\, K = 0$ and $\mathrm{Char}\, L = p$.

II-5) Let $K = \mathbb{C}(X)$. Prove that, for any $a \in \mathbb{C}$, the mapping $\pi_a: \tilde{K} \to \tilde{\mathbb{C}}$ defined by $\pi_a(F\cdot G^{-1}) = F(a)\cdot G(a)^{-1}$ (where F , G are relatively prime) and $\pi_a\infty = \infty$, is a place of K , and any place π of K such that $\mathbb{C}[X] \subseteq A_\pi$ is equivalent to π_a for exactly one $a \in \mathbb{C}$.

II-6) Let A and A' be valuation rings of K such that $A \subseteq A'$ and let v be a Krull valuation corresponding to A with value group Γ . Prove that there is a Krull valuation w of $\mathcal{K}' = A'/\mathfrak{M}_{A'}$, corresponding to the valuation ring $A/\mathfrak{M}_{A'}$ of $\mathcal{K}'$, with value group $\Phi_{A'} = v(U_{A'})$ and such that $w(x + \mathfrak{M}_{A'}) = vx$ for all $x \in U_{A'}$.

II-7) Let K be a field of characteristic zero and of transcendence degree d over its prime field, and let p be any prime number. Prove the existence of a valuation ring A of K , $A \neq K$, such that the transcendence degree of $A/\mathfrak{M}_A$ over its prime field is equal to d . What can be said about the number of valuation rings A of K such that $A/\mathfrak{M}_A$ has a transcendence degree less than d ?

II-8) Let R be a subring of K . Give a direct proof (using only (9.1)) of the following statement: For any proper ideal $\mathfrak{A}$ (resp. prime ideal $\mathfrak{P}$) of R there is a valuation ring A of K such that $R \subseteq A$ and $\mathfrak{A} \subseteq$ (resp. $\mathfrak{P} =$) $\mathfrak{M}_A \cap R$.

II-9) Let $K_o(z) \subseteq A \subseteq K$ where A is a valuation ring of K and z is transcendental over K_o . Prove that there are infinitely many valuation rings A' of K such that $K_o \subseteq A' \subseteq A \subseteq K$.

II-10) Let R , S be integral domains, $R \subseteq S$, and let M be a multiplicatively closed subset of R . Prove that if S is integral over R then the ring of fractions S_M is integral over the ring of fractions R_M .

II-11) Let R be a subring of a field L and $Q(R)$ be the quotient field of R contained in L . Prove that $Q(I_L(R)) = I_L(Q(R)) = (I_L(R))_{R^*}$, where R^* is the multiplicative closed set of all non-zero elements of R .

II-12) Let R be a subring of K and A be a valuation ring of K which contains R . Prove that for any prime ideal $\mathfrak{P}'$ of R which contains $\mathfrak{M}_A \cap R$ there exists a valuation ring A' of K such that $R \subseteq A' \subseteq A$ and $\mathfrak{M}_{A'} \cap R = \mathfrak{P}'$.

II-13) Let $\mathfrak{A}$ be a set of valuation rings of K which is totally ordered with respect to inclusion. Prove that $\bigcap_{A \in \mathfrak{A}} A$ is a valuation ring of K .

II-14) Let R be a subring of K , $\mathfrak{A}$ be the set of all valuation rings of K which contain R , and $\mathfrak{A}'$ be the set of all minimal elements of $\mathfrak{A}$. Prove:

a) For any $A \in \mathfrak{A}$ there is an $A' \in \mathfrak{A}'$ such that $A' \subseteq A$.

b) $I_K(R) = \bigcap_{A' \in \mathfrak{A}'} A'$.

c) For any $A' \in \mathfrak{A}'$, $\mathfrak{M}_{A'} \cap R$ is a maximal ideal of R .

II-15) Let $R = \mathbb{Q}[X,Y]$ and $K = \mathbb{Q}(X,Y)$, where X and Y are indeterminates over $\mathbb{Q}$. Prove:

a) Any valuation ring A of K such that $R \subseteq A$ and $\mathfrak{M}_A \cap R$ is a principal ideal is essential for R and has rank 1.

b) There exist valuation rings A of K of rank 1 and rank 2 such that $R \subseteq A$ and $\mathfrak{M}_A \cap R = R\cdot X + R\cdot Y$; they are not essential for R .

II-16) Let R be a subring of K . Prove that the following conditions are equivalent:

(i) For any maximal ideal $\mathfrak{M}$ of R , $R_{\mathfrak{M}}$ is a noetherian valuation ring of K .

(ii) Any valuation ring A of K such that $R \subseteq A \subset K$ is essential for R and is discrete.

In this case R is called an almost Dedekind ring of K .

II-17) Let R be an almost Dedekind ring of K . Prove:

a) R is a Prüfer ring of K and any non-zero prime ideal of R is maximal.

b) Any subring of K which contains R is an almost Dedekind ring of K .

II-18) Let A_i be a valuation ring of K and v_i a corresponding Krull valuation with value group Γ_i $(i = 1,2)$. Prove: If $A_1 \cdot A_2 \neq K$ then, for any $\alpha \in \Gamma_1$, there is a $\beta \in \Gamma_2$ such that $(\alpha,\beta) \neq (v_1x, v_2x)$ for all $x \in K$.

Chapter III

III-1) Let $L|K$ be a field extension, $\mathfrak{A}$ be a set of valuation rings of K , $\mathfrak{B}(A)$ be the set of all valuation rings of L which lie over $A \in \mathfrak{A}$, and $\mathfrak{B} = \bigcup_{A\in\mathfrak{A}} \mathfrak{B}(A)$. Prove:

$$I_L\Big(\bigcap_{A\in\mathfrak{A}} A\Big) = \bigcap_{A\in\mathfrak{A}} I_L(A) = \bigcap_{B\in\mathfrak{B}} B .$$

III-2) Let $L|K$ be an algebraic extension and R be a Prüfer ring of K .

a) Prove that $I_L(R)$ is a Prüfer ring of L .

b) Study the relationship between the prime ideals of $I_L(R)$ and those of R .

c) Assuming $[L:K] < \infty$ prove that if R is an almost Dedekind ring of K then $I_L(R)$ is an almost Dedekind ring of L .

III-3) Let $L|K$ be a finite field extension. Prove that if R is a Krull ring of K then $I_L(R)$ is a Krull ring of L . Prove also similar statements for generalized Krull rings, Dedekind rings, and generalized Dedekind rings.

III-4) Let $L|K$ be an algebraic extension, A a valuation ring of K , $D = I_L(A)$, $\mathcal{P}_D$ (resp. $\mathcal{P}_A$) the set of all prime ideals of D (resp. A), and $\mathcal{C}$ (resp. $\mathcal{A}$) the set of all valuation rings of L (resp. K) which contain D (resp. A). Prove:

a) The mappings $\mathcal{C} \to \mathcal{A}$ and $\mathcal{P}_D \to \mathcal{P}_A$ defined by $C \mapsto C \cap K$ and $\mathfrak{P} \mapsto \mathfrak{P} \cap A$, respectively, are surjective and commute with the bijective mappings $\mathcal{A} \to \mathcal{P}_A$ and $\mathcal{C} \to \mathcal{P}_D$ defined in (6.6) and (13.4).

b) For any maximal ideal $\mathfrak{M}$ of D , the mapping $\mathcal{P}_D \to \mathcal{P}_A$ induces a bijective mapping $\{\mathfrak{P} \in \mathcal{P}_D \mid \mathfrak{P} \subseteq \mathfrak{M}\} \to \mathcal{P}_A$.

c) For any valuation ring B of L which lies over A , the mapping $\mathcal{C} \to \mathcal{A}$ induces a bijective mapping $\{C \in \mathcal{C} \mid B \subseteq C\} \to \mathcal{A}$. (Hint: Use the going down theorem, Zariski & Samuel [36] Chap. V, §3).

III-5) Give an example of a purely inseparable field extension $L|K$ and a valuation ring B of L such that $e_{B|K} = [L:K] < \infty$ (resp. $f_{B|K} = [L:K] < \infty$).

III-6) Let K be a field of characteristic $p \neq 0$, $L|K$ be a purely inseparable field extension, B be a valuation ring of L, and w (resp. ζ) be a Krull valuation (resp. place) corresponding to B. Moreover, let Δ (resp. Γ) be the value group of w (resp. $w|K$) and $\mathfrak{L}$ (resp. $\mathcal{K}$) be the residue field of ζ (resp. $\zeta|K$). Prove:

a) Δ/Γ is a p-group and $\mathfrak{L}|\mathcal{K}$ is a purely inseparable extension.

b) If $[L:K] = p^m$ then $e_{B|K}\cdot f_{B|K} = p^n$ for some $n \le m$.

III-7) Let $L|K$ be a finite field extension, v be a Krull valuation of K with value group Γ, and $w_1,\dots,w_r$ be pairwise non-equivalent valuations of L which extend v and whose value groups $\Delta_1,\dots,\Delta_r$ are contained in the divisible closure Γ_c of Γ. Prove: For any $\delta_1 \in \Delta_1$ there exist $\delta_2 \in \Delta_2,\dots,\delta_r \in \Delta_r$ such that $\delta_1 < \min\{\delta_2,\dots,\delta_r\}$ and $(\delta_1,\dots,\delta_r)$ is compatible (cf end of §11). (Hint: Use induction on r and the fact that, for $i \neq j$, the isolated subgroups Φ_{ij} of Δ_i and Φ_{ji} of Δ_j which correspond to the product of the valuation rings of w_i and w_j, by (7.4), generate the same isolated subgroup of Γ_c.)

III-8) Let $N|K$ be a normal field extension with Galois group G, B be a valuation ring of N, and w be a Krull valuation corresponding to B. Prove:

a) $G^Z(B|K) = \{\sigma \in G \mid w\circ\sigma = w\} = \{\sigma \in G \mid \frac{\sigma x}{x} \in U_B \text{ for all } x \in N^*\}$.

b) For any ideal $\mathfrak{A}$ of B and any $\sigma \in G^Z(B|K)$ we have $\sigma\mathfrak{A} = \mathfrak{A}$.

III-9) Let $N|K$ be a finite normal field extension, A be a valuation ring of K which is indecomposed in N, and v be a Krull valuation corresponding to A with value group Γ. Prove that the unique Krull valuation w of N which extends v and whose value group is contained in Γ_c, satisfies $[N:K]\cdot wy = v(\mathfrak{N}_{N|K}y)$ for all $y \in N$, where $\mathfrak{N}_{N|K}$ is the norm with respect to $N|K$.

III-10) a) Give a direct proof of the implication (vii) ⇒ (i) of (16.3).

b) Show that in condition (viii) of (16.3), the inclusion $\subseteq A$ may be replaced by $\subseteq \mathfrak{M}_A$.

c) Prove: If A is indecomposed in N then, for any prime ideal $\mathfrak{P}$ of A and any $y \in N$, $c(y) \in \mathfrak{P}$ implies $\{c_1(y),\ldots,c_{n(y)}(y)\} \subseteq \mathfrak{P}$.

III-11) Prove that $\mathbb{Q}$ is the only henselian valuation ring of $\mathbb{Q}$.

III-12) Prove that for distinct prime numbers p , q there is no monomorphism $\hat{\mathbb{Q}}_p \to \hat{\mathbb{Q}}_q$. (Hint: Construct a polynomial in $\mathbb{Z}[X]$ which is irreducible in $\hat{\mathbb{Q}}_q$ and is reducible in $\hat{\mathbb{Q}}_p$.)

III-13) Let $L|K$ be a finite field extension, $n = [L:K]$, A be a discrete valuation ring of K , and $y \in L$ a root of some polynomial $F = X^n + a_1 \cdot X^{n-1} + \ldots + a_n \in A[X]$ such that $a_1,\ldots,a_n \in \mathfrak{M}_A$ and $a_n \notin \mathfrak{M}_A^2$. (Such a polynomial is called eisensteinian.) Prove: $L = K(y)$, $F = P_{y|K}$, A is indecomposed in L , and $e_{B|K} = n$, $f_{B|K} = 1$, where $B = I_L(A)$.

III-14) Using the notation of (17.4), we say that a henselian extension $(\bar{K},\bar{A})$ of (K,A) is L-<u>distinguished</u> if, for any $\lambda \in \mathrm{Mon}(L|K)$, $(\lambda L \cdot \bar{K}, \bar{D} \cap L \cdot \bar{K})$ is an immediate extension of $(\lambda L, \lambda B_\lambda)$.

a) Prove that any henselization of (K,A) is L-distinguished for any separable extension $L|K$.

b) Prove that if A has rank 1 then the completion $(\hat{K},\hat{A})$ of (K,A) is L-distinguished for any finite extension $L|K$.

c) Characterize those henselian extensions $(\bar{K},\bar{A})$ of (K,A) which are K-distinguished and those which are $sc(K)$- (resp. $ac(K)$-) distinguished.

III-15) a) Prove that in (18.2) the henselization $(\bar{K},\bar{A})$ can be replaced by any immediate L-distinguished and L-allowable henselian extension of (K,A) .

b) Let $L|K$ be a finite field extension and A a valuation ring of K which is defectless in L . Prove that any immediate henselian extension of (K,A) is L-allowable and L-distinguished.

III-16) Let $L|K$ be a finite field extension and B be a valuation ring of L . Prove that $e_{B|K}$ is a multiple of the initial index $e_{B|K}$.

III-17) Assume that the following diagram, consisting of groups and group homomorphisms, is commutative and has exact rows and columns (E denoting the trivial group).

$$\begin{array}{ccccccc} & & & & E & & \\ & & & & \downarrow & & \\ & & G_1 & \rightarrow & G_1' & & \\ & & \downarrow & & \downarrow & & \\ G_2'' & \rightarrow & G_2 & \rightarrow & G_2' & \rightarrow & E \\ & & \downarrow & & \downarrow & & \\ G_3'' & \rightarrow & G_3 & \rightarrow & G_3' & & \\ \downarrow & & & & & & \\ E & & & & & & \end{array}$$

Prove that the homomorphism $G_1 \to G_1'$ is surjective.

III-18) Let $N|K$ be a finite normal extension, B be a valuation ring of N , and q be a prime number which divides $e_{B|K}$ and is different from Char $B/\mathfrak{M}_B$. Prove that $B/\mathfrak{M}_B$ has a primitive q-th root of unity.

III-19) Let $N|K$ be a finite Galois field extension, B be a valuation ring of N , ζ be a place corresponding to B , and $\mathfrak{h}$ (resp. $\mathfrak{K}$) the residue field of ζ (resp. $\zeta|K$). Prove:

a) N is a solvable extension of $K^T(B|K)$.

b) N is a solvable extension of $K^Z(B|K)$ if and only if the separable closure of $\mathfrak{K}$ in $\mathfrak{h}$ is a solvable extension of $\mathfrak{K}$.

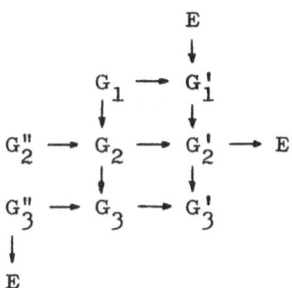

III-20) Using the notation of §21, show that the correspondence $\mathfrak{G}^V \leftrightarrow \mathfrak{J}^V$ is induced by a Galois connexion between the set of all subgroups of $G^V(B|K)$, ordered by inclusion, and the set of all proper ideals of B, ordered by inverse inclusion.

III-21) Using the notation of §21, let $G^T(\mathfrak{A}) = \{\sigma \in G \mid \sigma x - x \in \mathfrak{A}$ for all $x \in B\}$, for any $\mathfrak{A} \in \mathfrak{J}$.

a) Prove that, for any $\mathfrak{A} \in \mathfrak{J}$, $G^T(\mathfrak{A})$ is a normal subgroup of $G^Z(B|K)$.

b) Which relationship holds between the groups $G^T(\mathfrak{A})$ and $G^V(\mathfrak{A})$?

c) Prove a theorem similar to (21.5), replacing $G^V(s\cdot B)$ by $G^T(s\cdot B)$ and replacing $H(N^*/K^*, \mathfrak{h}^+)$ by the additive abelian group of all derivatives $\delta: B \to \mathfrak{h}$ such that $\delta(B \cap K) = \{0\}$.

III-22) Using the notation of (21.5), endow $Hom(N^*/K^*, \mathfrak{h}+)$ with a topology such that the homomorphism χ_s becomes continuous.

Chapter IV

IV-1) Let K be an algebraically closed field and φ be a valuation of K. Let $\mu(H;z)$ be the multiplicity of $z \in K$ as a root of $H \in K[X]$. (Note that $\mu(H;z) = 0$ if and only if z is not a root of H.) Prove: For any $F \in \mathfrak{m}_n$ and any $\varepsilon > 0$ there is a $\delta > 0$ such that, for any $G \in \mathfrak{U}_\delta(F)$ and any root $x \in K$ of F, we have

$$\sum_{y \in \mathfrak{V}_\varepsilon(x)} \mu(G;y) = \mu(F;x) \quad \text{(where } \mathfrak{V}_\varepsilon(x) \text{ is as in §1).}$$

IV-2) Let $\varphi^1,\dots,\varphi^k$ be pairwise non-equivalent non-trivial valuations of K and let $\mathfrak{m}_n$ be endowed with the intersection of the topologies defined by $\varphi^1,\dots,\varphi^k$ (cf §24). Prove that $\mathfrak{S}_n$ is dense in $\mathfrak{M}_n$.

IV-3) Let y be a 0-solution of the φ-prescription $\mathfrak{y} \in \mathfrak{T}_n$ (i.e., $\rho\mathfrak{y} = P_{y|K}$). Prove: For any $\epsilon > 0$ there is a $\delta > 0$ such that $P_{z|K} \in \mathfrak{U}_\epsilon(P_{y|K})$ for any δ-solution z of $\mathfrak{y}$.

IV-4) Let φ be a valuation of K and $K(y)$ be a finite separable extension of K . Let $\psi_1,\ldots,\psi_r$ be all extensions of φ to $K(y)$ and, for any $i \in \{1,\ldots,r\}$, let $\lambda_i\colon (K(y),\psi_i) \to (\hat{L}_i,\hat{\varphi})$ be a completion with $\lambda_i|K = \iota_K$. Prove: For any $\epsilon > 0$ there is a $\delta > 0$ such that, for any root z of any $F \in \mathfrak{J}_n \cap \mathfrak{U}_\delta(P_{y|K})$, φ has exactly r extensions $\chi_1,\ldots,\chi_r$ to $K(z)$ and there are completions $\mu_i\colon (K(z),\chi_i) \to (\hat{L}_i,\hat{\varphi})$ such that $\mu_i|K = \iota_K$ and $\hat{\varphi}(\lambda_i y - \mu_i z) \le \epsilon$ $(i = 1,\ldots,r)$. Moreover, prove that $\mathfrak{J}_n \cap \mathfrak{U}_\delta(P_{y|K})$ may be replaced by $\mathfrak{U}_\delta(P_{y|K})$ whenever $r = 1$.

IV-5) Prove (24.4), (24.5), and (24.7), replacing φ by an arbitrary Krull valuation v . Try to extend §23, §24, and the first part of §25 up to (25.3), to Krull valuations v , replacing the completion $(\hat{K},\hat{\varphi})$ by a henselization of (K,A_v) .

IV-6) Show that in (25.5) and (27.5) $\mathfrak{M}_n$ and $\mathfrak{J}_n$ may be replaced by $\mathfrak{S}_n$ and $\mathfrak{J}_n \cap \mathfrak{S}_n$, respectively. (Hint: Use Exercise IV-2).

IV-7) Prove the following generalization of Theorem (26.7): For any non-trivial valuation φ of K and any $n > 1$ the following conditions are equivalent:

(i) φ is m-antihenselian for all $m \in \{2,\ldots,n\}$.

(ii) There is no separable extension $\hat{L}$ of $\hat{K}$ such that $1 < [\hat{L}:\hat{K}] \le n$.

(iii) For any $m \in \{2,\ldots,n\}$ we have $\mathfrak{T}_m \subseteq \hat{K} \times\ldots\times \hat{K}$ (m times).

(iv) For any $m \in \{2,\ldots,n\}$ we have $\mathfrak{S}_m \subseteq \hat{K} \times\ldots\times \hat{K}$ (m times).

(v) Any separable extension L of K such that $[L:K] \le n$ has exactly $[L:K]$ valuations which extend φ .

IV-8) Let $\varphi^1,\dots,\varphi^k$ be pairwise non-equivalent non-trivial valuations of K . Prove that if φ^1 is henselian then, for any $n \geq 1$, we have $\overline{\mathfrak{S}}_n = \mathfrak{O}_n^1 \times \mathfrak{R}_n^2 \times\dots\times \mathfrak{R}_n^k$, where $\mathfrak{R}_n^j = \hat{K}^j \times\dots\times \hat{K}^j$ (n times). Prove a similar statement under the weaker hypothesis that φ^1 be m-henselian for all $m \in \{1,\dots,n\}$.

IV-9) Let K be any subfield of $\mathbb{R}$. Prove that K has no 2-henselian valuation.

IV-10) Let $L|K$ be an algebraic extension, ψ a henselian non-archimedean valuation of L , and $\varphi = \psi|K$. Prove that φ is henselian if

a) $L|K$ is normal and L is not separably closed; or

b) $[L:K]_{sep} < \infty$ and L is not separably closed; or

c) $[L:K]_{sep} < \infty$ and K is not real closed.

(Hint: Use (14.3) and Artin-Schreier's theorem.)

IV-11) Define "saturated" and "antihenselian" for valuation rings of arbitrary rank in such a way that the conditions of (17.15) become equivalent to each of the following conditions:

(i) A is saturated and is defectless in any separable extension.

(ii) A is antihenselian.

IV-12) Let $\xi = (\Delta_1,\dots,\Delta_r, \mathfrak{L}_1,\dots,\mathfrak{L}_r)$ be a Krull φ-prescription of degree n , where φ is a non-archimedean non-trivial valuation of K . Let L be a separable extension of K of degree $[L:K] \leq n$ such that there exist at least r extensions $\psi_1,\dots,\psi_r$ of φ to L and, for any $i \in \{1,\dots,r\}$, a Γ_φ-monomorphism $\Delta_i \to \Gamma_{\psi_i}$ and a $\mathcal{K}_\varphi$-monomorphism $\mathfrak{L}_i \to \mathcal{K}_{\psi_i}$. Prove: Any primitive element of $L|K$ is a solution of ξ .

IV-13) Let φ be a non-archimedean non-trivial valuation whose residue field is finitely generated over its prime field. Prove that φ is non-n-saturated and non-n-antihenselian for any

$n > 1$.

IV-14) Is it true that any saturated non-archimedean valuation is antihenselian? (We do not know the answer.)

IV-15) It is known that the Steinitz numbers (i.e. formal products $\prod_{p\in\mathbb{P}} p^{n_p}$, where $\mathbb{P}$ is the set of all prime numbers and $n_p \in \mathbb{N} \cup \{\infty\}$ for all $p \in \mathbb{P}$) are in 1-1 correspondence with:

a) those subgroups of the additive group $\mathbb{Q}$ which contain $\mathbb{Z}$;

b) the subfields of the algebraic closure of any finite prime field.

For any non-archimedean non-trivial valuation φ of any algebraic extension K of $\mathbb{Q}$, let $e(\varphi)$ (resp. $f(\varphi)$) be the Steinitz number corresponding to the value group (resp. residue field) of φ . Prove: φ is saturated if and only if φ is n-saturated for almost all $n \geq 1$, if and only if $e(\varphi) = f(\varphi) = \prod_{p\in\mathbb{P}} p^{\infty}$.

IV-16) Generalize Theorems (27.5), (27.6), and (27.7), by prescribing, in addition to a Krull $(\varphi^1,\ldots,\varphi^k)$-prescription $\bar{\xi}$ of degree n , the number of extensions for finitely many pairwise non-equivalent real-archimedean valuations of K . (Note that this number must be between $\frac{1}{2}n$ and n , because of (2.13).)

IV-17) Let $K|\mathbb{Q}$ be a finite extension of degree k , q be the number of extensions of φ_∞ (the absolute value of $\mathbb{Q}$) to K , and let $n \geq 1$. Prove:

a) For any extension L of K of degree n , the number r of extensions of φ_∞ to L satisfies

$$n\cdot q \geq r \geq \begin{cases} k\cdot \frac{n}{2} & \text{if } n \text{ is even,} \\ k\cdot \frac{n-1}{2} + q & \text{if } n \text{ is odd.} \end{cases}$$

b) For any integer r satisfying these inequalities, there exists an extension L of K of degree n which has exactly r extensions of φ_∞ .

IV-18) Let K , A , v , π , Γ , $\mathcal{K}$ be as in Theorem (27.1). Let Δ be a torsion extension of Γ and $\mathcal{L}$ an algebraic extension of $\mathcal{K}$. Prove the existence of a separable extension L of K with the following properties:

a) A is indecomposed in L .

b) There is a Krull valuation w of L lying over v with value group Δ .

c) There is a place ρ of L lying over π with residue field $\mathcal{L}$.

(Obviously, w and ρ correspond to the only valuation ring $I_L(A)$ of L which lies over A.)

IV-19) Prove that in (28.3) the hypothesis of existence of a valuation φ^o may be replaced by other hypotheses; for example:

I) $r^1 = 1$ and $(\Delta_1^1:\Gamma^1)\cdot[\mathcal{L}_1^1:\mathcal{K}^1) \geq \max\{r^2,\dots,r^k\}$.

II) $k \geq 3$, $\bigcap_{i=1}^{r^1} \Delta_i^1 \neq \Gamma^1$ or $\bigcap_{i=1}^{r^1} \mathcal{L}_i^1 \neq \mathcal{K}^1$, and φ^2 , φ^3 are not saturated.

IV-20) Let K be an algebraic extension of $\mathbb{Q}$, let $\varphi^1,\dots,\varphi^k$ be pairwise non-equivalent non-archimedean non-trivial valuations of K , and (e_i^j,f_i^j) pairs of Steinitz numbers such that $e(\varphi^j)$ divides e_i^j and $f(\varphi^j)$ divides f_i^j $(i = 1,\dots,r^j;\ j = 1,\dots,k$; cf Exercise IV-15). Assume that $r^1 = \dots = r^k = 1$ or that K has a non-trivial non-archimedean valuation which is non-equivalent to $\varphi^1,\dots,\varphi^k$ and is not saturated.

Prove the existence of an algebraic extension L of K with the following properties:

a) For any $j \in \{1,\dots,k\}$, φ^j has at least r^j extensions ψ_i^j

to L , and $e(\psi_i^j) = e_i^j$, $f(\psi_i^j) = f_i^j$ $(i = 1,\ldots,r^j \; ; \; j = 1,\ldots,k)$.

b) Any non-trivial non-archimedean valuation of L which is non-equivalent to all valuations ψ_i^j $(i = 1,\ldots,r^j \; ; \; j = 1,\ldots,k)$ is saturated.

Indicate other hypotheses under which this statement holds.

Bibliography

[1] - BACHMAN, G.: Introduction to p-Adic Numbers and Valuation Theory. Academic Press, New York and London 1964.

[2] - BASTOS, G.G.: Sobre um Problema de Existência na Teoria das Valorizações. Ph.D. Thesis, IMPA, Rio de Janeiro, 1972.

[3] - BOREVICH, Z.I., and SHAFAREVICH, I.R.: Number Theory. Academic Press, New York and London 1966.

[4] - BOURBAKI, N.: Topologie Générale. 3rd Ed. Hermann, Paris 1960- .

[5] - BOURBAKI, N.: Algèbre Commutative. Hermann, Paris 1961-1965.

[6] - ENDLER, O.: Bewertungstheorie. Unter Benutzung einer Vorlesung von W. Krull. Bonner math. Schriften Nr. 15, I,II. Bonn 1963.

[7] - ENDLER, O.: Über einen Existenzsatz der Bewertungstheorie. Math. Ann. 150 (1963), 54-65.

[8] - ENDLER, O.: Sobre Corpos Valorizados Completos com Corpos de Restos Perfeitos. See Ribenboim [32], p. 121-165.

[9] - ENDLER, O.: Endlich separable Körpererweiterungen mit vorgeschriebenen Bewertungsfortsetzungen. I. Abh. Math. Sem. Hamburg 33 (1969), 80-101.

[10]- GILMER, R.W.: Multiplicative Ideal Theory. I, II. Queen's Papers on Pure and Applied Mathematics No. 12. Kingston, Ontario 1968.

[11]- GOLDHABER J.K., and EHRLICH, G.: Algebra. Macmillan, London 1970.

[12]- GRIFFIN, M.: Some Results on v-Multiplication Rings. Can. J. Math. 19 (1967), 710-722.

[13]- GRIFFIN, M.: Families of Finite Character and Essential Valuations. Trans. Amer. Math. Soc. 130 (1968), 75-85.

[14]- GRIFFIN, M.: Rings of Krull Type. J. reine angew. Math. 229 (1968), 1-27.

[15]- HASSE, H.: Zwei Existenzsätze über algebraische Zahlkörper. Math. Ann. 95 (1925), 229-238.

[16]- HASSE, H.: Zahlentheorie. 2nd Ed. Akademie-Verlag, Berlin 1963.

[17]- HENSEL, K.: Theorie der algebraischen Zahlen. Teubner, Leipzig 1908.

[18]- HILL, L.C.: On Prescribing Henselizations for Valued Fields. Ph. D. Thesis, University of Rochester, Rochester, N.Y. 1972.

[19]- JACOBSON, N.: Lectures in Abstract Algebra. III. Van Nostrand, Princeton 1964.

[20]- KRULL, W.: Allgemeine Bewertungstheorie. J. reine angew. Math. 167 (1931), 160-196.

[21]- KRULL, W.: Idealtheorie. Springer-Verlag, Berlin 1935. Reprint: Chelsea, New York 1948.

[22] - KRULL, W.: Über einen Existenzsatz der Bewertungstheorie. Abh. Math. Sem. Hamburg 23 (1959), 29-35.

[23] - KRULL, W.: Eine Bemerkung zur Bewertungstheorie. Anais Acad. Brasil. Ci. 35 (1963), 457-481.

[24] - LANG, S.: Algebra. Addison-Wesley, Reading, Mass. 1965.

[25] - NORTHCOTT, D.G.: Ideal Theory. Cambridge University Press, Cambridge 1953.

[26] - OHM, J.: Some Counterexamples Related to Integral Closure in D[[X]]. Trans. Amer. Math. Soc. 122 (1966), 321-333.

[27] - RIBENBOIM, P.: Anneaux Normaux Réels à Caractère Fini. Summa Brasil. Math. 3 (1956), 213-253.

[28] - RIBENBOIM, P.: Remarques sur le Prolongement des Valuations de Krull. Rend. Circ. Mat. Palermo, ser. II, 8 (1959), 1-8.

[29] - RIBENBOIM, P.: An Existence Theorem for Fields with Krull Valuations. Trans. Amer. Math. Soc. 105 (1962), 278-294.

[30] - RIBENBOIM, P.: Théorie des Valuations. Les Presses de l'Université de Montréal, 1964.

[31] - RIBENBOIM, P.: On the Existence of Fields with Few Discrete Valuations. J. reine angew. Math. 216 (1964), 43-49.

[32] - RIBENBOIM, P.: Tópicos de Teoria dos Números. Notas de Matemática No. 35, IMPA, Rio de Janeiro 1966.

[33] - SERRE, J.P.: Corps Locaux. Hermann, Paris 1962.

[34] - SERRE, J.P.: Cohomologie Galoisienne. Lecture Notes in Mathematics. Springer-Verlag, Berlin-Heidelberg-New York 1964.

[35] - WEISS, E.: Algebraic Number Theory. McGraw-Hill, New York 1963.

[36] - ZARISKI, O., and SAMUEL, P.: Commutative Algebra. I, II. Van Nostrand, Princeton 1958-1960.

Index